VOLUME ONE HUNDRED AND THREE

VITAMINS AND HORMONES

Anxiety

VOLUME ONE HUNDRED AND THREE

VITAMINS AND HORMONES

Anxiety

Series Editor

GERALD LITWACK, PhD

Toluca Lake, North Hollywood, California

ELSEVIER

ACADEMIC PRESS

An imprint of Elsevier

elsevier.com

Cover photo credit:
Meyer, T., Herrmann-Lingen, C.
Natriuretic Peptides in Anxiety and Panic Disorder
Vitamins and Hormones (2017) **103**, pp. 131-146.

Academic Press is an imprint of Elsevier
50 Hampshire Street, 5th Floor, Cambridge, MA 02139, United States
525 B Street, Suite 1800, San Diego, CA 92101-4495, United States
The Boulevard, Langford Lane, Kidlington, Oxford OX5 1GB, United Kingdom
125 London Wall, London, EC2Y 5AS, United Kingdom

First edition 2017

ISBN: 978-0-12-811914-3
ISSN: 0083-6729

For information on all Academic Press publications
visit our website at https://www.elsevier.com/

Publisher: Zoe Kruze
Acquisition Editor: Alex White
Editorial Project Manager: Helene Kabes
Production Project Manager: Vignesh Tamil
Cover Designer: Miles Hitchen

Typeset by SPi Global, India

CONTENTS

CONTRIBUTORS

D.C. Aguiar
Institute of Biological Sciences, Universidade Federal de Minas Gerais, Belo Horizonte, Brazil

B. Bonaz
Clinique Universitaire d'Hépato-Gastroentérologie, CHU de Grenoble; Université Grenoble Alpes, Grenoble Institut des Neurosciences, Fonctions Cérébrales et Neuromodulation, INSERM, Grenoble 09, France

A.P. Borrow
Colorado State University, Fort Collins, CO, United States

M.A. Briscione
Trauma Recovery Program, Mental Health Service Line, Atlanta Veterans Affairs Medical Center, Decatur; Emory University School of Medicine, Atlanta, GA, United States

P. Chanana
University Institute of Pharmaceutical Sciences, UGC Centre of Advanced Study, Panjab University, Chandigarh, India

U. Doboszewska
Faculty of Pharmacy, Jagiellonian University Medical College, Krakow, Poland

M. Gaweł
Faculty of Pharmacy, Jagiellonian University Medical College, Krakow, Poland

F.V. Gomes
University of Pittsburgh, Pittsburgh, PA, United States

F.S. Guimarães
Medical School of Ribeirão Preto, University of São Paulo; Center for Interdisciplinary Research on Applied Neurosciences (NAPNA), University of São Paulo (USP), Ribeirão Preto, Brazil

K. Gulati
Vallabhbhai Patel Chest Institute, University of Delhi, Delhi, India

R.J. Handa
Colorado State University, Fort Collins, CO, United States

S.M.J. Hemmings
Faculty of Medicine and Health Sciences, Department of Psychiatry, Stellenbosch University, Cape Town, South Africa

C. Herrmann-Lingen
University of Göttingen Medical Centre; German Centre for Cardiovascular Research, University of Göttingen, Göttingen, Germany

F. Huang
Yiyang Medical College, Yiyang, Hunan, PR China

T. Jovanovic
Emory University School of Medicine, Atlanta, GA, United States

A. Kumar
University Institute of Pharmaceutical Sciences, UGC Centre of Advanced Study, Panjab University, Chandigarh, India

C.-Q. Li
School of Basic Medical Sciences, Central South University, Changsha, Hunan, PR China

S.F. Lisboa
Medical School of Ribeirão Preto, University of São Paulo; Center for Interdisciplinary Research on Applied Neurosciences (NAPNA), University of São Paulo (USP), Ribeirão Preto, Brazil

S. Malan-Müller
Faculty of Medicine and Health Sciences, Department of Psychiatry, Stellenbosch University, Cape Town, South Africa

T. Meyer
University of Göttingen Medical Centre; German Centre for Cardiovascular Research, University of Göttingen, Göttingen, Germany

V. Michopoulos
Emory University School of Medicine; Yerkes National Primate Research Center, Atlanta, GA, United States

K. Młyniec
Faculty of Pharmacy, Jagiellonian University Medical College, Krakow, Poland

F.A. Moreira
Institute of Biological Sciences, Universidade Federal de Minas Gerais, Belo Horizonte, Brazil

S.D. Norrholm
Trauma Recovery Program, Mental Health Service Line, Atlanta Veterans Affairs Medical Center, Decatur; Emory University School of Medicine, Atlanta, GA, United States

G. Nowak
Faculty of Pharmacy, Jagiellonian University Medical College, Krakow, Poland

S. Pellissier
Laboratoire Interuniversitaire de Psychologie, Personnalité, Cognition, Changement Social, Université Savoie Mont-Blanc, Chambéry, France

N. Rai
Vallabhbhai Patel Chest Institute, University of Delhi, Delhi, India

A. Ray
Vallabhbhai Patel Chest Institute, University of Delhi, Delhi, India

L.B.M. Resstel
Medical School of Ribeirão Preto, University of São Paulo; Center for Interdisciplinary Research on Applied Neurosciences (NAPNA), University of São Paulo (USP), Ribeirão Preto, Brazil

G. Starowicz
Faculty of Pharmacy, Jagiellonian University Medical College, Krakow, Poland

A.L.B. Terzian
Medical School of Ribeirão Preto, University of São Paulo; Center for Interdisciplinary Research on Applied Neurosciences (NAPNA), University of São Paulo (USP), Ribeirão Preto, Brazil

Z. Yang
Yiyang Medical College, Yiyang, Hunan, PR China

PREFACE

While occasional episodes of anxiety are normal life experiences, anxiety disorder is chronic and seems to be inherited. Panic attacks and social anxiety are two constituents of anxiety disorder. Some disorder that includes anxiety in the Unites States alone may involve about 40 million persons. The probability of developing any anxiety disorder approaches 30%, and females are 60% more likely to have anxiety disorders than males. The non-Hispanic white population is more susceptible to having anxiety disorder than any other group.

Anxiety disorder can be dissected into at least five subgroups. Thus, there are about 6.8 million persons in the United States with generalized anxiety disorder, 2.2 million persons have obsessive compulsive disorder, 7.7 million persons have posttraumatic stress disorder, 15 million persons have social phobia, and 19 million persons have a specific other type of phobia. Obviously, these statistics equate to a major health problem, requiring investigations into the mechanisms involved. Neuroendocrine involvement is clearly indicated. This volume reviews the roles that hormones and other factors play in anxiety and stress.

In the opening chapter, A. Ray, K. Gulati, and N. Rai examine "Stress, anxiety, and immunomodulation: a pharmacological analysis." Next, A.P. Borrow and R.J. Handa report on "Estrogen receptors and modulation of anxiety-like behavior." Following this, "Neuroendocrine underpinnings of increased risk for posttraumatic stress disorder in women" is the topic reviewed by M.A. Briscione, V. Michopoulos, T. Jovanovic, and S.D. Norrholm. The ever-enlarging role of RNAs in disease is exemplified by the chapter by S. Malan-Müller and S.M.J. Hemmings on "The big role of small RNAs in anxiety and stress-related disorders." T. Meyer and C. Herrmann-Lingen then describe "Natriuretic peptides in anxiety and panic disorder." Two chapters deal with the relatively new gaseous hormone, nitric oxide. The first is "The role of nitric oxide in stress-induced anxiety: from pathophysiology to therapeutic target" by A. Kumar and P. Chanana. The second is "Nitric oxide and anxiety" by K. Gulati, N. Rai, and A. Ray. S.F. Lisboa, F.V. Gomes, A.L.B. Terzian, D.C. Aguiar, F.A. Moreira, L.B.M. Resstel, and F.S. Guimarães report on "The endocannabinoid system and anxiety." "The melatonergic system in anxiety disorders and the role of melatonin in conditional fear" is the subject of F. Huang, Z. Yang, and C.-Q. Li.

"The role of elements in anxiety" is examined by K. Młynec, M. Gaweł, U. Doboszewska, G. Starowicz, and G. Nowak. In the final chapter, S. Pellissier and B. Bonaz describe "The place of stress and emotions in the irritable bowel syndrome."

The illustration on the cover is a reproduction of Fig. 3 from Chapter 5 by T. Meyer and C. Herrmann-Lingen entitled: "Natriuretic peptides in anxiety and panic disorder." (A) Crystal structure of the extracellular ligand-binding domain of the atrial natriuretic (ANP) receptor dimer complexed with rat ANP7-27. The receptor structure is shown as a ribbon model with carbohydrate structures depicted in magenta, while the receptor-bound hormone is depicted in a space-filling model with nitrogen atoms marked in blue, oxygen atoms in red, and sulfur atoms in yellow. The figure was drawn with PyMOL (DeLano Scientific) using data from the Protein Database file 1T34 (Ogawa et al., 2004). (B) Overview of the pleiotropic functions of cGMP-mediated ANP signaling including its effect on anxiety.

Publication of this book was facilitated by Helene Kabes of Elsevier, Oxford, UK, and by Vignesh Tamilselvvan, Elsevier, S&T Book Production, Chennai, India.

Gerald Litwack
July 18, 2016
North Hollywood, California

CHAPTER ONE

Stress, Anxiety, and Immunomodulation: A Pharmacological Analysis

A. Ray[1], K. Gulati, N. Rai
Vallabhbhai Patel Chest Institute, University of Delhi, Delhi, India
[1]Corresponding author: e-mail address: arunabha14@yahoo.co.in

Contents

Abstract

Stress and stressful events are common occurrences in our daily lives and such aversive situations bring about complex changes in the biological system. Such stress responses influence the brain and behavior, neuroendocrine and immune systems, and these responses orchestrate to increase or decrease the ability of the organism to cope with such stressors. The brain via expression of complex behavioral paradigms controls peripheral responses to stress and a bidirectional link exists in the modulation of stress effects. Anxiety is a common neurobehavioral correlate of a variety of stressors, and both acute and chronic stress exposure could precipitate anxiety disorders. Psychoneuroimmunology involves interactions between the brain and the immune system, and it is now being increasingly recognized that the immune system could contribute to the neurobehavioral responses to stress. Studies have shown that the brain and its complex neurotransmitter networks could influence immune function, and there could be a possible link between anxiogenesis and immunomodulation during stress. Physiological and pharmacological data have highlighted this concept, and the present review gives an overview of the relationship between stress, anxiety, and immune responsiveness.

Vitamins and Hormones, Volume 103
ISSN 0083-6729
http://dx.doi.org/10.1016/bs.vh.2016.09.007

1. INTRODUCTION

Stress is a response to aversive stimuli, external or internal, capable of altering the physiological homeostasis and the ability to cope with such stressful situation is a crucial determinant of health and disease. Exposure to adverse conditions initiates a series of adaptive response organized to defend the stability of the internal environment and enhance an organism's survival. This orchestrated process usually referred to as "stress response" involves various mechanisms that allow the body to make necessary physiological and metabolic adjustments required to cope with the demands of homeostatic challenge (Gold & Chrousos, 2002). Such changes may occur on the physiological (emotional and cognitive), behavioral, and biological level (altered autonomic and neuroendocrine function). The central nervous system (CNS) plays a crucial role in the regulation of stress responses, and complex neurochemical pathways have been proposed (Carrasco & Van de Kar, 2003).

In the early 19th century, the French physiologist Claude Bernard introduced a theory suggesting that, as organisms become more independent of their surroundings, they develop more complex ways of stabilizing their internal environments to counter the changes in their external environment. The importance of adaptive mechanisms was thus recognized as the constancy of "milieu interieur," which would be the condition of a free and independent existence (Goldstein & Kopin, 2007). In the early 1900, Walter Cannon expanded this theory and coined the term "homeostasis." He demonstrated in several seminal experiments that the sympathoadrenal system was responsible for coordinating the "fight and flight" response necessary to meet external challenges (Cannon, 1929; LeMoal, 2007). Cannon was able to show that both emotional and physical disturbances triggered the same response from the organism. In addition, he proposed that there was a "critical" level of stress, in terms of magnitude and duration, against which the homeostatic mechanisms fail and the organism perishes. Cannon believed that an individual organism's susceptibility to this critical stress varied under different general conditions and during the normal and pathologic ups and down of existence in an ordinary life style (LeMoal, 2007). In 1930s Hans Selye began research on stress and thus helped in understanding the connection between stress and health (Selye, 1936). His definition of stress as being the "nonspecific response of the body to any demand upon it" was so persuasive that it persisted and remains widely used even today. By "nonspecific," Selye meant a set of shared elements of responses—regardless of the nature of the causative agent, or stressor. Selye proposed three universal stages of coping with a stressor—the "General Adaptation Syndrome"—an initial "alarm reaction," analogous to Cannon's "fight or flight" response, a stage of adaptation associated with resistance to the stressor, and eventually a stage of exhaustion and organismic death. It was later demonstrated that these changes are associated with, and to some extent result from, activation of the hypothalamic–pituitary–adrenocortical (HPA) axis. Steroids released into the circulation from adrenal cortex contribute to resistance but may also be responsible for pathological changes. Selye's concept that prolonged stress can produce physical disease and mental disorders is now widely accepted (Perdrizet, 1997). Mason criticized Selye's theory of nonspecificity. Anxiety and fear were understood as the important factors that contribute to nonspecific responses on exposure to several stressors (Mason, 1971).

More modern concepts view stress as a consciously or unconsciously sensed threat to homeostasis, in which the response has a degree of specificity, depending among other things, on the particular challenge to homeostasis, the organism's perception of the stressor and the perceived ability to cope with it. Although homeostasis suggests constancy of values for variables, ranges of acceptable values are now recognized to be decidedly inconstant (Korte, Koolhaas, Wingfield, & McEwen, 2005). There are diurnal variations in body temperature, heart rate, and blood pressure. In addition, adaptations to different stressors include alterations in acceptable levels for monitored variables. Sterling and Eyer (1988) introduced the term, "allostasis," to describe the attainment of stability by natural alterations in acceptable ranges of variables. The brain is the site at which effects of stressors are sensed and appropriate coordinated behavioral and neuroendocrine responses initiated. Adaptations involving allostasis to cope with real, simulated, or imagined challenges are determined by genetic, developmental, and previous experimental factors. While they may be effective for a short interval, over time, the alterations may have cumulative adverse effects. For instance, chronic elevation of blood pressure ensures adequate blood flow to the brain eventually lead to atherosclerosis and stroke or coronary occlusion. Risk of such adverse effects is termed "allostatic load" (McEwen, 1998). The emerging concept of allostatic load provides a means of augmenting the concept of stress as a conceptual and methodological basis for elucidating the behavioral and physiological mechanisms by which genes, early life experiences, living and working environment, interpersonal relationships, diet, exercise, sleep, and other lifestyle factors all converge to affect body chemistry, structure, and function over a life time (McEwen, 2000). Stress is an interdisciplinary topic, and understanding health consequences of stress requires an integrative approach. Merging of the homeostatic definitions of stress and distress with the concept of allostasis should provide a better understanding of the roles of stress and distress in chronic diseases and also provide a conceptual basis for the further development of scientific integrative medicine (Levine, 2005).

2. MECHANISM OF STRESS RESPONSE

Exposure to hostile conditions (usually referred to as stressors) results in a series of coordinated responses organized to enhance the probability of survival. These coordinated responses, often referred to as "stress responses," are composed of alterations in behavior, autonomic function, and the secretion of multiple hormones including adrenocorticotropin hormone (ACTH) and cortisol/corticosterone, adrenal catecholamines, oxytocin, prolactin, and renin (Gold & Chrousos, 2002). Some of the physiological changes associated with the stress response include: (i) mobilization of energy to maintain brain and muscle function; (ii) sharpened and focused attention on the perceived threat; (iii) increased cerebral perfusion rates and local cerebral glucose utilization; (iv) enhanced cardiovascular output and respiration, and redistribution of blood flow, increasing substrate and energy delivery to the brain and muscles; (v) modulation of immune function; (vi) inhibition of reproductive physiology and sexual behavior; (vii) decreased feeding and appetite. These orchestrated responses are geared to alter the internal milieu in a way that increases the probability of survival (Goligorsky, 2001).

In general, stressors can be grouped into four broad categories: (a) psychological stressors based on a learned response to the threat of an impending adverse condition (fear, anxiety, exposure to a novel or uncontrollable environment); (b) stressors that consist of a physical stimulus and have a strong psychological component (pain, foot shock, immobilization); (c) stressors which challenge cardiovascular homeostasis (hemorrhage, orthostatic stress/upright tilt, exercise, heat exposure); (d) social stressors that distributed interactions among individuals (death of partner, marital separation, unemployment, etc.). The neuroendocrine responses to stressors are considered important survival mechanisms during exposure to life threatening stimuli (Dayas, Buller, Crane, Xu, & Day, 2001; Newport & Nemeroff,

2002). In terms of duration, stressors may be either (a) acute stressors (single, intermittent, time-dependent exposure); (b) chronic or repeated stressors (continuous long-term prolonged exposure and intermittent long-term exposure).

3. STRESS, CNS, AND BEHAVIOR

The brain is the key to interpreting and responding to potentially stressful events; it is also target for the actions of stress hormones particularly glucocorticoids. Acute and traumatic inescapable stressors can have lasting consequences on brain and behavior (Joca, Ferreira, & Guimaraes, 2007). The neural correlates of the behavioral deficits have been documented in the hippocampus, amygdala, and prefrontal cortex and include aspects of structural remodeling and cell proliferation, as well as changes in amine, neuropeptide, and corticosteroid systems (Herman et al., 2003).

Exposure to stress precipitates a coordinated series of responses, which may lead to changes in brain functions. Most of the changes in brain function are mediated by stress-induced activation of neurotransmitter systems in the brain, which include neurotransmitter receptors-mediated signal transduction pathways. The signal transduction pathways are ultimately responsible for stress-induced changes in neuronal functions. Some of these changes are beneficial, however, depending upon the nature and duration of stressful stimuli; the changes in neuronal functions may be harmful, which may lead to many disorders.

The response of the brain to both acute and chronic stress must be regarded in terms of its capacity to show adaptive plasticity. Hippocampal neurons express receptors for circulating adrenal steroids and work in many laboratories has shown that hippocampus has both types of adrenal steroid receptors which mediate a variety of effects on neuronal excitability, neurochemistry, and structural plasticity (McLaughlin, Gomez, Baran, & Conrad, 2007). These effects that involve hormone-mediated effects on gene expression include the regulation of branching and length of dendrites in the pyramidal cells of Ammon's horn and the replacement of nerve cells in dentate gyrus (Sandi et al., 2003). Many of these hormone effects do not occur alone but rather in the context of ongoing neuronal activity. In particular, excitatory amino acids and NMDA receptors as well as serotonin play an important role in structural and functional changes produced in the hippocampal formation by steroid hormones. The hippocampus is also sensitive to gonadal hormones and expresses both intracellular androgen and estrogen receptors (Karst & Joels, 2003). Gonadal and adrenal hormones participate in functional and structural changes in adult life, as well as developmental events, which include sexual differentiation and influences of early stressful life experiences (Adamec, 2003).

Severe stress can produce lasting changes in emotional behavior in animals and humans. Interestingly, lasting emotional changes like those produced by severe stress are also produced by repeated seizure activity invading limbic system circuitry in humans and animals, e.g., kindling and behavioral change (Adamec & Young, 2000). Associated with these behavioral changes are decreases in mRNA expression for glucocorticoid receptors in the CA1 subfield of the dorsal hippocampus.

A large body of evidence has demonstrated that amygdala, in addition to playing a critical role in emotional learning, is central to the neural circuitry underlying stress and anxiety. Recent reports have also raised the possibility that cellular substrates for stress-induced changes in emotional behavior may reside in amygdala (Rademacher et al., 2008). The serine protease tissue plasminogen activator in the amygdala has been identified as a critical component in the sequence of molecular events linking repeated restraint stress (RS)-induced plasticity with development of anxiety-like behavior (Bennur et al., 2007). Studies indicate that the same chronic immobilization stress that causes dendritic atrophy in hippocampal CA3 pyramidal neurons can elicit dendritic hypertrophy in the basolateral amygdala. More importantly, this form of dendritic remodeling in the basolateral amygdala depends on the type of stressor: chronic immobilization stress, but not chronic unpredictable stress elicited dendritic remodeling of spiny neurons in the basolateral amygdala (Vyas, Mitra, Shankaranarayana Rao, & Chattarji, 2002). Pawlak et al. (2005) has suggested that acute episode of stress can trigger, within hours, a sequence molecular events related to structural modeling in the amygdala, which, in turn, can lead to the eventual development of anxiety. Depression is a complex disease that involves clusters of clinical symptoms. Anatomical studies have shown that in some patients clinical symptoms are accompanied by reversible changes in hippocampal volume. Such volume changes were suspected to be related to stress-induced reductions in adult proliferation and neurogenesis of dentate gyrus. Interestingly, stress-induced suppression of proliferation is prevented by antidepressant tianeptine (Czeh et al., 2001).

At present, it is unclear, whether the effects of lasting or repeated stressors on hippocampal remodeling proceed directly through MR- (mineralocorticoid receptor) and GR (glucocorticoid receptor)-mediated actions on structural and cell cycle genes in neurons, as indicated by the identification of responsive target genes. However, the corticosteroid signaling cascade is complex and involves numerous chaperones, accessory proteins, coregulators, and interacting transcription factors that permit differentiation between GR- and MR-mediated actions. This might be altered after long-lasting stress (Garcia, Steiner, Kronenberg, Bick-Sander, & Kempermann, 2004).

In addition to the corticosteroid receptors, other mediators, such as growth factors or the central monoaminergic system, have also been implicated in altered brain function after chronic stress. During chronic stress, noradrenaline and CRH mutually enhance the activity of one another in the amygdala–locus coeruleus circuitry. Chronic stress, particularly when coping is impossible, attenuates the rewarding value of stimulating the mesocortical dopaminergic system and inhibits dopamine release in various terminal areas, including the hypothalamus (Holsboer, 2000).

The monoaminergic system and the corticosteroid receptors also interact to establish changes in the brain. This probably occurs predominantly through brain MRs, as GRs are relatively resistant to the effects of chronic stress. Moreover, the manipulation of 5-HT, noradrenergic, and peptidergic inputs to the hippocampus enhances the expression of MR in particular. This might have profound implications for the integrity of the dentate gyrus, as several approaches have identified the MR as a crucial factor in the control of dentate gyrus neurogenesis and cell death. Therefore, induction of MRs by antidepressants might mediate the recently reported drug effects on neurogenesis (Sabban & Kvetnansky, 2001). Although most studies use animal models of depression and behavioral response to antidepressants, few existing studies, including female subjects, demonstrate that males and females can respond either differently or even in opposite directions (McEwen, 2003). However, more studies are required to understand the behavioral differences between males and females in response to acute and chronic stressors.

4. STRESS AND ANXIETY

Clinical and experimental studies have shown that a clear association exists between stress and anxiety. In fact, a variety of stressful situations, both emotional and environmental, can precipitate anxiety disorders, and animal experiments have shown that a positive correlation actually exists. Anxiety is also sometimes referred to as the psychophysiological signal of the stress response. For example, neuroanatomical data have shown that the limbic brain areas like amygdala, prefrontal cortex, hippocampus, bed nucleus of stria terminalis, etc., through their complex neural networking, regulate both stress-related responses including anxiety. Anxiety and depression are by far the commonest psychopathologies seen after chronic stress exposures. Anxiety disorders like posttraumatic stress disorder (PTSD) are precipitated by stressful and traumatic life events. Neuropharmacological data have also indicated that the neurochemical mechanisms involved in both situations are very similar (Lupien, Maheu, Tu, Fiocco, & Schramek, 2007; Shin & Liberzon, 2010).

Experimental studies have shown that a variety of stressors are known to elicit anxiogenic responses which depend on the duration and intensity of the aversive stimuli. Such stress responses impact the behavioral, endocrinological, and immune system and are differentially influenced by the classical antistress agents or adaptogens. Benzodiazepine drugs (which are the prototype adaptogens) such as diazepam, chlordiazepoxide, alprazolam, and midazolam have been used for many years as anxiolytic drugs in humans and experimental animals. These drugs act by enhancing GABA-ergic neurotransmission through an allosteric interaction at the benzodiazepine–$GABA_A$–barbiturate chloride ionophore receptor complex and are effective in inhibiting the secretion of ACTH to several stressors (Groenink et al., 1996). However, chlordiazepoxide and midazolam fail to inhibit the effects of conditioned fear stress on the secretion of renin (Van de Kar et al., 1985), suggesting that central GABA-ergic neurons do not play a significant role in the neural circuits that mediate the effects of stressors on the secretion of renin. Diazepam inhibits the ACTH but not plasma catecholamine responses to nonhypotensive head-up tilt (Matzen, Secher, Knigge, Bach, & Warberg, 1993). Furthermore, diazepam does not inhibit any hormone responses to hypotensive head-up tilt (Matzen et al., 1993). Interestingly, benzodiazepines do not affect renin secretion, although they inhibit ACTH, oxytocin, and prolactin secretion. The ineffective role of benzodiazepines in reducing the renin response to stress further confirms that brain circuit and/or receptor mechanisms that mediate the effects of stressors on the secretion of ACTH, oxytocin, and prolactin are different from those who mediated the secretion of renin. It is also becoming apparent that stress-induced anxiety-related responses are also not totally attributable dysfunctions in the BZD–GABA receptor moiety and that other complexly interactive mechanisms could be involved.

5. STRESS AND VARIOUS ANXIOLYTIC DRUGS

The effects of anxiolytic drugs, CRF receptor antagonists, agonists and antagonists for serotonin, opioid, histamine, and adrenergic receptors on the neuroendocrine responses to stressors. The majority of the cited studies are preclinical studies that focus on psychological stressors and their effects on laboratory animals, and there is some information about human studies. New information is emerging about drugs affecting peptide neurotransmitters (CRF, CCK, VIP, neuropeptide Y, substance P, and others) (Gully et al., 2002). However, their role in stress-induced secretion of hormones remains to be examined. Most of these new studies only determined the ACTH or corticosterone response to stress, and there is no information about other stress hormones. Among the various classes of drugs those interacting with CRF1, neuropeptides, 5-HT_{1A}, $GABA_A$, histaminergic H_2 receptors, and α_1-adrenoceptors appear to interact directly with the neural circuits involved in stress-induced hormone release (Van de Kar & Blair, 1999). Promising results have been obtained using the new class of peptide and nonpeptide antagonists of CRF receptors. The neuroendocrine information for these drugs is still incomplete and most of the behavioral studies just used acute stressors. However, although these drugs will require a thorough examination, they are a new class of antidepressant and anxiolytic drugs that offer new targets on specific receptors involved in the onset of the stress response. All the antagonists of the CRF1 receptor inhibit the ACTH/corticosterone response to stressors and have a good anxiolytic score (Maciag et al., 2002). 5-HT_{1A} receptor agonists, buspirone and ipsapirone, inhibit the effects of stressors on the secretion of prolactin and corticosterone (Matzen, 1995) and also inhibit the effect of conditioned fear stress on the secretion of renin (Matheson et al., 1997).

Studies employing serotonergic receptor antagonists indicate that 5-HT_2 and 5-HT_3 receptors may also mediate the stress-induced increases in hormone release (Van de Kar et al., 2001). Central histaminergic neurons participate in mediating the effects of stressors on the secretion of ACTH, prolactin, adrenal catecholamines, and renin (Matheson et al., 1997). Ranitidine and cimetidine (histamine H_2 receptor antagonists) inhibit the effects of immobilization on prolactin and renin release, suggesting that H_2 receptors mediate the effects of immobilization stress on these hormones (Van de Kar & Blair, 1999). In general, the influence of histamine is indirect and is mediated by intrahypothalamic neurons (such as inhibition of tuberoinfundibular dopaminergic neurons) or by altering the function of incoming nerve terminals, such as serotonergic inputs (Knigge & Warberg, 1991).

During exposure to stress, the release of catecholamines in the hypothalamus increases ACTH secretion by activation of α_1-adrenoceptors, while activation of α_2-adrenoceptors inhibits this effect (Grino, Paulmyer-Lacroix, Faudon, Renard, & Anglade, 1994). Injection of the α_1-adrenoceptor antagonist prazosin into the central amygdala reduces the corticosterone response to a stressful combination of ether and light flashes (Feldman & Weidenfeld, 1996), suggesting that catecholamines in other parts of the limbic circuit also play a role. These data are consistent with the fact that catecholaminergic neurons send collaterals to both the central amygdala and paraventricular hypothalamic nucleus (Petrov, Krukoff, & Jhamandas, 1993) and that catecholaminergic innervation of the paraventricular nucleus is required for full expression of the ACTH response to several stressors (Gaillet, Lachuer, Malaval, Assenmacher, & Szafarczyk, 1991).

Antagonists for β-adrenoceptors reduce basal plasma renin levels (Blair & Gengo, 1995) and attenuate the renin response to a wide range of stressors both in humans and in experimental animals (Keeton & Campbell, 1980) and do not reduce ACTH stress response in hypoglycemic stress and conditioned stress (Van de Kar et al., 1985). The β-adrenoceptors that mediate the renin response to conditioned fear are proposed to reside within the hypothalamus (Morton, Johnson, & Van de Kar, 1995). The stressors that stimulate the secretion of renin include conditioned fear (Van de Kar et al., 1985), immobilization stress (Matzen, Knigge, & Warberg, 1990), head-up tilt (Golin, Gotoh, Said, & Ganong, 1988), hemorrhage (Blair, Hisa, Sladek, Radke, & Gengo, 1991), water deprivation (Chen et al., 1988), and heat exposure (Escourrou, Freund, Rowell, & Johnson, 1982). β-Adrenoceptor antagonists are less effective in attenuating the renin response to hemorrhage (Roarty & Raff, 1988), water deprivation (Blair, Woolf, & Felten, 1997), or heat exposure (Groza, Ardeleanu, & Lazar, 1981) in rats than in human subjects, nonhuman primates, or dogs. On the other hand, direct injection of propranolol or sotalol into the hypothalamic paraventricular nucleus reduces the effect of ether and conditioned fear, respectively, on the secretion of corticosterone (Richardson Morton et al., 1990). While the local effects of these drugs may be due to blockade of β-adrenoceptors, the receptor specificity of their effects on the paraventricular nucleus requires confirmation.

Recent studies have shown that nitric oxide (NO) may play a critical role in the regulation of anxiety-like behavior. NO is a unique gasotransmitter widely distributed in the brain, and its diverse role in CNS pathophysiology has been suggested. Recent studies have shown that NO and its interactions with oxidative stress markers may play a crucial role in the regulation of stress effects. Our studies, showing that both acute and chronic stress-induced anxiogenesis, were attenuated by NO mimetics and aggravated by NO synthase inhibitors. These effects were accompanied by predictable changes in brain NO metabolites and oxidative stress markers (Gulati, Ray, Masood, & Vijayan, 2006). Cyclic GMP-mediated pathways are probably not involved as the phosphodiesterase-5 inhibitor, sildenafil, was without any significant effect on stress-induced anxiety-like behavior. Studies also showed that NO regulated gender-based differences in stress-induced anxiogenesis with male rats having lower brain NO and higher anxiety scores as compared to females. Blockade of estrogen pathways in female rats resulted in similar anxiety-like behavior as seen in males.

Centrally acting drugs like diazepam and morphine, which are known for their antianxiety effects, also elevated brain levels of NO metabolites in stressed rats—suggesting the involvement of NO in the anxiolytic effects of these therapeutically useful agents. More recently, a link between heat shock proteins and antistress effects of NO has also been proposed (Gulati, Joshi, & Ray, 2015).

6. STRESS AND IMMUNITY

Studies on stress and immunity in both experimental animals and humans suggest that psychological challenges are capable of modifying various features of immune response (MacQueen & Bienenstock, 2006; Padgett & Glaser, 2003). Studies using animal models have shown that stress diminishes vaccine responses, exacerbates viral and bacterial pathogenesis, and slows wound healing and alters autoimmune diseases (Yang & Glaser, 2000). These studies have demonstrated that stress hormones inhibit the trafficking of neutrophils, macrophages, antigen-presenting cells, natural killer (NK) cells, T and B lymphocytes, downregulate the production of cytokines necessary for the generation of adaptive immune responses, and impair effector functions of macrophages, NK cells, and lymphocytes (Glaser, Rabin, Chesney, Cohen, & Natelson, 1999; Yang & Glaser, 2002a).

Acute vs chronic stressors may have differential effects on immunity. Dhabhar and McEwen (1997) have observed that the effects of acute vs chronic-RS of rats on delayed type hypersensitivity (DTH) responses appear to be in opposite directions. That is, following a single session of restraint prior to challenge, DTH responses in skin were enhanced, but following chronic-restraint sessions, DTH responses were suppressed compared to control mice. The adrenal hormones corticosterone and epinephrine may mediate these effects. It has been proposed that acute stress may induce "redeployment" of peripheral blood lymphocytes to the skin, and this would be an adaptive response during the "flight and fight" response (Glaser et al., 1999; Yang & Glaser, 2002b). Mitogen responses by rat lymphocytes were also suppressed following a single session of intermittent but not continuous footshock; this suppression was dependent upon production of beta-endorphin and was independent of elevated levels of corticosterone, which were observed with both intermittent and continuous shock. Plaque forming cell responses and serum antibody titers to sheep red blood cell immunization was found to be suppressed in mice following a single footshock session. In addition, a number of investigators have examined the effects of acute stressors on NK cell cytotoxicity. NK cells spontaneously lyse virally infected cells and tumor cells, and their function can be augmented by IFN-γ, and the effects of acute stress on NK cell function are almost uniformly suppressive (Dhabhar, Miller, McEwen, & Spencer, 1995; Petticrew & Hunter, 1999).

Generally, chronic stress is associated with suppression of NK cell responses and antigen-specific immune responses in rodent and nonhuman primate models, but also in humans. Stress-related impairments in immune function would be expected to place individuals at risk for decreased immune responses following viral infections (Brosschot et al., 1998). It has been shown that following laboratory infection with rhinovirus, individuals with higher levels of perceived stress had increased rates of infection. More recently, it has been reported that individuals with low numbers of NK cells were more likely to experience self-reported upper respiratory infection during high stress rather than low stress periods (Keeney et al., 2006).

The balance between Th1 and Th2 cell function or cytokine production may be altered by stressor administration (Kim & Maes, 2003). Exposure to the odors produced by stressed BALB/c mice results in a highly significant elevation of keyhole limpet hemocyanin (KLH)-specific production of the Th2 cytokine IL-4, with no change in production of IL-2 or IFN-γ. Interestingly, this increase in IL-4 production was associated with increased serum IgM and IgG anti-KLH antibody (Calcagni & Elenkov, 2006). In addition, lymphocytes from rats subjected to inescapable tail shock produced less IFN-γ following stimulation with T-cell mitogen. In contrast, however, in another study it was observed that chronic-RS significantly diminished both the Th1- (IL-2 and IFN-γ) and the Th2-derived (IL-10) cytokine responses following infection of C57B1/6 mice with influenza virus. Some of the observed differential effects of stressors on immune parameters may be due to differential neuroendocrine responses that trigger changes in one, but not necessarily both, types of Th cells (Kiank et al., 2006; Reiche, Nunes, & Morimoto, 2004).

Glucocorticoids appear to be crucial in controlling the individual repertoire of immunological responses, including the direction and magnitude of immune reactions (Wilckens & De Rijk, 1997). If optimally balanced, glucocorticoid-dependent functions will contribute to a resolution of infection, trauma, or other immunologically related stressors (Jessop & Harbuz, 2003). However, disruption or malfunction of these dynamic interactions may result in fatal outcome of acute inflammation or may predispose for autoimmunity or atopic reactions. Glucocorticoid receptors expressed on a variety of immune cells bind cortisol and interfere with the function of NF-κB, which regulates the activity of cytokine producing immune cells (Riccardi, Bruscoli, & Migliorati, 2002). Adrenergic receptors bind epinephrine and norepinephrine and activate the c-AMP response element binding protein, inducing transcription of genes encoding for a variety of cytokines. The changes in gene expression mediated by glucocorticoid hormones and catecholamines can dysregulate immune function (Leonard, 2006).

The existence of an association between the brain and immunity has been documented. Data show that the nervous and immune systems communicate with one another to maintain immune homeostasis. A bidirectional circuit exists between the CNS and the immune system, since activation of the immune system results in the elaboration of cytokines and inflammatory mediators; these mediators induce hypothalamic CRF, which stimulates the release of the same immunosuppressive molecules that mediate the response to stress (Haddad, Saade, & Safieh-Garabedian, 2002). It has been reported that the antibody response arouses the metabolic variations of ACh, NA, 5-HT, and DA in the central and peripheral nervous systems, and then, these alterations can in turn influence immune function through neurotransmitter relevant receptors present on the immunocytes (Huang, Pang, Karalis, & Theoharides, 2003).

7. STRESS, ANXIETY, AND IMMUNOMODULATION

The past 20 years has witnessed the emergence of the field of psychoneuroimmunology. This field deals with the influence of the CNS on the immune system, or more specifically, whether and how thoughts and emotions affect immune function. Studies have concentrated, for the most part, on the effects of stress on the immune system. Stress is defined as a state of disharmony or threatened homeostasis provoked by a psychological, environmental, or physiologic stressor. It has also become apparent from these studies that the immune system can influence the CNS, and thus, a circuit exists between these two systems. Regulatory molecules or cytokines elaborated from activated immune cells evoke a CNS response which, in turn, affects the immune system. It is likely, therefore, that the brain is normally part of the immunoregulatory network. Investigators postulate that the events which occur in the brain, specifically at the HPA axis, in response to stress are similar to those which occur in response to internal immune system events; in both instances, the effect of HPA axis stimulation is a downregulation of immune system function.

Evidence for CNS–immune system interactions has been derived from several studies. Electrolytic lesions of certain areas of the hypothalamus produce either enhancement or inhibition of various immune functions. These changes, both facilitatory and inhibitory, are prevented by hypophysectomy, indicating that pituitary function mediates the hypothalamic effect. Lesioning of the pituitary gland can also inhibit or stimulate certain immune system functions. The facts that both the hypothalamus and the pituitary gland can produce both immune-enhancing and immune suppressive effects and, furthermore, that the pituitary is under hypothalamic control provide a mechanism for tightly regulating the immune response in both magnitude and duration. Further evidence for a CNS–immune system interaction is derived from the observation that certain neurotransmitters, neuropeptides, and neurohormones affect immune function both in vivo and in vitro; receptors for these molecules are present on lymphocytes and/or macrophages. Cytokines, in turn, elaborated from activated immune cells can alter HPA axis function, thereby closing a negative-feedback loop. In other studies, the sympathetic nervous system (SNS) has been found to innervate both primary (thymus, bone marrow) and secondary (spleen, lymph node, Peyer's patches) lymphoid tissues; this indicates that states of arousal can be transmitted neurogenically to lymphoid tissue. In addition, a number of studies indicate that the immune system can be conditioned. For example, conditioned immunosuppression occurs when a sensory stimulus is paired with an immunosuppressive drug; subsequent exposure to the sensory (conditioned) stimulus results in immunosuppression. Thus, a substantial body of evidence exists for bidirectional interactions between the CNS and the immune system (Black, 1994).

Psychoneuroimmunology (PNI), the study of interactions among behavioral, neural, and endocrine, and immune processes, coalesced as an interdisciplinary field of study in the late 1970s. Some of the early research that was critical in establishing neuroanatomical, neurochemical, and neuroendocrine pathways and functional relationships between the brain and the immune system are outlined here. These and subsequent studies led to the general acknowledgment that the nervous and immune systems are components of an integrated system of adaptive processes and that immunoregulatory processes can no longer be studied as the independent activity of an autonomous immune system. This paradigm shift in the study of immunoregulatory processes and the elaboration of the mechanisms underlying behaviorally induced alterations of immune function promise a better understanding and a new appreciation of the multidetermined etiology of pathophysiological states (Ader, 2000).

Psychological and social factors are thought to influence disease processes via two main mechanisms, psychosocial processes and health behaviors. Psychosocial processes include factors that affect interpretation of and response to life events and stressors, such as mental health and mood factors, personality characteristics, and resources such as social relationships. Health behaviors such as exercise, nutrition, and smoking serve as indirect pathways by which psychosocial processes can influence health, as they may be strongly influenced by factors such as mood.

Interactions between psychosocial and immunologic factors are relevant to a variety of diseases including inflammatory diseases, cardiovascular disease, infectious diseases, cancer, diabetes, osteoporosis, muscle wasting, and multiple sclerosis, and processes such as wound healing, surgical recovery, and efficacy of vaccination. There are many nonimmune interactive pathways as well; however, we will limit this review to pathways mediated by immune and neuroendocrine processes (Lutgendorf & Costanzo, 2003).

PNI has thus emerged as the scientific field of study investigating the link between bidirectional communications among the nervous system, the endocrine and immune system as well as the implications of these linkages for physical and mental health, thereby focusing on the measurable interaction between psychological and physiological processes. The CNS and the immune system are the two major adaptive entities of the body. Although the immune system has often been regarded as autonomous, the last two to three decades have provided strong evidence that the CNS receives messages from the immune system. On the other hand, there is substantial evidence for a brain-derived information outflow to the immune system.

Thus, the brain and the immune system are involved in functionally relevant cross talk, with homeostasis being the main function. These two systems communicate through intricate chemical messengers that are able to breach their independent and often sequestered anatomical locations. The CNS is without "classical" lymphatic drainage, meaning that in some ways it is devoid of the immune surveillance that is available for the rest of the body. In health there are mechanisms to exclude potentially destructive lymphoid cells from the brain, spinal cord, and peripheral nerves. However, multiple anatomic and physiological connections exist between the CNS and the immune system, including "hardwiring" of the autonomic nervous system. Multiple chemical messengers, ranging from classical substances like biogenic amines (noradrenaline, serotonin, histamine, etc.), neuropeptides like corticotrophin-releasing hormone (CRH), endogenous opioids, TRH, neurotensin, and also large proteins including cytokines, have been implicated. Most recently, our studies have shown that the atypical yet unique gasotransmitter, NO may play a crucial role in such CNS–immune interactions. Thus, these two elaborate systems relevant for sensing danger and mounting a counter attack to these threats are inextricably linked (Ziemssen & Kern, 2007).

Mast cells are a heterogeneous population of granulocytic cells of the immune system. They contain numerous mediators, including neurotransmitters, cytokines, chemokines, and lipid-derived factors. Mast cells in the brain are constitutively active, releasing their contents by means of piecemeal or anaphylactic degranulation. Additionally, their activity is increased by a wide range of stimuli including immune and nonimmune signals, such as hormones like corticotrophin-releasing hormone, and various neuropeptides like Substance P and neurotensin. Of mast cells over 50 mediators, some are synthesized upon activation (e.g., substance P, somatostatin, cytokines), while others are preformed and stored in granules, allow for very fast release (e.g., serotonin and histamine). Due to their ability to migrate, they can serve as "single-cell glands" delivering mediators "on demand" and influencing neuronal activity. Mast cells can act via autocrine and paracrine mechanisms, and their secretions can reach a large spatial volume. Their granule remnants can even be acquired by neurons through endocytosis. The residence of mast cells in meninges and perivascular locations on the brain side of the blood–brain barrier, primarily in thalamic and hippocampal regions, indicates that they are strategically situated to initiate neural and vascular responses. However, the function of mast cells in the brain is mostly unknown. A role in normal physiology and behavior was suggested as the brain population of mast cells fluctuated with endocrine status and changes after stress and handling. Not surprisingly, there were individual differences in the number of brain mast cells within species, perhaps associated with behavioral and/or experiential differences. These properties indicated that mast cells could regulate neural networks responsible for behavior.

Studies showed that mast cell-deficient mice exhibited greater anxiety-related behavior as compared to nondeficient mice in the elevated plus maze and open field tests. Additionally, it was also shown that blockade of brain (and not peripheral) mast cells increased anxiety-like behavior. These experiments suggested that mast cells could play a crucial role in the modulation of anxiety-like behavior and give further evidence for the behavioral importance of neuroimmune links (Nautiyal, Ribeiro, Pfaff, & Silver, 2008).

PTSD is a severe anxiety disorder that develops after a stressful/traumatic event. Clinical and experimental data have shown that it is associated with neurodegenerative pathology and shrinkage of the hippocampus. The anatomical changes correlate well with cognitive deficits observed in such cases. Stress-induced sustained activation of the HPA axis has been implicated. Stress and glucocorticoids increase glutamate release, which is recognized as an important mediator of glucocorticoid-induced neurotoxicity.

Recent preclinical studies indicate that glutamate and NO play a causal role in such anxiety-related behaviors. Because of the prominent role of NO in neuronal toxicity, cellular memory processes, and as a neuromodulator, nitrergic pathways may have an important role in stress-related hippocampal degenerative pathology and cognitive deficits seen in patients with PTSD (Oosthuizen, Wegener, & Harvey, 2005).

Functional interactions between the SNS and HPA axis are required for appropriate responses to stress, anxiety, and fear (Vermetten & Bremner, 2002), and in PTSD, there is an evidence for dysregulation of both of these homeostasis-directed systems.

Several sequelae of PTSD are partially attributed to glucocorticoid-induced neuronal loss in the hippocampus and amygdala. Glucocorticoids and adrenergic agents cause both immediate and late sequelae and are considered from the perspective of their actions on the expression of cytokines as well as some of their physiological and psychological effects. A shift in immune system balance from Th1 to Th2 dominance is thought to result from the actions of both molecular groups. The secretion of glucocorticoids and adrenergic agents is commonly induced by trauma or stress, and synergy between these two parallel but separate pathways can produce long- and short-term sequelae in individuals with PTSD. Potential therapies are suggested, and older therapies that involve the early effects of adrenergics or glucocorticoids are reviewed for their control of acute symptoms. These therapies may also be useful for acute flashback therapy. Timely and more precise glucocorticoid and adrenergic control is recommended for maintaining these molecular groups within acceptable homeostatic limits and thus managing immune and brain sequelae.

Autologous adoptive immune therapy may also offer utility. In the traumatized PTSD brain, neurons in the hippocampus, amygdala, and other parts of the brain are destroyed by glucocorticoids, and neurochemical dysregulation is reported. Adrenergic agents and glucocorticoids, along with serotonin and other moieties, affect immune, chemical, and structural responses seen after PTSD. Interactions between the immune system, the HPA axis, and other related molecules have been implicated in this stress disorder (Griffin, Charron, & Al-Daccak, 2014).

A clinical study demonstrated that gender differences in PTSD were also linked to immune mechanisms. This study used gene microarray technology to determine which genes were activated in the subjects' monocytes, which are immune cells that regularly cross the barrier between the bloodstream and the brain, and thus give a broad picture of immune reaction in both the body and brain. A differential gender-based pattern in monocyte gene expression was observed, and where a suppressed monocyte gene expression was seen in males, females showed relatively balanced pattern of expression. It was inferred that men and women had markedly different immune system responses to PTSD (Neylan et al., 2011).

Recent experimental studies from our laboratory have investigated the neuropharmacological basis of brain–immune interactions during stress-induced anxiety. RS, which is a well-documented emotional stressor, resulted in anxiety modulation and also induced differential effects of adaptive immune responses. These effects were dependent on the duration/intensity and frequency of the stressor. In immunized rats, brain catecholamine depletion resulted in an exaggerated anxiogenic response with concurrent humoral immune suppression. On the other hand, drugs with well-known anxiolytic effects like benzodiazepines and morphine, enhanced humoral immune responses to the antigen, and drugs like flumazenil and naltrexone attenuated these effects. These studies showed that centrally acting agents were modulating behavior and immune responses in a receptor-specific manner (Ray, Gulati, Mediratta, & Sen, 1993; Ray, Mediratta, Puri, & Sen, 1991; Ray, Mediratta, & Sen, 1992). Our studies further showed that, the amygdala, which plays a crucial role in anxiety, also had a regulatory influence on immune responses. Electrolytic lesions in the amygdala resulted in anxiolysis, reduced gastric ulcerogenesis and immune modulation in response to stress. Intraamygdalar injections of DSP-4, a specific brain norepinephrine depletor, facilitated the stress-induced anxiogenic response, while suppressing the adaptive immune responses to the antigen. Intraamygdala applications of the benzodiazepine, midazolam, showed opposite effects on both stress-induced anxiety and immunomodulation (unpublished data). In another study to assess neuroimmune interactions during stress, RS-induced immune suppression also studied as a function of gastric ulcerogenesis (which is also an anxiety-related response and attenuated by anxiolytics), and drugs like benzodiazepines and morphine attenuated both responses. Modulatory effects on both stress responses were also seen with neuropeptides like TRH and neurotensin, and complex neurotransmitter interactions in such brain–immune interactions were proposed (Ray, 1994, PhD thesis). In a recent study, it was shown that such neuroimmune interactions during anxiety were also under the regulatory influence of the gasotransmitter, NO. Stress-induced anxiogenesis and immune suppression were associated with lowered levels of brain and blood NO metabolites. Further, prior administration of NO mimetics attenuated both behavioral and immune responses, whereas, the NO synthase inhibitor further aggravated both responses. Additionally, these stress responses were gender specific, with male rats showing greater intensity of both responses to the stressor. Interactions of NO with reactive oxygen species were also proposed (Chakraboti, 2007). All these data (preclinical and clinical) are strongly suggestive of a bidirectional interaction between brain and immune systems during stress and anxiety.

Taken together, anxiety is a complex psychosomatic disorder with both peripheral and central components playing significant roles in its pathogenesis. Though a variety of pharmacological agents are available for treatment, most of them have their shortcomings. The concept of the involvement of the brain–immune axis during anxiogenesis may provide new leads for drug development in this neuropsychiatric disorder.

REFERENCES

Adamec, R. E. (2003). Stress effects on limbic function and behavior. *Progress in Neuro-Psychopharmacology and Biological Psychiatry*, *27*, 1173–1175.

Adamec, R., & Young, B. (2000). Neuroplasticity in specific limbic system circuits may mediate specific kindling induced changes in animal affect-implications for understanding anxiety associated with epilepsy. *Neuroscience and Biobehavioral Reviews*, *24*, 705–723.

Ader, R. (2000). On the development of psychoneuroimmunology. *European Journal of Pharmacology*, *405*, 167–176.

Bennur, S., Shankaranarayana Rao, B. S., Pawlak, R., Strickland, S., McEwen, B. S., & Chattarji, S. (2007). Stress-induced spine loss in the medial amygdala is mediated by tissue-plasminogen activator. *Neuroscience*, *144*, 8–16.

Black, P. H. (1994). Central nervous system-immune system interactions: Psychoneuroendocrinology of stress and its immune consequences. *Antimicrobial Agents and Chemotherapy*, *38*, 1–6.

Blair, M. L., & Gengo, F. M. (1995). Beta-adrenergic control of renin in sodium-deprived conscious dogs: Renal versus extrarenal location. *Canadian Journal of Physiology and Pharmacology*, *73*, 1198–1202.

Blair, M. L., Hisa, H., Sladek, C. D., Radke, K. J., & Gengo, F. M. (1991). Dual adrenergic control of renin during nonhypotensive hemorrhage in conscious dogs. *The American Journal of Physiology*, *260*, E910–E919.

Blair, M. L., Woolf, P. D., & Felten, S. Y. (1997). Sympathetic activation cannot fully account for increased plasma renin levels during water deprivation. *The American Journal of Physiology*, *272*, R1197–R1203.

Brosschot, J. F., Godaert, G. L., Benschop, R. J., Olff, M., Ballieux, R. E., & Heijnen, C. J. (1998). Experimental stress and immunological reactivity: A closer look at perceived uncontrollability. *Psychosomatic Medicine*, *60*, 359–361.

Calcagni, E., & Elenkov, I. (2006). Stress system activity, innate and T helper cytokines, and susceptibility to immune-related diseases. *Annals of the New York Academy of Sciences*, *1069*, 62–76.

Cannon, W. B. (1929). Organization for physiological homeostasis. *Physiological Reviews*, *9*, 399–431.

Carrasco, G. A., & Van de Kar, L. D. (2003). Neuroendocrine pharmacology of stress. *European Journal of Pharmacology*, *463*, 235–272.

Chakraboti, A. (2007). *Studies to explore gender related differences in stress responses with special emphasis on the role of nitric oxide.* Ph.D (Pharmacology) thesis University of Delhi.

Chen, Y. H., Hisa, H., Radke, K. J., Izzo, J. L., Jr., Sladek, C. D., & Blair, M. L. (1988). Adrenergic control of renin in euhydrated and water-deprived conscious dogs. *The American Journal of Physiology*, *255*, E793–E800.

Czeh, B., Michaelis, T., Watanabe, T., Frahm, J., de Biurrun, G., van Kampen, M., et al. (2001). Stress-induced changes in cerebral metabolites, hippocampal volume, and cell proliferation are prevented by antidepressant treatment with tianeptine. *Proceedings of the National Academy of Sciences of the United States of America*, *98*, 12796–12801.

Dayas, C. V., Buller, K. M., Crane, J. W., Xu, Y., & Day, T. A. (2001). Stressor categorization: Acute physical and psychological stressors elicit distinctive recruitment patterns in the amygdala and in medullary noradrenergic cell groups. *The European Journal of Neuroscience*, *14*, 1143–1152.

Dhabhar, F. S., & McEwen, B. S. (1997). Acute stress enhances while chronic stress suppresses cell-mediated immunity in vivo: A potential role for leukocyte trafficking. *Brain, Behavior, and Immunity*, *11*, 286–306.

Dhabhar, F. S., Miller, A. H., McEwen, B. S., & Spencer, R. L. (1995). Effects of stress on immune cell distribution. Dynamics and hormonal mechanisms. *Journal of Immunology, 154*, 5511–5527.

Escourrou, P., Freund, P. R., Rowell, L. B., & Johnson, D. G. (1982). Splanchnic vasoconstriction in heat-stressed men: Role of renin-angiotensin system. *Journal of Applied Physiology: Respiratory, Environmental and Exercise Physiology, 52*, 1438–1443.

Feldman, S., & Weidenfeld, J. (1996). Involvement of amygdalar alpha adrenoceptors in hypothalamo-pituitary-adrenocortical responses. *Neuroreport, 7*, 3055–3057.

Gaillet, S., Lachuer, J., Malaval, F., Assenmacher, I., & Szafarczyk, A. (1991). The involvement of noradrenergic ascending pathways in the stress-induced activation of ACTH and corticosterone secretions is dependent on the nature of stressors. *Experimental Brain Research, 87*, 173–180.

Garcia, A., Steiner, B., Kronenberg, G., Bick-Sander, A., & Kempermann, G. (2004). Age-dependent expression of glucocorticoid- and mineralocorticoid receptors on neural precursor cell populations in the adult murine hippocampus. *Aging Cell, 3*, 363–371.

Glaser, R., Rabin, B., Chesney, M., Cohen, S., & Natelson, B. (1999). Stress-induced immunomodulation: Implications for infectious diseases? *JAMA, 281*, 2268–2270.

Gold, P. W., & Chrousos, G. P. (2002). Organization of the stress system and its dysregulation in melancholic and atypical depression: High vs low CRH/NE states. *Molecular Psychiatry, 7*, 254–275.

Goldstein, D. S., & Kopin, I. J. (2007). Evolution of concepts of stress. *Stress, 10*, 109–120.

Goligorsky, M. S. (2001). The concept of cellular "fight-or-flight" reaction to stress. *American Journal of Physiology. Renal Physiology, 280*, F551–F561.

Golin, R. M., Gotoh, E., Said, S. I., & Ganong, W. F. (1988). Pharmacological evidence that the sympathetic nervous system mediates the increase in renin secretion produced by immobilization and head-up tilt in rats. *Neuropharmacology, 27*, 1209–1213.

Griffin, G. D., Charron, D., & Al-Daccak, R. (2014). Post-traumatic stress disorder: Revisiting adrenergics, glucocorticoids, immune system effects and homeostasis. *Clinical and Translational Immunology, 3*, e27.

Grino, M., Paulmyer-Lacroix, O., Faudon, M., Renard, M., & Anglade, G. (1994). Blockade of alpha 2-adrenoceptors stimulates basal and stress-induced adrenocorticotropin secretion in the developing rat through a central mechanism independent from corticotropin-releasing factor and arginine vasopressin. *Endocrinology, 135*, 2549–2557.

Groenink, J., Spijker, J., van den Herik-Oudijk, I. E., Boeije, L., Rook, G., Aarden, L., et al. (1996). On the interaction between agalactosyl IgG and Fc gamma receptors. *European Journal of Immunology, 26*, 1404–1407.

Groza, P., Ardeleanu, S., & Lazar, I. (1981). Plasma renin activity in hyperthermic rats. *Physiologie, 18*, 83–88.

Gulati, K., Joshi, J. C., & Ray, A. (2015). Recent advances in stress research: Focus on nitric oxide. *European Journal of Pharmacology, 765*, 406–414.

Gulati, K., Ray, A., Masood, A., & Vijayan, V. K. (2006). Involvement of nitric oxide (NO) in the regulation of stress susceptibility and adaptation in rats. *Indian Journal of Experimental Biology, 44*, 816–820.

Gully, D., Geslin, M., Serva, L., Fontaine, E., Roger, P., Lair, C., et al. (2002). 4-(2-Chloro-4-methoxy-5-methylphenyl)-N-[(1S)-2-cyclopropyl-1-(3-fluoro-4-methylphenyl)ethyl] 5-methyl-N-(2-propynyl)-1,3-thiazol-2-amine hydrochloride (SSR125543A): A potent and selective corticotrophin-releasing factor(1) receptor antagonist. I. Biochemical and pharmacological characterization. *The Journal of Pharmacology and Experimental Therapeutics, 301*, 322–332.

Haddad, J. J., Saade, N. E., & Safieh-Garabedian, B. (2002). Cytokines and neuro-immune-endocrine interactions: A role for the hypothalamic-pituitary-adrenal revolving axis. *Journal of Neuroimmunology, 133*, 1–19.
Herman, J. P., Figueiredo, H., Mueller, N. K., Ulrich-Lai, Y., Ostrander, M. M., Choi, D. C., et al. (2003). Central mechanisms of stress integration: Hierarchical circuitry controlling hypothalamo-pituitary-adrenocortical responsiveness. *Frontiers in Neuroendocrinology, 24*, 151–180.
Holsboer, F. (2000). The corticosteroid receptor hypothesis of depression. *Neuropsychopharmacology, 23*, 477–501.
Huang, M., Pang, X., Karalis, K., & Theoharides, T. C. (2003). Stress-induced interleukin-6 release in mice is mast cell-dependent and more pronounced in Apolipoprotein E knockout mice. *Cardiovascular Research, 59*, 241–249.
Jessop, D. S., & Harbuz, M. S. (2003). Brain-immune interactions in stress—The impact of hormones on disease. *Stress, 6*, 71–72.
Joca, S. R., Ferreira, F. R., & Guimaraes, F. S. (2007). Modulation of stress consequences by hippocampal monoaminergic, glutamatergic and nitrergic neurotransmitter systems. *Stress, 10*, 227–249.
Karst, H., & Joels, M. (2003). Effect of chronic stress on synaptic currents in rat hippocampal dentate gyrus neurons. *Journal of Neurophysiology, 89*, 625–633.
Keeney, A., Jessop, D. S., Harbuz, M. S., Marsden, C. A., Hogg, S., & Blackburn-Munro, R. E. (2006). Differential effects of acute and chronic social defeat stress on hypothalamic-pituitary-adrenal axis function and hippocampal serotonin release in mice. *Journal of Neuroendocrinology, 18*, 330–338.
Keeton, T. K., & Campbell, W. B. (1980). The pharmacologic alteration of renin release. *Pharmacological Reviews, 32*, 81–227.
Kiank, C., Holtfreter, B., Starke, A., Mundt, A., Wilke, C., & Schutt, C. (2006). Stress susceptibility predicts the severity of immune depression and the failure to combat bacterial infections in chronically stressed mice. *Brain, Behavior, and Immunity, 20*, 359–368.
Kim, Y. K., & Maes, M. (2003). The role of the cytokine network in psychological stress. *Acta Neuropsychiatrica, 15*, 148–155.
Knigge, U., & Warberg, J. (1991). Neuroendocrine functions of histamine. *Agents and Actions. Supplements, 33*, 29–53.
Korte, S. M., Koolhaas, J. M., Wingfield, J. C., & McEwen, B. S. (2005). The Darwinian concept of stress: Benefits of allostasis and costs of allostatic load and the trade-offs in health and disease. *Neuroscience and Biobehavioral Reviews, 29*, 3–38.
LeMoal, M. (2007). Historical approach and evolution of the stress concept: A personal account. *Psychoneuroendocrinology, 32*, S3–S9.
Leonard, B. E. (2006). HPA and immune axes in stress: Involvement of the serotonergic system. *Neuroimmunomodulation, 13*, 268–276.
Levine, S. (2005). Stress: An historical perspective. In J. Steckler, N. H. Kalin, & J. M. H. M. Reul (Eds.), *Handbook of stress and the brain* (pp. 3–24). Amsterdam: Elsevier Inc.
Lupien, S. J., Maheu, F., Tu, M., Fiocco, A., & Schramek, T. E. (2007). The effects of stress and stress hormones on human cognition: Implications for the field of brain and cognition. *Brain and Cognition, 65*, 209–237.
Lutgendorf, S. K., & Costanzo, E. S. (2003). Psychoneuroimmunology and health psychology: An integrative model. *Brain, Behavior, and Immunity, 17*, 225–232.
Maciag, C. M., Dent, G., Gilligan, P., He, L., Dowling, K., Ko, T., et al. (2002). Effects of a non-peptide CRF antagonist (DMP696) on the behavioral and endocrine sequelae of maternal separation. *Neuropsychopharmacology, 26*, 574–582.
MacQueen, G. M., & Bienenstock, J. (2006). Stress and immune regulation. *Clinical and Experimental Allergy, 36*, 969–971.

Mason, J. W. (1971). A re-evaluation of the concept of "non-specificity" in stress theory. *Journal of Psychiatric Research, 8*, 323–333.
Matheson, G. K., Knowles, A., Guthrie, D., Gage, D., Weinzapfel, D., & Blackbourne, J. (1997). Actions of serotonergic agents on hypothalamic-pituitary-adrenal axis activity in the rat. *General Pharmacology, 29*, 823–828.
Matzen, S. H. (1995). Neuroendocrine mechanisms during reversible hypovolaemic shock in humans with emphasis on the histaminergic and serotonergic system. *Acta Physiologica Scandinavica, 628*(Suppl.), 1–31.
Matzen, S., Knigge, U., & Warberg, J. (1990). Brain regulation of renin secretion involves central histaminergic neurons. *Neuroendocrinology, 52*, 175–180.
Matzen, S., Secher, N. H., Knigge, U., Bach, F. W., & Warberg, J. (1993). Effect of diazepam on endocrine and cardiovascular responses to head-up tilt in humans. *Acta Physiologica Scandinavica, 148*, 143–151.
McEwen, B. S. (1998). Stress, adaptation, and disease. Allostasis and allostatic load. *Annals of the New York Academy of Sciences, 840*, 33–44.
McEwen, B. S. (2000). Allostasis and allostatic load: Implications for neuropsychopharmacology. *Neuropsychopharmacology, 22*, 108–124.
McEwen, B. S. (2003). Mood disorders and allostatic load. *Biological Psychiatry, 54*, 200–207.
McLaughlin, K. J., Gomez, J. L., Baran, S. E., & Conrad, C. D. (2007). The effects of chronic stress on hippocampal morphology and function: An evaluation of chronic restraint paradigms. *Brain Research, 1161*, 56–64.
Morton, K. D., Johnson, M. D., & Van de Kar, L. D. (1995). Serotonin and stress-induced increases in renin secretion are not blocked by sympathectomy/adrenal medullectomy but are blocked by beta antagonists. *Brain Research, 698*, 185–192.
Nautiyal, K. M., Ribeiro, A. C., Pfaff, D. W., & Silver, R. (2008). Brain mast cells link the immune system to anxiety-like behavior. *Proceedings of the National Academy of Sciences of the United States of America, 105*, 18053–18057.
Newport, D. J., & Nemeroff, C. B. (2002). Stress. In V. S. Ramachandran (Ed.), *Encyclopedia of the human brain* (pp. 449–462). Amsterdam: Elsevier Inc.
Neylan, T. C., Sun, B., Rempel, H., Ross, J., Lenoci, M., O'Donovan, A., et al. (2011). Suppressed monocyte gene expression profile in men versus women with PTSD. *Brain, Behavior, and Immunity, 25*, 524–531.
Oosthuizen, F., Wegener, G., & Harvey, B. H. (2005). Nitric oxide as inflammatory mediator in post-traumatic stress disorder (PTSD): Evidence from an animal model. *Neuropsychiatric Disease and Treatment, 1*, 109–123.
Padgett, D. A., & Glaser, R. (2003). How stress influences the immune response. *Trends in Immunology, 24*, 444–448.
Pawlak, R., Rao, B. S., Melchor, J. P., Chattarji, S., McEwen, B., & Strickland, S. (2005). Tissue plasminogen activator and plasminogen mediate stress-induced decline of neuronal and cognitive functions in the mouse hippocampus. *Proceedings of the National Academy of Sciences of the United States of America, 102*, 18201–18206.
Perdrizet, G. A. (1997). Hans Selye and beyond: Responses to stress. *Cell Stress and Chaperones, 2*, 214–219.
Petrov, T., Krukoff, T. L., & Jhamandas, J. H. (1993). Branching projections of catecholaminergic brainstem neurons to the paraventricular hypothalamic nucleus and the central nucleus of the amygdala in the rat. *Brain Research, 609*, 81–92.
Petticrew, M., & Hunter, D. (1999). Stress-induced immunomodulation. *JAMA, 282*, 2209–2210.
Rademacher, D. J., Meier, S. E., Shi, L., Ho, W. S., Jarrahian, A., & Hillard, C. J. (2008). Effects of acute and repeated restraint stress on endocannabinoid content in the amygdala, ventral striatum, and medial prefrontal cortex in mice. *Neuropharmacology, 54*, 108–116.

Ray, A. (1994). *Neuropeptidergic regulation of some psychoneuroimmunologica; aspects of stress*. Ph.D (Pharmacology) thesis University of Delhi.
Ray, A., Gulati, K., Mediratta, P. K., & Sen, P. (1993). Opioidergic regulation of gastric ulcerogenesis and immune responses during stress in rats. *Experimental and Clinical Gastroenterology*, *3*, 7–11.
Ray, A., Mediratta, P. K., Puri, S., & Sen, P. (1991). Effects of stress on immune responsiveness, gastric ulcerogenesis and plasma corticosterone in rats: Modulation by diazepam and naltrexone. *Indian Journal of Experimental Biology*, *29*, 233–236.
Ray, A., Mediratta, P. K., & Sen, P. (1992). Modulation by naltrexone of stress-induced changes in humoral immune responsiveness and gastric mucosal integrity in rats. *Physiology and Behavior*, *51*, 293–296.
Reiche, E. M., Nunes, S. O., & Morimoto, H. K. (2004). Stress, depression, the immune system, and cancer. *The Lancet Oncology*, *5*, 617–625.
Riccardi, C., Bruscoli, S., & Migliorati, G. (2002). Molecular mechanisms of immunomodulatory activity of glucocorticoids. *Pharmacological Research*, *45*, 361–368.
Richardson Morton, K. D., Van de Kar, L. D., Brownfield, M. S., Lorens, S. A., Napier, T. C., & Urban, J. H. (1990). Stress-induced renin and corticosterone secretion is mediated by catecholaminergic nerve terminals in the hypothalamic paraventricular nucleus. *Neuroendocrinology*, *51*, 320–327.
Roarty, T. P., & Raff, H. (1988). Renin response to graded haemorrhage in conscious rats. *Clinical and Experimental Pharmacology and Physiology*, *15*, 373–378.
Sabban, E. L., & Kvetnansky, R. (2001). Stress-triggered activation of gene expression in catecholaminergic systems: Dynamics of transcriptional events. *Trends in Neurosciences*, *24*, 91–98.
Sandi, C., Davies, H. A., Cordero, M. I., Rodriguez, J. J., Popov, V. I., & Stewart, M. G. (2003). Rapid reversal of stress induced loss of synapses in CA3 of rat hippocampus following water maze training. *The European Journal of Neuroscience*, *17*, 2447–2456.
Selye, H. (1936). A syndrome produced by diverse nocuous agents. *Nature*, *38*, 32.
Shin, L. M., & Liberzon, I. (2010). The neurocircuitry of fear, stress, and anxiety disorders. *Neuropsychopharmacology*, *35*, 169–191.
Sterling, P., & Eyer, J. (1988). Allostasis: A new paradigm to explain arousal pathology. In J. Fisher & J. Reason (Eds.), *Handbook of life stress, cognition, and health* (pp. 629–649). New York: Johns Wiley & Sons Inc.
Van de Kar, L. D., & Blair, M. L. (1999). Forebrain pathways mediating stress-induced hormone secretion. *Frontiers in Neuroendocrinology*, *20*, 1–48.
Van de Kar, L. D., Javed, A., Zhang, Y., Serres, F., Raap, D. K., & Gray, T. S. (2001). 5-HT2A receptors stimulate ACTH, corticosterone, oxytocin, renin, and prolactin release and activate hypothalamic CRF and oxytocin-expressing cells. *The Journal of Neuroscience*, *21*, 3572–3579.
Van de Kar, L. D., Lorens, S. A., Urban, J. H., Richardson, K. D., Paris, J., & Bethea, C. L. (1985). Pharmacological studies on stress-induced renin and prolactin secretion: Effects of benzodiazepines, naloxone, propranolol and diisopropyl fluorophosphate. *Brain Research*, *345*, 257–263.
Vermetten, E., & Bremner, J. D. (2002). Circuits and systems in stress. I. Preclinical studies. *Depression and Anxiety*, *15*, 126–147.
Vyas, A., Mitra, R., Shankaranarayana Rao, B. S., & Chattarji, S. (2002). Chronic stress induces contrasting patterns of dendritic remodeling in hippocampal and amygdaloid neurons. *The Journal of Neuroscience*, *22*, 6810–6818.
Wilckens, T., & De Rijk, R. (1997). Glucocorticoids and immune function: Unknown dimensions and new frontiers. *Immunology Today*, *18*, 418–424.
Yang, E. V., & Glaser, R. (2000). Stress-induced immunomodulation: Impact on immune defenses against infectious disease. *Biomedicine and Pharmacotherapy*, *54*, 245–250.

Yang, E. V., & Glaser, R. (2002a). Stress-associated immunomodulation and its implications for responses to vaccination. *Expert Review of Vaccines*, *1*, 453–459.
Yang, E. V., & Glaser, R. (2002b). Stress-induced immunomodulation and the implications for health. *International Immunopharmacology*, *2*, 315–324.
Ziemssen, T., & Kern, S. (2007). Psychoneuroimmunology—Cross-talk between the immune and nervous systems. *Journal of Neurology*, *254*, II8–II11.

CHAPTER TWO

Estrogen Receptors Modulation of Anxiety-Like Behavior

A.P. Borrow, R.J. Handa[1]
Colorado State University, Fort Collins, CO, United States
[1]Corresponding author: e-mail address: bhanda@colostate.edu

Contents

Abstract

Estrogens exert profound effects on the expression of anxiety in humans and rodents; however, the directionality of these effects varies considerably within both clinical and preclinical literature. It is believed that discrepancies regarding the nature of estrogens' effects on anxiety are attributable to the differential effects of specific estrogen receptor (ER) subtypes. In this chapter we will discuss the relative impact on anxiety and anxiety-like behavior of each of the three main ERs: ERα, which has a generally anxiogenic effect, ERβ, which has a generally anxiolytic effect, and the G-protein-coupled ER known as GPR30, which has been found to both increase and decrease anxiety-like behavior. In addition, we will describe the known mechanisms by which these receptor subtypes exert their influence on emotional responses, focusing on the hypothalamic–pituitary–adrenal axis and the oxytocinergic and serotonergic systems. The impact of estrogens on the expression of anxiety is likely the result of their combined effects on all of these neurobiological systems.

Vitamins and Hormones, Volume 103
ISSN 0083-6729
http://dx.doi.org/10.1016/bs.vh.2016.08.004

1. INTRODUCTION

Changes in circulating estrogen levels across the reproductive lifespan have long been associated with changes in the incidence of anxiety in women. The risk of developing an anxiety disorder is elevated at menarche (Patton et al., 1996), a developmental period characterized by an increase in circulating estradiol from prepubertal to adult levels (Ojeda & Bilger, 2000). By contrast, an increase in anxiety symptoms has also been noted when estradiol levels drop, such as following surgical menopause (Rocca et al., 2008) and in postmenopausal women (Sahingoz, Uguz, & Gezginc, 2011). Moreover, toward the end of the luteal phase of the menstrual cycle, which is characterized by a dramatic decline in circulating estradiol levels, there is an increase in symptoms of anxiety in patients with anxiety disorders (Cameron, Kuttesch, McPhee, & Curtis, 1988), as well as in patients with premenstrual disorders (Yonkers, O'Brien, & Eriksson, 2008).

While the association between anxiety symptoms and low endogenous estradiol levels might suggest a therapeutic effect of estrogens, both clinical and preclinical studies have reported that treatment with estradiol yields contradictory results. For example, in postmenopausal women, anxiety levels have been reported to either decrease (Gleason et al., 2015) or remain unaffected (Demetrio et al., 2011) following estrogen therapy. Using rodent models, studies have shown that the effect of estradiol on anxiety-like behavior may be dependent on both the dose administered and the behavioral testing paradigm. A recent study by Kastenberger, Lutsch, and Schwarzer (2012) conducted in ovariectomized mice found that a high dose of estradiol (0.25 mg/kg) but not a low dose (0.025 mg/kg) decreased anxiety-like behavior in the elevated plus maze. In contrast, only the low dose of estradiol increased anxiety-like behavior in the open field test (Kastenberger et al., 2012). A different group also noted that a low dose of estradiol (2.0 μg/day) increased anxiety in ovariectomized mice in the light/dark test, while a dose that was even lower (0.2 μg/day) instead decreased anxiety-like behavior (Tomihara et al., 2009). Other studies have reported either anxiogenic (Mora, Dussaubat, & Diaz-Véliz, 1996), anxiolytic (Tian et al., 2013), or null (Walf & Frye, 2008) effects of estradiol in a variety of rodent models. Consequently, the conflicting effects of estradiol on anxiety and anxiety-like behavior have been proposed to be the result of diverging roles for estrogen receptor subtypes (Kastenberger et al., 2012).

Currently, researchers are aware of at least three types of estrogen receptors. The classic estrogen receptors, ERα and ERβ, are highly homologous in structure and belong to one subclass of a large superfamily of nuclear hormone receptors (Burris et al., 2013) which also include the receptors for androgens, glucocorticoids, mineralocorticoids, thyroid hormone, and retinoic acid, to name but a few. A more complete examination of these receptor proteins can be found in several recent reviews (Huang, Chandra, & Rastinejad, 2010; Morrill, Kostellow, & Gupta, 2015), and their distributions into the various subclasses of receptors, specific receptor-related transcriptomics, and reagents are now cataloged online at the nuclear receptor signaling atlas website (www.nursa.org). It is classically thought that the nuclear receptors influence gene expression both directly, as ligand-activated transcription factors, and perhaps indirectly, through the membrane localization of the classical receptors which can rapidly impact intracellular signaling cascades (Edwards, 2005). A third, more recently identified receptor, the G-protein-coupled estrogen receptor (GPR30 or GPER1), is not related to the nuclear hormone receptors and exerts its effects through both rapid signaling events and, to a lesser extent, direct transcriptional activation (Prossnitz & Arterburn, 2015). Additionally, receptors such as the STX-sensitive receptor, a membrane-associated receptor which can mediate some of the actions of estradiol (Smith, Bosch, Wagner, Rønnekleiv, & Kelly, 2013), and ER-X, a putative membrane receptor for 17α estradiol (Toran-Allerand et al., 2002), have recently been described, although our understanding of their effects is currently, at best, very limited. In this chapter, we will discuss the contribution of each of the major estrogen receptors to anxiety and anxiety-like behavior, and their potential roles as therapeutic targets for anxiety disorders.

2. NUCLEAR ESTROGEN RECEPTOR FORM AND FUNCTION

ERα and ERβ show a number of similarities, both in their modular structure (Ascenzi, Bocedi, & Marino, 2006) as well as in homology within the central DNA binding domain (Pettersson, Grandien, Kuiper, & Gustafsson, 1997). The two ERs arose from a single duplication event approximately 450 million years ago (Kelley & Thackray, 1999). Since then, they have undergone a parallel evolution resulting in substantial sequence homology, yet they exhibit unique roles in regulated mammalian physiology. The generation of transgenic mouse lines that are null for

ERα (Antonson, Omoto, Humire, & Gustafsson, 2012; Dupont et al., 2000), ERβ (Dupont et al., 2000; Krege et al., 1998), or both (Dupont et al., 2000; Kudwa & Rissman, 2003), has helped to determine the function of the nuclear estrogen receptors. These receptors influence expression of target genes through both genomic and nongenomic mechanisms (Björnström & Sjöberg, 2005) (Fig. 1). Classically, it has been shown that ligand binding to nuclear ERs results in receptor dimerization, followed by binding to promoter regions of target genes at sites known as estrogen response elements, ultimately attracting coregulatory proteins to these areas which can positively or negatively influence gene transcription. In addition, nuclear ERs are able to indirectly influence gene expression via tethering to other transcription factors, as well as through modulating the synthesis of various second messenger proteins (Björnström & Sjöberg, 2005). Despite

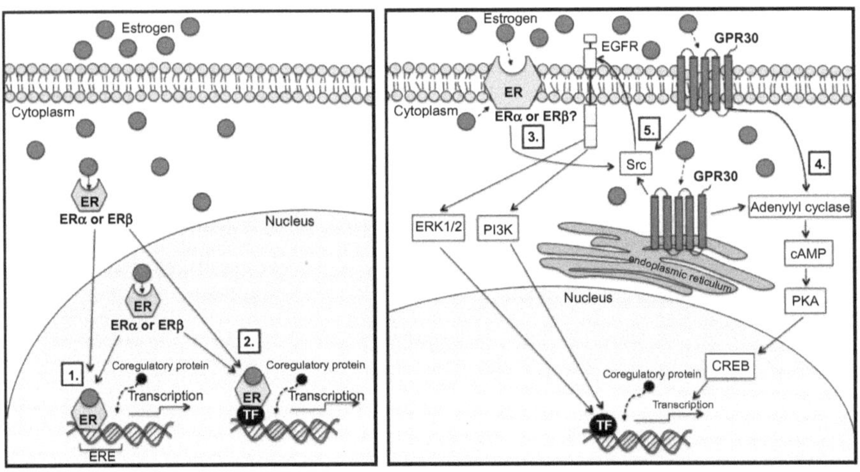

Fig. 1 A simplified description of known estrogen receptor signaling pathways. Classically, ligand-activated estrogen receptors (ER) α and β bind directly to promoter regions at estrogen response element (ERE) sites, attracting coregulatory proteins, which influence gene transcription (1). ERα and β can also indirectly affect gene transcription by influencing the activity of other transcription factors (TF) (2) and by mediating the synthesis of second messenger proteins such as Src kinase (3). Ligand binding to GPR30 initiates several second messenger signaling cascades, including the activation of adenylyl cyclase which increases the production of cAMP, ultimately activating the transcription factor cAMP response element-binding protein (CREB) (4), and the recruitment of Src kinase, which activates epidermal growth factor receptors (EGFRs), resulting in the activation of mitogen-activated protein kinases (MAPKs) ERK1/2 and phosphatidylinositol 3-kinases (PI3Ks), ultimately leading to the activation of proteins that regulate transcription factors (5). See Björnström and Sjöbergg (2005), Maggiolini and Picard (2010), and Prossnitz et al. (2008) for review.

the marked similarities between ERα and ERβ, these receptors differ significantly in tissue distribution (Shughrue, Lane, & Merchenthaler, 1997) and function, particularly in brain.

3. ESTROGEN RECEPTOR ALPHA

ERα is the original ER to be cloned and characterized and has long been thought of as the ER that is essential for regulating reproductive physiology and behaviors. Knockout models of ERα consistently show impaired reproductive hormone secretion, physiology, and behaviors (Antonson et al., 2012; Dupont et al., 2000; McDevitt et al., 2007). In regard to the regulation of nonreproductive behaviors, the activation of ERα is largely associated with an increase in anxiety-like behavior in the rodent. For example, downregulation of ERα within the medial preoptic area and the posterodorsal amygdala of ovariectomized rats following infusion of an adeno-associated viral vector encoding for a small hairpin RNA (shRNA) targeting the ERα gene produced a decrease in anxiety-like behavior in the light–dark box and open field test (Spiteri et al., 2010; Spiteri, Ogawa, Musatov, Pfaff, & Ågmo, 2012). It is important to note that not all studies have observed effects of ERα on anxiety-like behavior. In ERα knockout mice, there were no observable changes in anxiety-like behavior in the elevated plus maze or open field tests (Krężel, Dupont, Krust, Chambon, & Chapman, 2001). Additionally, some researchers have observed no effects (Jacome et al., 2010; Walf & Frye, 2005) or mild effects (Lund, Rovis, Chung, & Handa, 2005) of the selective ERα agonist propylpyrazole triol (PPT) on anxiety-like behaviors when tested in the open field and elevated plus maze in rat models. Thus, it is possible that the effects of ERα on anxiety are context or species dependent, and further work is required to tease out the subtleties of this response.

While ERα's effects appear mostly anxiogenic in the rodent, recent research has suggested an opposing effect of ERα on anxiety-like behavior in female animals following reproductive experience. Expression of ERα in the medial amygdala may change during the postpartum period, demonstrating a downregulation at 3 weeks postpartum, followed by an upregulation at 5 weeks (Furuta et al., 2013). Studies of primiparous rats have shown a decrease in anxiety-like behavior in the elevated plus maze following treatment with PPT at 3 weeks (albeit, not as effectively as estradiol treatment) (Furuta et al., 2013) or at 7 weeks (Byrnes, Casey, & Bridges, 2012; Byrnes, Casey, Carini, & Bridges, 2013) postpartum. It is unknown whether this temporal change in ERα's effects is the result of hormonal changes during pregnancy and the postnatal period, maternal experience, or some combination of these two factors.

To date, only a small number of studies have considered a link between single nucleotide polymorphisms (SNPs) of the ERα gene (ESR1) and the occurrence of anxiety disorders. A study of adult men in treatment for substance abuse noted an association between symptoms of anxiety and dinucleotide polymorphisms (Comings, Muhleman, Johnson, & MacMurray, 1999). In an elderly human population, a study investigating SNPs found that, in women only, ESR1 haplotypes consisting of the T allele of rs2234693 and the A allele of rs9340799 were correlated with an increased likelihood of anxiety disorders (Tiemeier et al., 2005). More recently, a study by Ryan et al. noted that these same haplotypes were linked to an increase in panic disorder in elderly women, and that this association was mediated by the use of hormone therapy (Ryan et al., 2011). While further verification in a younger female population is still needed, these studies suggest that specific ESR1 polymorphisms may result in an increased vulnerability to the development of anxiety disorders.

4. ESTROGEN RECEPTOR BETA

ERβ was first described in 1996, after being cloned from a rat prostate cDNA library (Kuiper, Enmark, Pelto-Huikko, Nilsson, & Gustafsson, 1996). While studies in the prostate have demonstrated a role for ERβ in activating apoptosis and limiting prostate growth, studies utilizing ERβ knockout models have suggested that in brain, this receptor is a major component of estradiol's anxiolytic-like effects in the rodent. A number of studies have reported increased anxiety-like behaviors in female ERβ knockout mice (Imwalle, Gustafsson, & Rissman, 2005; Krężel et al., 2001; Walf, Koonce, & Frye, 2008), as well as higher levels of plasma corticosterone (Walf, Koonce, Manley, & Frye, 2009). Further evidence supporting an anxiolytic-like effect of ERβ can be found in pharmacological studies utilizing selective ERβ agonists. The most commonly used ERβ agonist, the selective estrogen receptor modulator (SERM) diarylpropionitrile (DPN), causes decreased anxiety-like behavior in rats and mice in a number of tests, including the open field test (Lund et al., 2005; Oyola et al., 2012; Walf et al., 2008), elevated zero maze test (Walf et al., 2008), elevated plus maze test (Lund et al., 2005; Oyola et al., 2012), and light–dark box test (Lund et al., 2005; Oyola et al., 2012). Administration of a different ERβ agonist,

WAY-200070, increased anxiolytic-like responding in the open field, the elevated plus maze (Weiser, Wu, & Handa, 2009), the four-plate test, which "punishes" exploration of a novel environment with mild foot shock, and in a test of stress-induced hypothermia (Hughes et al., 2008). Furthermore, DPN administration to the paraventricular nucleus (PVN) decreased restraint stress-induced c-fos mRNA levels within the PVN and decreased the plasma adrenocorticotropic hormone (ACTH) response to in the male rat (Lund, Hinds, & Handa, 2006), suggesting a central site for ERβ action. This mechanism also holds true for mice, since peripheral DPN administration decreased stress-induced plasma corticosterone and ACTH in ovariectomized mice (Oyola et al., 2012). Despite all of the evidence suggesting an anxiolytic-like effect of ERβ activation, an influence of this receptor on anxiety-like behavior has not always been observed (Jacome et al., 2010; Patisaul, Burke, Hinkle, Adewale, & Shea, 2009). The absence of effects of ERβ agonists on anxiolytic behaviors in some studies may be due to dose or timing phenomena since the selectivity of ERβ agonists is rather modest and, therefore, too high or too low a dose may negate its actions by interactions with ERα (at higher doses) or by failing to activate ERβ (at lower doses). Nonetheless, there remains strong support for ERβ as an inhibitor of rodent anxiety.

Of importance, the activation of ERβ is not restricted to estrogens or selective ERβ agonists. Naturally occurring metabolites of androgens have been recently described that can also bind and activate ERβ and consequently can modulate physiology and behaviors. The two best described androgen metabolites that functionally impact ERβ signaling pathways are 5-androst*en*-3β,17β-diol (ADIOL), a metabolite of the adrenal androgen dehydroepiandrosterone (Saijo, Collier, Li, Katzenellenbogen, & Glass, 2011), and 5α-androstane-3β,17β-diol (3β-diol) (Weihua, Lathe, Warner, & Gustafsson, 2002), a metabolite of the potent nonaromatizable androgen, dihydrotestosterone. Both of these androgens have been shown to be antiinflammatory (Saijo et al., 2011; Zuloaga, O'Connor, Handa, & Gonzales, 2012), and the effects of 3β-diol have been described in regulating anxiety-like behaviors. Treatment of rats with 3β-diol dipropionate reduces anxiety in the elevated plus maze (Handa, Pak, Kudwa, Lund, & Hinds, 2008) and reduces the acute neuroendocrine response to restraint stress (Handa et al., 2008; Lund et al., 2006). Moreover, 3β-diol has been shown to act through ERβ to regulate promoter activity of several neuropeptides, including oxytocin and vasopressin (Hiroi et al., 2013; Pak, Chung, Hinds, & Handa, 2007; Pak et al., 2005).

Unfortunately, despite the large body of literature describing ERβ's effects in the rodent, research regarding the influence of ERβ on anxiety in humans is currently very limited. A recent study noted that, for the ERβ gene ESR2, the SNP rs1256049 was associated with a greater incidence of generalized anxiety disorder in older women (Ryan et al., 2011). It is unknown if this correlation is observable in younger women or in men. Given the compelling data from preclinical models, further investigation into the role of ERβ in human anxiety disorders is critically needed.

5. G-PROTEIN-COUPLED ESTROGEN RECEPTOR

Recent research has identified the existence of a novel estrogen receptor known as GPR30 or GPER1 (Filardo, Quinn, Bland, & Frackelton, 2000; Revankar, Cimino, Sklar, Arterburn, & Prossnitz, 2005; Thomas, Pang, Filardo, & Dong, 2005). This receptor has been described as localized both at the membrane (Filardo et al., 2007; Funakoshi, Yanai, Shinoda, Kawano, & Mizukami, 2006; Thomas et al., 2005) as well as within several intracellular structures, including endoplasmic reticulum (Revankar et al., 2005) and cytokeratin intermediate filaments (Sandén et al., 2011). GPR30 exhibits a widespread distribution across the rodent brain, with high levels of immunoreactivity noted in regions implicated in affective behavior, such as the cortex, hippocampus, hypothalamus, and brainstem (Brailoiu et al., 2007; Hazell et al., 2009). In addition, GPR30 immunoreactivity is increased following acute stress in the basolateral amygdala of ovariectomized mice, which is accompanied by increased inhibitory synaptic transmission in the basolateral amygdala (Tian et al., 2013), suggesting a role in the rodent stress response.

GPR30 is capable of influencing gene expression through several known mechanisms, resulting in both short-term and long-term effects on transcription (see Fig. 1). Agonist binding to GPR30 can cause the activation of adenylyl cyclase, stimulating the production of cAMP, which ultimately activates the transcription factor cAMP response element-binding protein (CREB). Additionally, in vitro activation of GPR30 has been found to recruit second messengers such as Src kinase, producing downstream transactivation of epidermal growth factor receptors (EGFRs), leading to the activation of mitogen-activated protein kinases (MAPKs) ERK1/2, and the activation of phosphatidylinositol 3-kinases (PI3Ks), ultimately resulting in the activation of proteins that regulate transcription factors (for review, see Maggiolini & Picard, 2010; Prossnitz et al., 2008).

Although several recent studies have implicated GPR30 in modulating the expression of anxiety-like behaviors in the mouse, there are conflicting reports as to the nature of this receptor's effects. Importantly, the impact of this receptor on anxiety-like behavior may be sex dependent and may vary depending on the behavioral testing paradigm employed. For example, following an acute stressor, administration of the GPR30 agonist G-1 to the basolateral amygdala prevented an increase in anxiety-like behavior in ovariectomized mice in the open field test and elevated plus maze (Tian et al., 2013). Peripheral administration of G-1 produced a decrease in anxiety-like behavior in the elevated plus maze in gonadectomized male mice, while ovariectomized subjects were unaffected by treatment (Hart et al., 2014). Conversely, G-1 treatment led to an increase in anxiety-like behavior in ovariectomized mice and intact male mice in the elevated plus maze and open field test (Kastenberger et al., 2012). All told, it is possible that this discrepancy in the effects of G-1 on behavior may result from differences in dose of G-1, as well as differences in the timing of G-1 administration relative to behavioral testing.

While little is known about the nature of GPR30's influence on anxiety-like behavior, even less is known about the potential mechanisms responsible for GPR30's effects on this behavior. Activation of GPR30 increases inhibitory synaptic transmission within the basolateral amygdala of ovariectomized mice (Tian et al., 2013). Based on an observed high level of colocalization of GPR30 with both GABAergic interneurons and glutamatergic pyramidal neurons, coupled with GPR30's prevention of stress-induced downregulation of $GABA_A$ receptor expression and upregulation of glutamate receptors, Tian et al. (2013) have proposed that GPR30 mediates anxiety-like behavior by altering the balance between GABAergic and glutamatergic signaling within the basolateral amygdala (Tian et al., 2013). This receptor's effects may not be the result of activation of the ERK signaling pathway, as two studies investigating anxiety-like behavior following G-1 administration found no changes in phosphorylated ERK in the hippocampus of gonadectomized male (Hart et al., 2014) and female (Anchan, Clark, Pollard, & Vasudevan, 2014) mice. However, potential changes in phosphorylated ERK were not assessed in other regions implicated in regulating anxiety-like behavior that also express GPR30, thereby currently limiting conclusions about the ERK signaling pathway.

6. CROSS TALK BETWEEN ESTROGEN RECEPTORS

The above discussion has focused on the individual contribution of each estrogen receptor in anxiety and anxiety-like behavior. However, evidence suggests that these receptor subtypes are capable of interacting with one another. One such example is the relation between ERα and ERβ. Male ERα knockout mice show decreased expression of ERβ protein within the medial preoptic area and increased expression within the bed nucleus of the stria terminalis, suggesting that ERβ expression can be modulated by ERα. This observation can be coupled with a relative insensitivity to estradiol's effects on ERβ expression, as assessed through gonadectomy and estradiol replacement (Nomura, Korach, Pfaff, & Ogawa, 2003) although this lack of effect may be brain region dependent (Patisaul, Whitten, & Young, 1999). GPR30 and ERα have also shown some interdependence. A recent study by Hart et al. (2014) found that GPR30's agonist G-1 increased protein expression of hippocampal phosphorylated ERα in male mice (Hart et al., 2014). In addition, an in vitro study using ovarian cancer cells reported that both GPR30 and ERα were required for either estradiol or G-1 to induce an upregulation in expression of the oncogene *c-fos* (Albanito et al., 2007). The relation between ERβ and GPR30 has not yet been assessed. In sum, these findings demonstrate that estrogen receptors do not function entirely independently of one another, which likely has implications for estrogen receptors' mediation of anxiety.

7. MECHANISMS FOR ESTROGEN RECEPTOR EFFECTS ON ANXIETY

Researchers are continuing to investigate the pathways by which estrogen receptors influence anxiety and anxiety-like behavior. The most well-described mechanisms, discussed later, demonstrate that estrogen receptors are capable of mediating a diverse range of neural systems. Importantly, these systems do not operate in isolation; therefore, estrogen receptor effects on anxiety are likely the result of a number of interactions between pathways.

7.1 Effects on the Hypothalamo-Pituitary-Adrenal Axis

In response to a stressor, neurons within the PVN of the hypothalamus secrete the peptides corticotropin-releasing hormone (CRH) and vasopressin, which act in concert to stimulate the release of ACTH from the anterior pituitary into the peripheral circulation (Fig. 2). Moreover, ACTH acts upon the cortex of the adrenal gland, causing the upregulation of enzymes involved in steroidogenesis and ultimately the enhanced secretion of glucocorticoids. Glucocorticoids then act to prevent further activity of this system

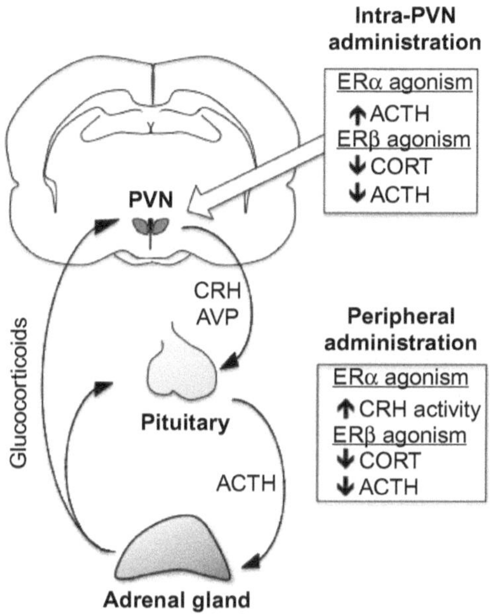

Fig. 2 Diagrammatic representation of the hypothalamo-pituitary-adrenal (HPA) axis and a summary of the effects of estrogen receptor agonists.

through a long-loop negative feedback mechanism that can involve the hippocampus, hypothalamus, and pituitary gland. This neuroendocrine system, known as the hypothalamo-pituitary-adrenal (HPA) axis, is one of the primary ways in which an organism is capable of reacting to a stressor (Herman & Cullinan, 1997) and maintaining homeostasis. Of importance for the current discussion, estrogen receptors have been shown to impact neuroendocrine stress responsivity at several different levels of this axis.

ERα may mediate anxiety-like behavior through its actions on HPA axis function. Glucocorticoids have long been known to activate brain regions that are involved in mediating anxiety and fear responses in rodents. In particular, the central nucleus of the amygdala has been shown to express glucocorticoid receptors, and neurons in this brain area respond to glucocorticoids to enhance anxiety-like behaviors. Thus, increases in glucocorticoid secretion can directly influence behaviors such as anxiety (Shepard, Barron, & Myers, 2003; Weiser, Foradori, & Handa, 2010) and correspondingly, factors that influence HPA activity can have similar actions on anxiety. Administration of estradiol, PPT, or the ERα-selective agonist, moxestrol, to the PVN augmented the ACTH response to restraint stress in the male rat (Lund et al., 2006), which correlated with an elevation in c-fos mRNA levels within this brain region (Lund et al., 2006). Interestingly, ERα is expressed at only low levels within the rat PVN (Shughrue et al., 1997), suggesting an alternative route for the effects of these ERα agonists on behavior. It has been proposed that the activation of the ERα-containing GABAergic neurons surrounding the PVN may inhibit GABAergic inhibition of the PVN, resulting in an increase in HPA activity. Correspondingly, increased stressor-induced *C-fos* immunoreactivity has been shown in CRH neurons following administration of PPT to ovariectomized rats (Thammacharoen, Geary, Lutz, Ogawa, & Asarian, 2009). Furthermore, as shown in vitro by Lalmansingh and Uht (2008) using an amygdaloid cell line, ERα may regulate CRH gene (*crh*) expression through histone acetylation in the cAMP regulatory element region of the *crh* promoter (Lalmansingh & Uht, 2008). This molecular regulation provides another pathway for ERα's anxiogenic effects.

Substantial evidence for ERβ influencing anxiety through HPA axis mediation has also been shown. Peripheral administration of the ERβ agonist DPN causes decreased stress-induced plasma corticosterone and ACTH levels in the ovariectomized mouse (Oyola et al., 2012). In contrast to ERα, ERβ is robustly expressed within both the rat and mouse (Fig. 3) PVN. A possible target for ERβ in the HPA axis is a direct action on a population of neurosecretory cells in the PVN that produce CRH. Neuroendocrine CRH neurons in the rat PVN express ERβ in a subregion-dependent fashion (13–60%) (Laflamme, Nappi, Drolet, Labrie, & Rivest, 1998; Suzuki & Handa, 2005). These CRH-expressing neurons initiate the HPA axis response to stress, ultimately increasing circulating glucocorticoid levels (Herman & Cullinan, 1997). Similar to peripheral DPN injection, implantation of a DPN-containing pellet near the male rat PVN has produced a decrease in stress-induced plasma corticosterone and ACTH (Lund et al., 2006). Thus, ERβ may mediate HPA axis activity by actions directly on PVN neurons. It has been proposed that this receptor subtype also regulates *crh* expression through histone acetylation in the cAMP regulatory element region of the *crh* promoter. Unlike ERα, ERβ may have a more indirect influence on expression, acting by attracting the cAMP-binding protein-binding protein (CBP) to the CRE element in the *crh* promoter (Lalmansingh & Uht, 2008).

7.2 Effects on the Serotonergic System

The serotonergic system has long been implicated in modulating anxiety in humans and rodent models (Handley, McBlane, Critchley, & Njung'e,

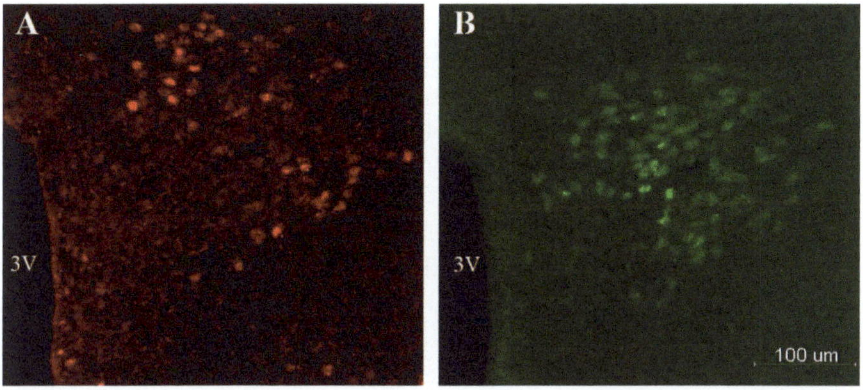

Fig. 3 Confocal microscopic image showing distribution of immunofluorescent labeling of estrogen receptors α (A) and β (B) in the mouse paraventricular nucleus. ERα was detected by standard immunofluorescent techniques using an AlexaFluor 594-labeled secondary antibody, and ERβ-expressing neurons were detected in the same brain section by examining endogenous ERβ-EGFP expression in a transgenic mouse model. *3V*, 3rd ventricle.

1993; Naughton, Mulrooney, & Leonard, 2000). Estrogens have potent effects on this system, causing changes in levels of serotonin, its metabolite 5-hydroxyindoleacetic acid (5-HIAA), the serotonin transporter, and serotonin receptors (Borrow & Cameron, 2014). These effects may be driven by the actions of ERβ and GPR30.

The estrogen receptor that has received the most attention as a facilitator of estradiol's effects on the serotonergic system is ERβ. Treatment of male mice with the ERβ agonist WAY-200070 has resulted in increased levels of serotonin within the striatum (Hughes et al., 2008), while DPN treatment caused increased levels of 5-HIAA in OVX rats (Jacome et al., 2010). Male ERβ knockout mice have been reported to have decreased levels of 5-hydroxytryptophan, the precursor of serotonin, within the frontal cortex (Hughes et al., 2008), while it has been shown that female ERβ knockout mice have lower levels of serotonin within the hippocampus, bed nucleus of the stria terminalis, and the preoptic area (Imwalle et al., 2005). These findings suggest that estrogens may mediate serotonin synthesis through ERβ.

One way in which ERβ may influence anxiety-like behavior through the serotonergic system is by mediating expression of tryptophan hydroxylase, the rate-limiting enzyme required for the synthesis of serotonin. Knockdown of this enzyme within the caudal dorsal raphe nucleus of OVX rats removes the anxiolytic-like effects of estradiol in the open field test (Hiroi, McDevitt, Morcos, Clark, & Neumaier, 2011). Correspondingly, both male and female ERβ-null mice have decreased expression of tryptophan hydroxylase. In addition, OVX decreased the number of tryptophan hydroxylase-positive neurons within the DRN, which was rescued following treatment with estradiol or the ERβ agonist LY3201 (Suzuki et al., 2013). Similarly, our laboratory has shown an upregulation in tryptophan hydroxylase 2 in OVX rats following systemic or local treatment with DPN (Donner & Handa, 2009). However, despite evidence implicating ERβ as a mediator of anxiety-like behavior through its effects on the serotonergic system, ERβ may not act directly on DRN neurons to influence anxiety-like behavior. We have previously reported that direct administration of DPN to this region did not alter anxiety-like behavior in the open field or elevated plus maze in OVX rats (Donner & Handa, 2009). Furthermore, rats may not express ERβ within the DRN (Sheng et al., 2004). This suggests an alternative site of action for ERβ's effects on anxiety through serotonergic modulation.

Another potential site for ERβ's effects on the serotonergic system is the amygdala. Krężel et al. (2001) reported that ERβ knockout mice showed an enhanced induction of long-term potentiation within the amygdala relative to wild-type controls. The authors also noted an increase in the receptor 5-HT1A within the medial amygdala of ERβ knockout female animals. When coupled with an observed potentiation of synaptic response following GABAergic inhibition and the known effects of serotonin on amygdala GABAergic tone (Koyama, Kubo, Rhee, & Akaike, 1999), this led them to hypothesize that enhanced 5-HT1A receptor activation might mediate long-term potentiation within the amygdala through the inhibition of GABA release (Krężel et al., 2001). Direct confirmation of changes in anxiety-like behavior as a result of ERβ's effects on the serotonergic system within the amygdala is still needed.

Despite the large body of data implicating ERβ in estrogens' mediation of anxiety-like behavior through the serotonergic system, it has recently been proposed that GPR30, not ERβ, is the estrogen receptor responsible for some of estrogens' effects. In the female rat PVN, a desensitization of 5-HT1A receptors following administration of estradiol was found to be the result of GPR30 expression, and not of ERβ (McAllister, Creech, Kimball, Muma, & Li, 2012). GPR30 is also colocalized with 5-HT1A receptors within the PVN, and administration of the agonist G-1 has attenuated 5-HT1A receptor signaling within this region (Xu et al., 2009). Thus, at least within the PVN, which shows an absence of ERα in the rat (Shughrue et al., 1997), estrogens may influence serotonergic signaling through GPR30 activation.

7.3 Effects on the Oxytocinergic System

Oxytocin (OT) is a nonapeptide produced primarily by magnocellular neurons within the PVN and the supraoptic nucleus of the hypothalamus. Its function has classically been associated with regulation of parturition and lactation, although more recent studies indicate that oxytocin can have a variety of effects on the central nervous system (CNS) to regulate such behaviors as reproductive and parental behavior, anxiety, social bonding, and social recognition (for review, see Borrow & Cameron, 2012; Lee, Macbeth, Pagani, & Young, 2009; Neumann, 2008). OT is secreted by three known methods: release into the peripheral circulation via activation of the neurohypophyseal system, local dendritic release followed by diffusion into local brain structures, and release into regions such as the hippocampus and amygdala by axonal projections from OT neurons within the PVN (Murgatroyd et al., 2004). Following its release, OT acts on its receptors (OTR), which are widely expressed throughout the brain and periphery. OT is an important mediator of anxiety symptoms in both humans and rodents. Plasma OT levels are negatively correlated with anxiety symptoms in depressed patients (Scantamburlo et al., 2007), and intranasal OT administration attenuates emotional fear reactivity in patients with social anxiety disorder (Labuschagne et al., 2010). Furthermore, an SNP within the oxytocin receptor gene, rs2254298, has been correlated with anxiety symptoms in women and adolescent girls (Chen & Johnson, 2012; Thompson, Parker, Hallmayer, Waugh, & Gotlib, 2011). In the rodent, central administration of OT or an OTR agonist decreases anxiety-like behavior and suppresses HPA axis activation in response to stress (Mak, Broussard, Vacy, & Broadbear, 2012; Sabihi, Durosko, Dong, & Leuner, 2014; Windle, Shanks, Lightman, & Ingram, 1997). Estradiol has been found to increase levels of plasma OT (Amico, Seif, & Robinson, 1981; Yamaguchi, Akaishi, & Negoro, 1979), OT mRNA (Hiroi et al., 2013; Nomura, McKenna, Korach, Pfaff, & Ogawa, 2002; Sharma, Handa, & Uht, 2012), and OTR binding (Krege et al., 1998), presenting an additional mechanism by which estrogens are capable of mediating anxiety.

The receptor primarily implicated in the effects of estradiol on OT is ERβ. OT and ERβ are highly colocalized within the PVN in the female rat (Hrabovszky et al., 2004; Suzuki & Handa, 2005), and colocalization of ERβ and OT has also been reported in the human PVN (Hrabovszky et al., 2004). ERβ knockout mice do not show the increase in OT mRNA following estradiol administration observed in wild-type controls (Nomura et al., 2002). Administration of the ERβ ligand 5α-androstane-3β,17β-diol (3β-diol) increased OT mRNA in a mouse hypothalamic cell line (Hiroi et al.,

2013; Sharma et al., 2012) and in ovariectomized rats (Hiroi et al., 2013). Finally, the anxiolytic effects of the ERβ agonist DPN are blocked in the rat following treatment with an OT antagonist (Kudwa, McGivern, & Handa, 2014). Recent in vitro evidence suggests that ERβ regulates OT expression by binding to the composite hormone response element of the *Ot* gene promoter, inducing the recruitment of the coactivators cAMP response element-binding protein (CREB) and steroid receptor coactivator (SRC-1) to the promoter site, resulting in an increase in histone H4 acetylation, ultimately leading to an increase in *Ot* gene transcription (Hiroi et al., 2013; Sharma et al., 2012).

Research investigating the regulation of OT by ERα and GPR30 is more limited, but bears mentioning. ERα knockout mice do not show an increase in OT receptor binding following estradiol treatment (Krege et al., 1998). Moreover, OT receptors have been shown to be estrogen regulated (de Kloet, Voorhuis, Boschma, & Elands, 1986). Such data indicate that the OT receptor is under estrogenic control, through ERα but not ERβ, whereas oxytocin may be differentially regulated by ERβ and not ERα. Finally, GPR30 is colocalized with OT neurons within the rat PVN (Sakamoto et al., 2007), but the influence of GPR30 on OT is currently unknown, and warrants further investigation.

8. CONCLUSIONS

There is compelling evidence demonstrating that estrogens influence anxiety and anxiety-like behaviors in both humans and rodents. The nature of this influence differs based on the subtype of estrogen receptor utilized (Table 1). While animal models suggest that activation of ERβ may serve to generate anxiolytic-like effects, ERα appears to have largely anxiogenic-like properties. The more recently discovered membrane estrogen receptor, GPR30, has also been implicated in the rodent anxiety response, although, at present, the extent of its participation is not clear. Given the importance of receptor subtype in modulating anxiety, SERMs demonstrate great potential as novel therapeutic agents in the treatment of anxiety disorders. SERMs that target ERβ may be particularly effective in patients with specific polymorphisms of ESR1 and ESR2, and in patients with anxiety symptoms temporally associated with declining estrogen levels, such as during the premenstrual period or menopause.

Table 1 The Effect of Estrogen Receptor Subtypes on Anxiety-Like Behavior in the Rodent

Receptor	Method	Species	References	Effect
ERα				
Agonists	PPT s.c., 1 mg/kg	Female[a] and male mice	Kastenberger et al. (2012)	No effect in EPM, LD, OFT
	PPT s.c., 1 mg/kg for 4 days	Female and male rats[a]	Lund et al. (2005)	↑ Anxiety in EPM
	PPT s.c., 10 μg	Female rats[a]	Walf and Frye (2005)	No effect in DFT, EPM, ET, OFT, VT
	PPT s.c., 3 or 5 mg/kg for 2 days	Female rats[a]	Jacome et al. (2010)	No effect in EPM
	PPT s.c., 1 mg/kg for 7 days	Female rats[a]	Weiser et al. (2009)	No effect in EPM, OFT
	PPT s.c., 1 mg/kg for 3 days	Male rats	Patisaul et al. (2009)	No effect in LD
	PPT s.c., 1 mg/kg for 4 days	Female rats[a]	Byrnes et al. (2012)	↓ Anxiety in primiparous females in EPM
	PPT s.c., 1 mg/kg	Female rats[a]	Byrnes et al. (2013)	↑ Anxiety in nulliparous females, ↓ anxiety in primiparous females in EPM
	PPT s.c., 1 mg/kg	Primiparous female rats	Furuta et al. (2013)	↓ Anxiety in EPM
Genetic models	shRNA targeting ERα intra-MPOA, estradiol s.c., 18 μg/kg	Female rats[a]	Spiteri et al. (2012)	↓ Anxiety in LD, OFT
	shRNA targeting ERα intra-meAMYG, estradiol s.c., 18 μg/kg	Female rats[a]	Spiteri et al. (2010)	↓ Anxiety in LD
	ERα knockout mice	Female and male mice	Krężel et al. (2001)	No effect in EPM or OFT

Continued

Table 1 The Effect of Estrogen Receptor Subtypes on Anxiety-Like Behavior in the Rodent—cont'd

Receptor	Method	Species	References	Effect
ERβ				
Agonists	DPN s.c., 1 mg/kg	Female[a] and male mice	Kastenberger et al. (2012)	No effect in EPM, LD, OFT
	DPN s.c., 1 mg/kg for 4 days	Female and male rats[a]	Lund et al. (2005)	↓ Anxiety in EPM, LD (only females tested), and OFT (only females tested)
	DPN s.c., 3 mg/kg for 2 days	Female rats[a]	Jacome et al. (2010)	No effect in EPM
	DPN s.c., 10 μg	Female rats[a]	Walf and Frye (2005)	↓ Anxiety in DFT, EPM, ET, OFT, VT
	DPN s.c., 1 mg/kg for 4 days	Female rats[a]	Byrnes et al. (2012)	No effect in EPM
	DPN s.c., 0.5 or 2 mg/kg for 4 days, 1 mg/kg for 3 days	Male rats	Patisaul et al. (2009)	No effect in EPM or LD
	DPN s.c., 1 mg/kg	Primiparous female rats	Furuta et al. (2013)	No effect in EPM
	DPN s.c., 0.1 mg/kg	Female mice[a]	Walf et al. (2008)	↓ Anxiety in EPM, EZM, OFT, SIT
	S-DPN s.c., 1 mg/kg	Female mice[a]	Oyola et al. (2012)	↓ Anxiety in EPM, LD, OFT, no effect in MBT
	DPN or S-DPN s.c., 2 mg/kg for 7 days	Females rats[a]	Weiser et al. (2009)	↓ Anxiety in EPM, OFT
	Coumestrol s.c., 10 μg	Female rats[a]	Walf and Frye (2005)	↓ Anxiety in DFT, EPM, ET, OFT, VT
	WAY-200070 s.c., 30 mg/kg	Male mice	Hughes et al. (2008)	↓ Anxiety in 4PT, SIH
	WAY-200070 s.c., 2 mg/kg for 7 days	Female rats[a]	Weiser et al. (2009)	↓ Anxiety in EPM, OFT

Table 1 The Effect of Estrogen Receptor Subtypes on Anxiety-Like Behavior in the Rodent—cont'd

Receptor	Method	Species	References	Effect
Genetic models	ERβ knockout mice	Female mice[a]	Oyola et al. (2012)	No effect in EPM, LD, MBT, OFT
	ERβ knockout mice	Female and male mice	Krężel et al. (2001)	↑ Anxiety in EPM, OFT (females only)
	ERβ knockout mice, s.c. estradiol (85–100 pg/mL)	Female mice[a]	Imwalle et al. (2005)	↑ Anxiety in EPM
	ERβ knockout mice, s.c. DPN 0.1 mg/kg	Female mice[a]	Walf et al. (2008)	↑ Anxiety in EPM, EZM, OFT, SIT
GPR30				
Agonists	(G-1) s.c., 0.3 μg	Female and male mice[a]	Hart et al. (2014)	↓ Anxiety in EPM (males only), no effect in OFT
	(G-1) s.c., 1 mg/kg	Female[a] and male mice	Kastenberger et al. (2012)	↑ Anxiety in EPM, LD (males only), OFT
	(G-1) intra-BLA, 0.5 μM	Female mice[a]	Tian et al. (2013)	↓ Anxiety in EPM, OFT
Genetic models	GPR30 knockout mice	Female and male mice	Kastenberger and Schwarzer (2014)	↓ Anxiety in EPM (males only) and LD (males only), no effect in OFT

[a]Gonadectomized subjects.

Note: *4PT*, four-plate test; *BLA*, basolateral amygdala; *DFT*, defensive freezing test; *EPM*, elevated plus maze test; *ET*, emergence test; *EZM*, elevated zero maze test; *LD*, light–dark box; *MBT*, marble burying test; *meAMYG*, medial amygdala; *MPOA*, medial preoptic area; *OFT*, open field test; *SIH*, stress-induced hypothermia; *SIT*, social interaction test; *VT*, Vogel test.

REFERENCES

Albanito, L., Madeo, A., Lappano, R., Vivacqua, A., Rago, V., Carpino, A., … Maggiolini, M. (2007). G protein-coupled receptor 30 (GPR30) mediates gene expression changes and growth response to 17β-estradiol and selective GPR30 ligand G-1 in ovarian cancer cells. *Cancer Research*, *67*(4), 1859–1866.

Amico, J. A., Seif, S. M., & Robinson, A. G. (1981). Oxytocin in human plasma: Correlation with neurophysin and stimulation with estrogen. *The Journal of Clinical Endocrinology and Metabolism*, *52*(5), 988–993.

Anchan, D., Clark, S., Pollard, K., & Vasudevan, N. (2014). GPR30 activation decreases anxiety in the open field test but not in the elevated plus maze test in female mice. *Brain and Behavior, 4*(1), 51–59.

Antonson, P., Omoto, Y., Humire, P., & Gustafsson, J. A. (2012). Generation of ERalpha-floxed and knockout mice using the Cre/LoxP system. *Biochemical and Biophysical Research Communications, 424*(4), 710–716.

Ascenzi, P., Bocedi, A., & Marino, M. (2006). Structure–function relationship of estrogen receptor α and β: Impact on human health. *Molecular Aspects of Medicine, 27*(4), 299–402.

Björnström, L., & Sjöberg, M. (2005). Mechanisms of estrogen receptor signaling: Convergence of genomic and nongenomic actions on target genes. *Molecular Endocrinology, 19*(4), 833–842.

Borrow, A. P., & Cameron, N. M. (2012). The role of oxytocin in mating and pregnancy. *Hormones and Behavior, 61*(3), 266–276.

Borrow, A. P., & Cameron, N. M. (2014). Estrogenic mediation of serotonergic and neurotrophic systems: Implications for female mood disorders. *Progress in Neuro-Psychopharmacology and Biological Psychiatry, 54*, 13–25.

Brailoiu, E., Dun, S. L., Brailoiu, G. C., Mizuo, K., Sklar, L. A., Oprea, T. I., … Dun, N. J. (2007). Distribution and characterization of estrogen receptor G protein-coupled receptor 30 in the rat central nervous system. *Journal of Endocrinology, 193*(2), 311–321.

Burris, T. P., Solt, L. A., Wang, Y., Crumbley, C., Banerjee, S., Griffett, K., … Kojetin, D. J. (2013). Nuclear receptors and their selective pharmacologic modulators. *Pharmacological Reviews, 65*(2), 710–778.

Byrnes, E. M., Casey, K., & Bridges, R. S. (2012). Reproductive experience modifies the effects of estrogen receptor alpha activity on anxiety-like behavior and corticotropin releasing hormone mRNA expression. *Hormones and Behavior, 61*(1), 44–49.

Byrnes, E. M., Casey, K., Carini, L. M., & Bridges, R. S. (2013). Reproductive experience alters neural and behavioural responses to acute oestrogen receptor alpha activation. *Journal of Neuroendocrinology, 25*(12), 1280–1289.

Cameron, O. G., Kuttesch, D., McPhee, K., & Curtis, G. C. (1988). Menstrual fluctuation in the symptoms of panic anxiety. *Journal of Affective Disorders, 15*(2), 169–174.

Chen, F. S., & Johnson, S. C. (2012). An oxytocin receptor gene variant predicts attachment anxiety in females and autism-spectrum traits in males. *Social Psychological and Personality Science, 3*, 93–99.

Comings, D. E., Muhleman, D., Johnson, P., & MacMurray, J. P. (1999). Potential role of the estrogen receptor gene (ESR1) in anxiety. *Molecular Psychiatry, 4*(4), 374–377.

de Kloet, E. R., Voorhuis, D. A., Boschma, Y., & Elands, J. (1986). Estradiol modulates density of putative 'oxytocin receptors' in discrete rat brain regions. *Neuroendocrinology, 44*(4), 415–421.

Demetrio, F. N., Rennó, J., Jr., Gianfaldoni, A., Gonçalves, M., Halbe, H. W., Filho, A. H., & Gorenstein, C. (2011). Effect of estrogen replacement therapy on symptoms of depression and anxiety in non-depressive menopausal women: A randomized double-blind, controlled study. *Archives of Women's Mental Health, 14*(6), 479–486.

Donner, N., & Handa, R. J. (2009). Estrogen receptor beta regulates the expression of tryptophan-hydroxylase 2 mRNA within serotonergic neurons of the rat dorsal raphe nuclei. *Neuroscience, 163*(2), 705–718.

Dupont, S., Krust, A., Gansmuller, A., Dierich, A., Chambon, P., & Mark, M. (2000). Effect of single and compound knockouts of estrogen receptors alpha (ERalpha) and beta (ERbeta) on mouse reproductive phenotypes. *Development, 127*(19), 4277–4291.

Edwards, D. P. (2005). Regulation of signal transduction pathways by estrogen and progesterone. *Annual Review of Physiology, 67*, 335–376.

Filardo, E. J., Quinn, J. A., Bland, K. I., & Frackelton, A. R., Jr. (2000). Estrogen-induced activation of Erk-1 and Erk-2 requires the G protein-coupled receptor homolog, GPR30, and occurs via trans-activation of the epidermal growth factor receptor through release of HB-EGF. *Molecular Endocrinology*, *14*(10), 1649–1660.

Filardo, E., Quinn, J., Pang, Y., Graeber, C., Shaw, S., Dong, J., & Thomas, P. (2007). Activation of the novel estrogen receptor G protein-coupled receptor 30 (GPR30) at the plasma membrane. *Endocrinology*, *148*(7), 3236–3245.

Funakoshi, T., Yanai, A., Shinoda, K., Kawano, M. M., & Mizukami, Y. (2006). G protein-coupled receptor 30 is an estrogen receptor in the plasma membrane. *Biochemical and Biophysical Research Communications*, *346*(3), 904–910.

Furuta, M., Numakawa, T., Chiba, S., Ninomiya, M., Kajiyama, Y., Adachi, N., … Kunugi, H. (2013). Estrogen, predominantly via estrogen receptor α, attenuates postpartum-induced anxiety- and depression-like behaviors in female rats. *Endocrinology*, *154*(10), 3807–3816.

Gleason, C. E., Dowling, N. M., Wharton, W., Manson, J. E., Miller, V. M., Atwood, C. S., … Asthana, S. (2015). Effects of hormone therapy on cognition and mood in recently postmenopausal women: Findings from the randomized, controlled KEEPS–cognitive and affective study. *PLoS Medicine*, *12*(6), e1001833.

Handa, R. J., Pak, T. R., Kudwa, A. E., Lund, T. D., & Hinds, L. (2008). An alternate pathway for androgen regulation of brain function: Activation of estrogen receptor beta by the metabolite of dihydrotestosterone, 5alpha-androstane-3beta,17beta-diol. *Hormones and Behavior*, *53*(5), 741–752.

Handley, S. L., McBlane, J. W., Critchley, M. A., & Njung'e, K. (1993). Multiple serotonin mechanisms in animal models of anxiety: Environmental, emotional and cognitive factors. *Behavioural Brain Research*, *58*(1–2), 203–210.

Hart, D., Nilges, M., Pollard, K., Lynn, T., Patsos, O., Shiel, C., … Vasudevan, N. (2014). Activation of the G-protein coupled receptor 30 (GPR30) has different effects on anxiety in male and female mice. *Steroids*, *81*, 49–56.

Hazell, G. G., Yao, S. T., Roper, J. A., Prossnitz, E. R., O'Carroll, A. M., & Lolait, S. J. (2009). Localisation of GPR30, a novel G protein-coupled oestrogen receptor, suggests multiple functions in rodent brain and peripheral tissues. *Journal of Endocrinology*, *202*(2), 223–236.

Herman, J. P., & Cullinan, W. E. (1997). Neurocircuitry of stress: Central control of the hypothalamo–pituitary–adrenocortical axis. *Trends in Neurosciences*, *20*(2), 78–84.

Hiroi, R., Lacagnina, A. F., Hinds, L. R., Carbone, D. G., Uht, R. M., & Handa, R. J. (2013). The androgen metabolite, 5alpha-androstane-3beta,17beta-diol (3beta-diol), activates the oxytocin promoter through an estrogen receptor-beta pathway. *Endocrinology*, *154*(5), 1802–1812.

Hiroi, R., McDevitt, R. A., Morcos, P. A., Clark, M. S., & Neumaier, J. F. (2011). Overexpression or knockdown of rat tryptophan hyroxylase-2 has opposing effects on anxiety behavior in an estrogen-dependent manner. *Neuroscience*, *176*, 120–131.

Hrabovszky, E., Kalló, I., Steinhauser, A., Merchenthaler, I., Coen, C. W., Petersen, S. L., & Liposits, Z. (2004). Estrogen receptor-beta in oxytocin and vasopressin neurons of the rat and human hypothalamus: Immunocytochemical and in situ hybridization studies. *The Journal of Comparative Neurology*, *473*(3), 315–333.

Huang, P., Chandra, V., & Rastinejad, F. (2010). Structural overview of the nuclear receptor superfamily: Insights into physiology and therapeutics. *Annual Review of Physiology*, *72*, 247–272.

Hughes, Z. A., Liu, F., Platt, B. J., Dwyer, J. M., Pulicicchio, C. M., Zhang, G., … Day, M. (2008). WAY-200070, a selective agonist of estrogen receptor beta as a potential novel anxiolytic/antidepressant agent. *Neuropharmacology*, *54*(7), 1136–1142.

Imwalle, D. B., Gustafsson, J.-Å., & Rissman, E. F. (2005). Lack of functional estrogen receptor β influences anxiety behavior and serotonin content in female mice. *Physiology & Behavior*, *84*(1), 157–163.

Jacome, L. F., Gautreaux, C., Inagaki, T., Mohan, G., Alves, S., Lubbers, L. S., & Luine, V. (2010). Estradiol and ERβ agonists enhance recognition memory, and DPN, an ERβ agonist, alters brain monoamines. *Neurobiology of Learning and Memory*, *94*(4), 488–498.

Kastenberger, I., Lutsch, C., & Schwarzer, C. (2012). Activation of the G-protein-coupled receptor GPR30 induces anxiogenic effects in mice, similar to oestradiol. *Psychopharmacology*, *221*(3), 527–535.

Kastenberger, I., & Schwarzer, C. (2014). GPER1 (GPR30) knockout mice display reduced anxiety and altered stress response in a sex and paradigm dependent manner. *Hormones and Behavior*, *66*, 628–636.

Kelley, S. T., & Thackray, V. G. (1999). Phylogenetic analyses reveal ancient duplication of estrogen receptor isoforms. *Journal of Molecular Evolution*, *49*(5), 609–614.

Koyama, S., Kubo, C., Rhee, J. S., & Akaike, N. (1999). Presynaptic serotonergic inhibition of GABAergic synaptic transmission in mechanically dissociated rat basolateral amygdala neurons. *The Journal of Physiology*, *518*(2), 525–538.

Krege, J. H., Hodgin, J. B., Couse, J. F., Enmark, E., Warner, M., Mahler, J. F., … Smithies, O. (1998). Generation and reproductive phenotypes of mice lacking estrogen receptor beta. *Proceedings of the National Academy of Sciences of the United States of America*, *95*(26), 15677–15682.

Krężel, W., Dupont, S., Krust, A., Chambon, P., & Chapman, P. F. (2001). Increased anxiety and synaptic plasticity in estrogen receptor β-deficient mice. *Proceedings of the National Academy of Sciences of the United States of America*, *98*(21), 12278–12282.

Kudwa, A. E., McGivern, R. F., & Handa, R. J. (2014). Estrogen receptor beta and oxytocin interact to modulate anxiety-like behavior and neuroendocrine stress reactivity in adult male and female rats. *Physiology and Behavior*, *129*, 287–296.

Kudwa, A. E., & Rissman, E. F. (2003). Double oestrogen receptor alpha and beta knockout mice reveal differences in neural oestrogen-mediated progestin receptor induction and female sexual behaviour. *Journal of Neuroendocrinology*, *15*(10), 978–983.

Kuiper, G. G., Enmark, E., Pelto-Huikko, M., Nilsson, S., & Gustafsson, J.-A. (1996). Cloning of a novel receptor expressed in rat prostate and ovary. *Proceedings of the National Academy of Sciences of the United States of America*, *93*(12), 5925–5930.

Labuschagne, I., Phan, K. L., Wood, A., Angstadt, M., Chua, P., Heinrichs, M., … Nathan, P. J. (2010). Oxytocin attenuates amygdala reactivity to fear in generalized social anxiety disorder. *Neuropsychopharmacology*, *35*(12), 2403–2413.

Laflamme, N., Nappi, R. E., Drolet, G., Labrie, C., & Rivest, S. (1998). Expression and neuropeptidergic characterization of estrogen receptors (ERalpha and ERbeta) throughout the rat brain: Anatomical evidence of distinct roles of each subtype. *Journal of Neurobiology*, *36*(3), 357–378.

Lalmansingh, A. S., & Uht, R. M. (2008). Estradiol regulates corticotropin-releasing hormone gene (crh) expression in a rapid and phasic manner that parallels estrogen receptor-α and -β recruitment to a 3′,5′-cyclic adenosine 5′-monophosphate regulatory region of the proximal crh promoter. *Endocrinology*, *149*(1), 346–357.

Lee, H. J., Macbeth, A. H., Pagani, J. H., & Young, W. S., 3rd. (2009). Oxytocin: The great facilitator of life. *Progress in Neurobiology*, *88*(2), 127–151.

Lund, T. D., Hinds, L. R., & Handa, R. J. (2006). The androgen 5α-dihydrotestosterone and its metabolite 5α-androstan-3β, 17β-diol inhibit the hypothalamo–pituitary–adrenal response to stress by acting through estrogen receptor β-expressing neurons in the hypothalamus. *The Journal of Neuroscience*, *26*(5), 1448–1456.

Lund, T. D., Rovis, T., Chung, W. C. J., & Handa, R. J. (2005). Novel actions of estrogen receptor-β on anxiety-related behaviors. *Endocrinology*, *146*(2), 797–807.

Maggiolini, M., & Picard, D. (2010). The unfolding stories of GPR30, a new membrane-bound estrogen receptor. *Journal of Endocrinology*, *204*(2), 105–114.

Mak, P., Broussard, C., Vacy, K., & Broadbear, J. H. (2012). Modulation of anxiety behavior in the elevated plus maze using peptidic oxytocin and vasopressin receptor ligands in the rat. *Journal of Psychopharmacology*, *26*(4), 532–542.

McAllister, C., Creech, R., Kimball, P., Muma, N., & Li, Q. (2012). GPR30 is necessary for estradiol-induced desensitization of 5-HT 1A receptor signaling in the paraventricular nucleus of the rat hypothalamus. *Psychoneuroendocrinology*, *37*(8), 1248–1260.

McDevitt, M. A., Glidewell-Kenney, C., Weiss, J., Chambon, P., Jameson, J. L., & Levine, J. E. (2007). Estrogen response element-independent estrogen receptor (ER)-alpha signaling does not rescue sexual behavior but restores normal testosterone secretion in male ERalpha knockout mice. *Endocrinology*, *148*(11), 5288–5294.

Mora, S., Dussaubat, N., & Diaz-Véliz, G. (1996). Effects of the estrous cycle and ovarian hormones on behavioral indices of anxiety in female rats. *Psychoneuroendocrinology*, *21*(7), 609–620.

Morrill, G. A., Kostellow, A. B., & Gupta, R. K. (2015). Transmembrane helices in "classical" nuclear reproductive steroid receptors: A perspective. *Nuclear Receptor Signaling*, *13*, e003.

Murgatroyd, C., Wigger, A., Frank, E., Singewald, N., Bunck, M., Holsboer, F., … Spengler, D. (2004). Impaired repression at a vasopressin promoter polymorphism underlies overexpression of vasopressin in a rat model of trait anxiety. *Journal of Neuroscience*, *24*(35), 7762–7770.

Naughton, M., Mulrooney, J. B., & Leonard, B. E. (2000). A review of the role of serotonin receptors in psychiatric disorders. *Human Psychopharmacology*, *15*(6), 397–415.

Neumann, I. D. (2008). Brain oxytocin: A key regulator of emotional and social behaviours in both females and males. *Journal of Neuroendocrinology*, *20*(6), 858–865.

Nomura, M., Korach, K. S., Pfaff, D. W., & Ogawa, S. (2003). Estrogen receptor beta (ERbeta) protein levels in neurons depend on estrogen receptor alpha (ERalpha) gene expression and on its ligand in a brain region-specific manner. *Molecular Brain Research*, *110*(1), 7–14.

Nomura, M., McKenna, E., Korach, K. S., Pfaff, D. W., & Ogawa, S. (2002). Estrogen receptor-beta regulates transcript levels for oxytocin and arginine vasopressin in the hypothalamic paraventricular nucleus of male mice. *Molecular Brain Research*, *109*(1–2), 84–94.

Ojeda, S. R., & Bilger, M. (2000). Neuroendocrine regulation of puberty. In P. M. Michael & M. E. Freeman (Eds.), *Neuroendocrinology in physiology and medicine* (pp. 197–224). New York: Springer Science+Business Media, LLC.

Oyola, M. G., Portillo, W., Reyna, A., Foradori, C. D., Kudwa, A., Hinds, L., … Mani, S. K. (2012). Anxiolytic effects and neuroanatomical targets of estrogen receptor-β (ERβ) activation by a selective ERβ agonist in female mice. *Endocrinology*, *153*(2), 837–846.

Pak, T. R., Chung, W. C., Hinds, L. R., & Handa, R. J. (2007). Estrogen receptor-beta mediates dihydrotestosterone-induced stimulation of the arginine vasopressin promoter in neuronal cells. *Endocrinology*, *148*(7), 3371–3382.

Pak, T. R., Chung, W. C., Lund, T. D., Hinds, L. R., Clay, C. M., & Handa, R. J. (2005). The androgen metabolite, 5alpha-androstane-3beta, 17beta-diol, is a potent modulator of estrogen receptor-beta1-mediated gene transcription in neuronal cells. *Endocrinology*, *146*(1), 147–155.

Patisaul, H. B., Burke, K. T., Hinkle, R. E., Adewale, H. B., & Shea, D. (2009). Systemic administration of diarylpropionitrile (DPN) or phytoestrogens does not affect anxiety-related behaviors in gonadally intact male rats. *Hormones and Behavior*, *55*(2), 319–328.

Patisaul, H. B., Whitten, P. L., & Young, L. J. (1999). Regulation of estrogen receptor beta mRNA in the brain: Opposite effects of 17beta-estradiol and the phytoestrogen, coumestrol. *Molecular Brain Research*, *67*(1), 165–171.

Patton, G., Hibbert, M., Carlin, J., Shao, Q., Rosier, M., Caust, J., & Bowes, G. (1996). Menarche and the onset of depression and anxiety in Victoria, Australia. *Journal of Epidemiology and Community Health*, *50*(6), 661–666.

Pettersson, K., Grandien, K., Kuiper, G. G., & Gustafsson, J.-A. (1997). Mouse estrogen receptor β forms estrogen response element-binding heterodimers with estrogen receptor α. *Molecular Endocrinology*, *11*(10), 1486–1496.

Prossnitz, E. R., & Arterburn, J. B. (2015). International union of basic and clinical pharmacology. XCVII. G protein-coupled estrogen receptor and its pharmacologic modulators. *Pharmacological Reviews*, *67*(3), 505–540.

Prossnitz, E. R., Arterburn, J. B., Smith, H. O., Oprea, T. I., Sklar, L. A., & Hathaway, H. J. (2008). Estrogen signaling through the transmembrane G protein-coupled receptor GPR30. *Annual Review of Physiology*, *70*, 165–190.

Revankar, C. M., Cimino, D. F., Sklar, L. A., Arterburn, J. B., & Prossnitz, E. R. (2005). A transmembrane intracellular estrogen receptor mediates rapid cell signaling. *Science*, *307*(5715), 1625–1630.

Rocca, W. A., Grossardt, B. R., Geda, Y. E., Gostout, B. S., Bower, J. H., Maraganore, D. M., … Melton, L. J., 3rd. (2008). Long-term risk of depressive and anxiety symptoms after early bilateral oophorectomy. *Menopause*, *15*(6), 1050–1059.

Ryan, J., Scali, J., Carrière, I., Scarabin, P. Y., Ritchie, K., & Ancelin, M. L. (2011). Estrogen receptor gene variants are associated with anxiety disorders in older women. *Psychoneuroendocrinology*, *36*(10), 1582–1586.

Sabihi, S., Durosko, N. E., Dong, S. M., & Leuner, B. (2014). Oxytocin in the prelimbic medial prefrontal cortex reduces anxiety-like behavior in female and male rats. *Psychoneuroendocrinology*, *45*, 31–42.

Sahingoz, M., Uguz, F., & Gezginc, K. (2011). Prevalence and related factors of mood and anxiety disorders in a clinical sample of postmenopausal women. *Perspectives in Psychiatric Care*, *47*(4), 213–219.

Saijo, K., Collier, J. G., Li, A. C., Katzenellenbogen, J. A., & Glass, C. K. (2011). An ADIOL-ERβ-CtBP transrepression pathway negatively regulates microglia-mediated inflammation. *Cell*, *145*(4), 584–595.

Sakamoto, H., Matsuda, K., Hosokawa, K., Nishi, M., Morris, J. F., Prossnitz, E. R., & Kawata, M. (2007). Expression of G protein-coupled receptor-30, a G protein-coupled membrane estrogen receptor, in oxytocin neurons of the rat paraventricular and supraoptic nuclei. *Endocrinology*, *148*(12), 5842–5850.

Sandén, C., Broselid, S., Cornmark, L., Andersson, K., Daszkiewicz-Nilsson, J., Martensson, U. E., … Leeb-Lundberg, L. M. (2011). G protein-coupled estrogen receptor 1/G protein-coupled receptor 30 localizes in the plasma membrane and traffics intracellularly on cytokeratin intermediate filaments. *Molecular Pharmacology*, *79*(3), 400–410.

Scantamburlo, G., Hansenne, M., Fuchs, S., Pitchot, W., Maréchal, P., Pequeux, C., … Legros, J. J. (2007). Plasma oxytocin levels and anxiety in patients with major depression. *Psychoneuroendocrinology*, *32*(4), 407–410.

Sharma, D., Handa, R. J., & Uht, R. M. (2012). The ERβ ligand 5α-androstane, 3β,17β-diol (3β-diol) regulates hypothalamic oxytocin (Oxt) gene expression. *Endocrinology*, *153*(5), 2353–2361.

Sheng, Z., Kawano, J., Yanai, A., Fujinaga, R., Tanaka, M., Watanabe, Y., & Shinoda, K. (2004). Expression of estrogen receptors (alpha, beta) and androgen receptor in serotonin neurons of the rat and mouse dorsal raphe nuclei; sex and species differences. *Neuroscience Research*, *49*(2), 185–196.

Shepard, J. D., Barron, K. W., & Myers, D. A. (2003). Stereotaxic localization of corticosterone to the amygdala enhances hypothalamo-pituitary-adrenal responses to behavioral stress. *Brain Research, 963*(1–2), 203–213.

Shughrue, P. J., Lane, M. V., & Merchenthaler, I. (1997). Comparative distribution of estrogen receptor-alpha and -beta mRNA in the rat central nervous system. *The Journal of Comparative Neurology, 388*(4), 507–525.

Smith, A. W., Bosch, M. A., Wagner, E. J., Rønnekleiv, O. K., & Kelly, M. J. (2013). The membrane estrogen receptor ligand STX rapidly enhances GABAergic signaling in NPY/AgRP neurons: Role in mediating the anorexigenic effects of 17beta-estradiol. *American Journal of Physiology. Endocrinology and Metabolism, 305*(5), E632–E640.

Spiteri, T., Musatov, S., Ogawa, S., Ribeiro, A., Pfaff, D. W., & Ågmo, A. (2010). The role of the estrogen receptor α in the medial amygdala and ventromedial nucleus of the hypothalamus in social recognition, anxiety and aggression. *Behavioural Brain Research, 210*(2), 211–220.

Spiteri, T., Ogawa, S., Musatov, S., Pfaff, D. W., & Ågmo, A. (2012). The role of the estrogen receptor α in the medial preoptic area in sexual incentive motivation, proceptivity and receptivity, anxiety, and wheel running in female rats. *Behavioural Brain Research, 230*(1), 11–20.

Suzuki, H., Barros, R. P., Sugiyama, N., Krishnan, V., Yaden, B. C., Kim, H. J., … Gustafsson, J. A. (2013). Involvement of estrogen receptor β in maintenance of serotonergic neurons of the dorsal raphe. *Molecular Psychiatry, 18*(6), 674–680.

Suzuki, S., & Handa, R. J. (2005). Estrogen receptor-β, but not estrogen receptor-α, is expressed in prolactin neurons of the female rat paraventricular and supraoptic nuclei: Comparison with other neuropeptides. *The Journal of Comparative Neurology, 484*(1), 28–42.

Thammacharoen, S., Geary, N., Lutz, T. A., Ogawa, S., & Asarian, L. (2009). Divergent effects of estradiol and the estrogen receptor-α agonist PPT on eating and activation of PVN CRH neurons in ovariectomized rats and mice. *Brain Research, 1268*, 88–96.

Thomas, P., Pang, Y., Filardo, E. J., & Dong, J. (2005). Identity of an estrogen membrane receptor coupled to a G protein in human breast cancer cells. *Endocrinology, 146*(2), 624–632.

Thompson, R. J., Parker, K. J., Hallmayer, J. F., Waugh, C. E., & Gotlib, I. H. (2011). Oxytocin receptor gene polymorphism (rs2254298) interacts with familial risk for psychopathology to predict symptoms of depression and anxiety in adolescent girls. *Psychoneuroendocrinology, 36*(1), 144–147.

Tian, Z., Wang, Y., Zhang, N., Guo, Y. Y., Feng, B., Liu, S. B., & Zhao, M. G. (2013). Estrogen receptor GPR30 exerts anxiolytic effects by maintaining the balance between GABAergic and glutamatergic transmission in the basolateral amygdala of ovariectomized mice after stress. *Psychoneuroendocrinology, 38*(10), 2218–2233.

Tiemeier, H., Schuit, S. C. E., den Heijer, T., van Meurs, J. B., van Tuijl, H. R., Hofman, A., … Uitterlinden, A. G. (2005). Estrogen receptor α gene polymorphisms and anxiety disorder in an elderly population. *Molecular Psychiatry, 10*(9), 806–807.

Tomihara, K., Soga, T., Nomura, M., Korach, K. S., Gustafsson, J.-Å., Pfaff, D. W., & Ogawa, S. (2009). Effect of ER-β gene disruption on estrogenic regulation of anxiety in female mice. *Physiology & Behavior, 96*(2), 300–306.

Toran-Allerand, C. D., Guan, X., MacLusky, N. J., Horvath, T. L., Diano, S., Singh, M., … Tinnikov, A. A. (2002). ER-X: A novel, plasma membrane-associated, putative estrogen receptor that is regulated during development and after ischemic brain injury. *Journal of Neuroscience, 22*(19), 8391–8401.

Walf, A. A., & Frye, C. A. (2005). ERβ-selective estrogen receptor modulators produce anti-anxiety behavior when administered systemically to ovariectomized rats. *Neuropsychopharmacology, 30*(9), 1598–1609.

Walf, A. A., & Frye, C. A. (2008). Parity and estrogen-administration alter affective behavior of ovariectomized rats. *Physiology & Behavior, 93*(1), 351–356.
Walf, A. A., Koonce, C. J., & Frye, C. A. (2008). Estradiol or diarylpropionitrile decrease anxiety-like behavior of wildtype, but not estrogen receptor beta knockout, mice. *Behavioral Neuroscience, 122*(5), 974–981.
Walf, A. A., Koonce, C., Manley, K., & Frye, C. A. (2009). Proestrous compared to diestrous wildtype, but not estrogen receptor beta knockout, mice have better performance in the spontaneous alternation and object recognition tasks and reduced anxiety-like behavior in the elevated plus and mirror maze. *Behavioural Brain Research, 196*(2), 254–260.
Weihua, Z., Lathe, R., Warner, M., & Gustafsson, J. A. (2002). An endocrine pathway in the prostate, ERbeta, AR, 5alpha-androstane-3beta,17beta-diol, and CYP7B1, regulates prostate growth. *Proceedings of the National Academy of Sciences of the United States of America, 99*(21), 13589–13594.
Weiser, M. J., Foradori, C. D., & Handa, R. J. (2010). Estrogen receptor beta activation prevents glucocorticoid receptor-dependent effects of the central nucleus of the amygdala on behavior and neuroendocrine function. *Brain Research, 1336*, 78–88.
Weiser, M. J., Wu, T. J., & Handa, R. J. (2009). Estrogen receptor-beta agonist diarylpropionitrile: Biological activities of R- and S- enantiomers on behavior and hormonal response to stress. *Endocrinology, 150*(4), 1817–1825.
Windle, R. J., Shanks, N., Lightman, S. L., & Ingram, C. D. (1997). Central oxytocin administration reduces stress-induced corticosterone release and anxiety behavior in rats. *Endocrinology, 138*(7), 2829–2834.
Xu, H., Qin, S., Carrasco, G. A., Dai, Y., Filardo, E. J., Prossnitz, E. R., … Muma, N. A. (2009). Extra-nuclear estrogen receptor GPR30 regulates serotonin function in rat hypothalamus. *Neuroscience, 158*(4), 1599–1607.
Yamaguchi, K., Akaishi, T., & Negoro, H. (1979). Effect of estrogen treatment on plasma oxytocin and vasopressin in ovariectomized rats. *Endocrinologia Japonica, 26*(2), 197–205.
Yonkers, K. A., O'Brien, P. M. S., & Eriksson, E. (2008). Premenstrual syndrome. *Lancet, 371*(9619), 1200–1210.
Zuloaga, K. L., O'Connor, D. T., Handa, R. J., & Gonzales, R. J. (2012). Estrogen receptor beta dependent attenuation of cytokine-induced cyclooxygenase-2 by androgens in human brain vascular smooth muscle cells and rat mesenteric arteries. *Steroids*, 77(8–9), 835–844.

CHAPTER THREE

Neuroendocrine Underpinnings of Increased Risk for Posttraumatic Stress Disorder in Women

M.A. Briscione*[,†], V. Michopoulos[†,‡], T. Jovanovic[†], S.D. Norrholm*[,†,1]

*Trauma Recovery Program, Mental Health Service Line, Atlanta Veterans Affairs Medical Center, Decatur, GA, United States

[†]Emory University School of Medicine, Atlanta, GA, United States

[‡]Yerkes National Primate Research Center, Atlanta, GA, United States

[1]Corresponding author: e-mail addresses: seth.norrholm@emoryhealthcare.org; snorrho@emory.edu

Contents

Abstract

Women are particularly vulnerable to the effects of psychological trauma and the development of trauma-, stressor-, and anxiety-related mental illnesses such as posttraumatic stress disorder (PTSD). In the current chapter, we examine the female hormonal systems that interact with psychobiological stress response systems to elicit maladaptive behavior and mental disease states in traumatized female populations. In addition, we provide a contemporary translational example of a stress vulnerability genomic profile (coding for pituitary adenylate cyclase-activating polypeptide) that may underlie the specific susceptibilities observed in women. Translational scientific investigations such as those described herein may lead to the identification of risk and resilience factors for PTSD as well as enhanced clinical interventions for treating excessive fear and anxiety.

1. INTRODUCTION

Posttraumatic stress disorder (PTSD) affects women disproportionately to men, occurring in 18% of all women and 10% of men who survive

Vitamins and Hormones, Volume 103
ISSN 0083-6729
http://dx.doi.org/10.1016/bs.vh.2016.08.003

a life-threatening event (Breslau, Davis, Andreski, Peterson, & Schultz, 1997; Resnick, Kilpatrick, Dansky, Saunders, & Best, 1993). In addition to being twice as likely to develop the disorder, women also have longer symptoms duration (Breslau et al., 1997), increased symptom severity (Seedat, Stein, & Carey, 2005), and lower quality of life as compared to men (Holbrook, Hoyt, Stein, & Sieber, 2002); yet, it remains unclear how high levels of traumatic stressors interact with one's sex to influence dysregulated fear responses. Further, PTSD is a heterogeneous disorder with symptoms ranging from hyperarousal, intrusive reminders of the trauma, avoidance of trauma-related cues, to negative cognitions, and depressed mood. The heterogeneity and clear sex differences observed in PTSD imply the involvement of multiple underlying neurobiological mechanisms (Norrholm & Jovanovic, 2010). The purpose of this review chapter is to address how ovarian hormones and neuroactive metabolites are important moderators of risk vs resiliency and maintenance of PTSD.

1.1 Women and PTSD

While women are at a greater risk for developing PTSD, the rates of PTSD are much higher in specific populations of women at risk of experiencing trauma; this includes military personnel and individuals living in low socioeconomic status urban environments (Keane, Marx, & Sloan, 2009; Kessler, Sonnega, Bromet, Hughes, & Nelson, 1995). With regards to the military, recent drawdowns of military forces have led to an increasing number of returning Service Members and Veterans seeking care for PTSD. Approximately one-fourth of returning Operation Iraqi Freedom, Operation Enduring Freedom, and Operation New Dawn (OIF/OEF/OND) Veterans received mental health diagnoses upon return, and the most common service-related mental health diagnosis was PTSD (Seal, Bertenthal, Miner, Sen, & Marmar, 2007). Additionally, an increasing number of women are serving in the military and now comprise approximately 8% of surviving Veterans (Street, Vogt, & Dutra, 2009). This is significant given that women in the military are more likely to develop PTSD (27%) compared to their civilian (10–12%) counterparts (Kessler et al., 1995; Resnick et al., 1993). This is due, in part, to the fact that they are more likely than civilians to have experienced assault, childhood sexual abuse, and the additional risks of combat exposure and military sexual trauma (MST, Zinzow, Grubaugh, Monnier, Suffoletta-Maierle, & Frueh, 2007). In fact, female Veterans with MST are nine times more likely than female Veterans without sexual assault histories to develop PTSD (Suris, Lind, Kashner, Borman, & Petty, 2004). Similarly, studies have shown the highest rates of PTSD are among sexually assaulted (Kessler et al., 1995; Norris, 1992; Resnick et al., 1993) and combat-exposed individuals (Kang, Dalager, Mahan, & Ishii, 2005).

There is growing evidence from both human and animal studies suggesting an important role of ovarian hormones in anxiety-related sex differences, with women being more likely to report symptoms of depression, anxiety, and elevated fear responding during periods of relatively low estrogen levels (Cloitre et al., 2004; Dean & Kendell, 1981; Douma, Husband, O'Donnell, Barwin, & Woodend, 2005; Epperson, Kim, & Bale, 2014; Pigott, 2003; Rapkin, Mikacich, Moatakef-Imani, & Rasgon, 2002; Wegerer, Kerschbaum, Blechert, & Wilhelm, 2014; Yonkers, 1997). Further, fluctuations in ovarian hormones over the course of the menstrual cycle (Bryant et al., 2011; Ertman, Andreano, & Cahill, 2011; Glover et al., 2013) and pregnancy (Michopoulos et al., 2014) are associated with changes in PTSD symptom severity and profiles; however, the mechanisms underlying these hormone-induced symptom changes remain poorly understood. While estradiol (E2) targets a myriad of neurochemical systems that regulate behavior and physiology (Michopoulos, Checchi, Sharpe, & Wilson, 2011; Young, Altemus, Parkison, & Shastry, 2001), including reproduction, it is likely the contribution of E2 (Glover et al., 2012), progesterone (P4, Soni, Curran, & Kamboj, 2013), and neuroactive metabolites such as allopregnanolone (ALLO, Rasmusson et al., 2006) that ultimately moderate these effects of sex steroids on PTSD symptomatology. Additionally, these hormones affect behavior and physiology via cyclical fluctuations across the menstrual cycle (Brambilla et al., 2003; Sigmon et al., 2000). However, despite the established sex differences observed in clinical populations, it remains unclear how high levels of trauma exposure interact with these hormones to ultimately moderate an individual's vulnerability to, and subsequent development and manifestation of, fear-related psychopathology.

Ovarian hormones have unique and diverse mechanisms of action, and understanding their role in the development and maintenance of PTSD may lead to novel pharmacological and behavioral treatments. For example, E2 has effects on serotonergic brain systems (for a review, see Bethea, Lu, Gundlah, & Streicher, 2002; Rubinow, Schmidt, & Roca, 1998), and ALLO is a potent type-A gamma-aminobutyric acid receptor ($GABA_A$) positive allosteric modulator and that can produce anxiolytic effects (Engin & Treit, 2007). Both serotonergic and GABAergic systems are therapeutic targets of PTSD (e.g., selective serotonin reuptake inhibitors and benzodiazepines, respectively) and the neurobiology of comorbid mental disorders; thus underscoring the clinical importance of disentangling the interaction between ovarian hormones and underlying fear circuitry relevant to PTSD.

While evidence indicates there are clear underlying sex differences in fear circuitry, it is unclear how high levels of traumatic stressors interact with sex to influence dysregulated fear responses. The newly announced National Institutes of Health (NIH) policies to promote the inclusion of female subjects in preclinical trials further highlight the need for translational methods of research that can more easily "bridge the gap" between meaningful clinical sex differences and underlying neurocircuitry. Thus, another the goal of the current chapter is to highlight how a multidimensional assessment of ovarian hormones across behavioral and neurobiological measures, or units of analysis, is necessary to uncover sex-specific differences in dysregulations of fear-related pathways in response to trauma. Finally, the identification of genes underlying PTSD-related symptomology and relevant neurobiology will be discussed.

1.2 Sex Differences in the Brain and in Behavior

Sexual differentiation begins with genetic sex or more specifically, the presence or absence of the sex-determining region of the Y chromosome, SRY. Presence of this region of the Y chromosome is necessary and sufficient to cause male development (Koopman, Gubbay, Vivian, Goodfellow, & Lovell-Badge, 1991), or the development of testes as opposed to ovaries. Once either the testes or ovaries develop, early secretion of androgens or estrogen, respectively, initiates the development of primary sex characteristics and, in turn these time-dependent *organizational effects* permanently alter brain structure and behavior (Christensen & Gorski, 1978). The *activational effects* of hormones, or those secreted directly by the gonads during puberty, will subsequently drive the remainder of sexual differentiation. The organizational–activational theory was first introduced experimentally in 1959 with guinea pigs as a means of assessing whether administration of testosterone during gestation would govern whether adult offspring would behave like males or females (e.g., with the absence or presence of lordosis, respectively, Phoenix, Goy, Gerall, & Young, 1959). Phoenix et al. (1959) found that adult female guinea pigs whose mothers had prenatal exposure to testosterone exhibited less lordosis and more mounting as adults as compared to control females; thus, fetal testosterone masculinized the female subjects

(Phoenix et al., 1959). This effect was permanent as it extended into adulthood. Phoenix and colleagues argued that the organizational actions of hormones permanently altered the substrate on which gonadal hormones act in adulthood.

More recent studies have shown that testosterone released from testes diffuses into the male brain where aromatase enzymes function to locally convert testosterone to estradiol, which then initiates the process of masculinization. Therefore, paradoxically, treating genetically female animals with estrogen will actually impair feminine-typical development and promote masculine-typical behavioral and brain development (Whalen & Olsen, 1981). Circulating estradiol is sequestered in the bloodstream via α-fetoprotein binding, which accounts for the lack of masculinization by placental estradiol in unaltered females (McCarthy, 2008; Naftolin et al., 1975). The resulting sexual dimorphisms in the brain can affect any number of neurobiological indices (e.g., neuroanatomical and neurochemical) including, but not limited to, the number or shape of synapses, length of dendrites, number of neurons, and neurotransmitter levels (for a review, see McCarthy, 2008).

1.2.1 Estrogen

The biosynthesis of estrogen is initiated by the synthesis of androstenedione from cholesterol. Aromatase, a cytochrome P450 enzyme, catalyzes the biosynthesis of estrogen from androgens. The most common (and potent) endogenous estrogen produced is 17B-estradiol (E2). In premenopausal women, estrogens are produced primarily in the ovaries and in smaller concentrations in extragonadal sites including adipose tissues, adrenal cortex, and across multiple regions in the brain (Simpson, 2003). Accordingly, the principal sites of aromatase expression are the ovarian granulosa cells in which androgen precursors are derived primarily from the adjacent theca cells (Simpson et al., 1994). In the brain, varying cell types (e.g., neurons and astrocytes) express aromatase, and testosterone has been shown to stimulate aromatase activity in brain tissue (Balthazart, Foidart, & Hendrick, 1990). In postmenopausal women (and men), the primary mechanism for estrogen formation is the aromatization of androgens in adipose cells and skin fibroblasts (Grodin, Siiteri, & MacDonald, 1973). Early studies in women demonstrated that the fractional conversion of androstenedione to estrone (one of three hormones in the estrogen family) increases as a function of obesity and age due to increased aromatase messenger RNA (mRNA) levels and increased aromatase activity (Edman & MacDonald, 1978; Grodin et al., 1973). Expression of E2 is regulated via aromatase expression, the catalyst for E2 synthesis. Hormones and second messengers regulate distinct promoters for the aromatase gene, *CYP19*, in a tissue-specific manner, which may ultimately allow for targeted modifications of E2 and subsequently a wide array of hormone-induced behaviors (for a review, see Simpson, 2004).

Circulating E2, and associated feedback mechanisms on the brain, result in reproductive cycles observed in women. Prior to ovulation, increased gonadotropin-releasing hormone (GnRH) secretion from the hypothalamus results in the release of luteinizing hormone (LH) and follicle-stimulating hormone (FSH) from the anterior pituitary gland in a pulsatile fashion. However, at initial low concentrations of E2, LH release is inhibited. FSH is released in response to low E2 levels, and as E2 levels rise, FSH hormones decrease. In the ovaries, FSH acts through cyclic AMP (cAMP) to regulate aromatase production via the proximal promoter II (Simpson, 2003). Resulting high levels of E2 exert positive feedback, causing a surge of LH, which triggers ovulation. The cellular mechanisms of E2-induced positive feedback on LH are not fully understood. After ovulation, progesterone is released and exerts negative feedback on the hypothalamus resulting in reduction of GnRH release and ultimately decreased LH and FSH. Steroid synthesis is primarily maintained by the corpus luteum; once it is degraded, steroid synthesis of E2 and P4 by the ovaries is reduced, resulting in disinhibition of GnRH negative feedback, and the cycle begins again.

E2 signaling occurs through interactions with two distinct estrogen receptors (ERs), ERα and ERβ. These receptors are both part of the nuclear receptor family of transcription factors (TFs) and have similar affinities for E2 and bind the same DNA response elements (for a review, see Heldring et al., 2007). The ERs are predominately nuclear localized and contain (1) a highly conserved DNA-binding domain, (2) a ligand-binding domain, located on the carboxyl-terminal, and (3) a variable amino-terminal domain. Transcriptional activation is achieved through distinct ligand-inducible activation functions (AF) and the constitutively active AF (Nilsson et al., 2001). Transcriptional activation requires ligand binding, receptor dimerization, receptor DNA binding, interaction with cofactors and other TFs, and formation of a preinitiation complex. A ligand-bound ER in the cell nucleus can bind directly to estrogen response elements (EREs) in target gene promoter regions (Nilsson et al., 2001). ERα and ERβ can also exert effects without direct binding to DNA and can thus influence the expression of genes that do not contain EREs; this is accomplished through interactions with other TF complexes (Kushner et al., 2000).

In addition to development and maintenance of sexual and reproductive function, E2 influences a wide array of nonreproductive behavior such as learning and memory (Kramar, Babayan, Gall, & Lynch, 2013; Sherwin, 1994), neuroprotection (for a review, see Brann, Dhandapani, Wakade, Mahesh, & Khan, 2007), pain perception (e.g., Hellstrom & Anderberg, 2003), aggression (Unger et al., 2015), and feeding (Dagnault & Richard, 1997; Santollo, Torregrossa, & Eckel, 2011). E2 exerts its effects on behavior through a number of mechanisms and interactions with other endocrine systems. For example, E2 affects neurotransmitter systems, including serotonin. Specifically, in the dorsal raphe of macaques, ERβ mRNA showed localization with the serotonin reuptake transporter gene marker in serotonin and nonserotonin neurons (Gundlah, Lu, Mirkes, & Bethea, 2001).

Women are more likely to report symptoms of PTSD, depression, and anxiety during premenstrual, postpartum, and peri- and postmenopausal periods, when E2 levels are relatively low (Cloitre et al., 2004; Dean & Kendell, 1981; Douma et al., 2005; Epperson et al., 2014; Pigott, 2003; Rapkin et al., 2002; Yonkers, 1997). Additionally, women recruited from a hospital who experienced a traumatic event during the luteal phase of their cycle, when E2 levels are higher, were almost five times more likely to have flashbacks of the traumatic event (Bryant et al., 2011). Menstrual cycle phase is also relevant to fear learning and inhibition of fear. One study found that women with PTSD showed impaired retention of fear extinction during the mid-luteal phase (Pineles et al., 2016). On the other hand, another study found that the follicular phase was associated with impairment of fear inhibition (Glover et al., 2013). Glover et al. (2012) similarly demonstrated low E2 levels were associated with impaired inhibition of fear and decreased extinction of fear in women with PTSD compared to trauma-exposed women without PTSD. These findings indicate that timing of trauma exposure and/or extinction-based treatment during the cycle may have different clinical outcomes dependent on circulating levels of E2.

In healthy women, low levels of E2 promote stronger and more extinction-resistant conditioned emotional reactions and stronger, subsequent intrusive memories (Wegerer et al., 2014). Similarly, a recent study demonstrated that low estradiol levels in women with increased sympathetic arousal (measured via pupil diameter changes) facilitated memory bias for negative events compared to neutral ones when the stressor was administered immediately prior to encoding (Nielsen, Barber, Chai, Clewett, & Mather, 2015). When a stressor was administered during encoding, levels of estradiol were positively related to intrusive memories of negative images, consistent with estrogen enhancement of learning and memory under conditions of stress (Cheung, Chervonsky, Felmingham, & Bryant, 2013). A study in healthy women showed that administration of synthetic E2 increased extinction recall (Graham & Milad, 2013). A recent study utilizing functional magnetic resonance imaging showed that adrenal and gonadal hormone abnormalities are associated with brain activity deficits in stress response circuits in a sex-dependent manner. That is, hypercortisolemia was present in both men and women with psychoses, but in men hypercortisolemia was associated with hyperactivity in stress response regions (right hypothalamus, left anterior cingulate cortex, medial prefrontal cortex, bilateral orbitofrontal cortex, right parahippocampal gyrus, and periaqueductal gray), and in women with psychoses, hypercortisolemia was associated with hyporeactivity in the medial prefrontal and orbitofrontal cortices (Goldstein et al., 2015).

1.2.2 Progesterone/Allopregnanolone

In nonpregnant women, P4 is primarily responsible for the release of mature oocytes and mediation of sexually responsive behavior (Graham & Clarke, 1997). P4 is released from the corpus luteum during the luteal phase on the menstrual cycle. Progesterone receptor (PR) expression is increased by estrogen (Leavitt, Chen, & Allen, 1977). The PR is also a TF that interacts with progestin response elements in target genes (Gronemeyer, 1991). PR expression is found in reproductive tissue and centrally in the pituitary, ventromedial hypothalamus, and preoptic areas (Kato, Onouchi, & Okinaga, 1978). Additionally, levels of the neuroactive steroid ALLO, a metabolite of P4, vary across the menstrual cycle with higher levels observed in the luteal compared to the follicular phase. ALLO is a potent agonist of the $GABA_A$ receptor with anxiolytic effects.

There are stress-induced changes in central levels of P4 and ALLO. Specifically, P4 (in addition to pregnenolone, 3α, 5α-TH PROG, and deoxycorticosterone) is increased in response to acute exposure to stressors such as carbon dioxide (Barbaccia, Roscetti, Bolacchi, et al., 1996; Barbaccia, Roscetti, Trabucchi, et al., 1996) and foot shock (Barbaccia, Roscetti, Bolacchi, et al., 1996; Barbaccia, Roscetti, Trabucchi, et al., 1996). P4 has also been shown to have anxiolytic effects (Fernandez-Guasti & Picazo, 1995); yet, blocking the conversion of P4 to ALLO precluded the previously observed anxiolytic effects, suggesting they are mediated through ALLO's actions at $GABA_A$ receptors (Bitran, Shiekh, & McLeod, 1995). Further,

ALLO administration in rats decreases stress-induced burying behavior (Fernandez-Guasti & Picazo, 1995).

While the effects of progesterone on fear physiology are less clear, recent evidence in rodents suggests P4 interacts with E2 in a biphasic and time-dependent manner to mediate the facilitating effects of E2 on extinction (Graham & Daher, 2016). Specifically, E2-dependent facilitation of extinction was potentiated by P4 when administered 6 h prior to extinction training, but was abolished when P4 was administered 24 h prior to extinction training. Similarly, pharmacological blockade of the P4 receptor during proestrus, when E2 and P4 levels peak, prevented impaired extinction during metestrus. These results suggest that the enhanced fear extinction observed in proestrus is mediated by E2, whereas, the impairments in extinction during metestrus are mediated by increased levels of P4 (Graham & Daher, 2016). In healthy women, the E2-to-P4 ratio, but not E2 or P4 alone, negatively correlated with greater spontaneous intrusions (e.g., visual image or verbal thought), reported 72 h after viewing a distressing film (Soni et al., 2013). Interestingly, Nielsen and colleagues also found that in women with increased sympathetic arousal low levels of P4 also facilitated a memory bias for negative events compared to neutral ones, and this may be due to the combined effects of low E2 and P4 (Nielsen et al., 2015).

Collectively, these studies suggest an important role of E2 and P4 in the vulnerability to, and development of, fear-related pathophysiology; however, this does not preclude the involvement of potent neuroactive metabolites such as ALLO. In regards to PTSD, studies have shown that central levels of ALLO are decreased in women with PTSD compared to healthy controls (Rasmusson et al., 2006). While the mechanisms underlying these effects are not entirely understood, studies suggest a role of neuroactive steroids in mediating other psychological disorders including major depressive disorder (MDD) and panic disorder. In patients with MDD, lower levels of ALLO were normalized after successful treatment with antidepressants (Romeo et al., 1998). Further, greater than normal plasma concentrations of ALLO have been observed in patients with panic disorder (Strohle et al., 2002).

Ovarian hormones likely affect behavior and physiology via cyclical fluctuations across the menstrual cycle (Brambilla et al., 2003; Sigmon et al., 2000). For example, in patients with panic disorder, the pattern of secretion of P4 and ALLO is altered. Specifically, patients with panic disorder displayed higher levels of P4 in the mid-luteal and premenstrual phase, and increased ALLO in the early follicular phase (Brambilla et al., 2003) compared to healthy controls. A compensatory mechanism to counteract underlying hypothalamic–pituitary–adrenal (HPA) axis hyperactivity and reduced $GABA_A$ receptor sensitivity has been proposed to explain the observed hypersecretion of neurosteroids. Another study determined that increased endogenous cycling hormones and exogenous hormones (i.e., oral contraceptives) were associated with decreased functional connectivity in the anterior portion of the default mode network and executive control network, which are associated with cognition and affect (Petersen, Kilpatrick, Goharzad, & Cahill, 2014). It is likely that the contribution of cyclical fluctuations of E2, P4, and ALLO across the menstrual cycle ultimately mediate the observed hormone-induced symptom changes in PTSD and fear-related pathologies.

Overall, these steroid hormones are complex; E2, P4, and ALLO are secreted cyclically in the periphery, de novo in the brain (Resko & Roselli, 1992), and are influenced by natural fluctuations or are maintained at constitutive levels due to exogenous hormone interventions (e.g., contraceptives and hormone replacement therapy). Further, target effects of these sex steroids are dependent on many factors, including rate of synthesis and degradation/clearance. However, these complex interactions represent potential avenues for targeted and personalized treatment and intervention strategies directed toward the growing population of women at risk for trauma (e.g., veterans) and the development of PTSD.

While the above-mentioned studies (summarized in Table 1) suggest that relative levels and cyclical secretions of ovarian hormones and metabolites across the reproductive cycle may play a critical role in women's increased vulnerability to, and severity of, PTSD symptoms, the mechanisms underlying these hormone-induced symptom changes are not well understood.

1.3 Reproductive Hormones Influence Stress Axis Function

Stress exposure activates the HPA axis via the actions of corticotropin-releasing hormone (CRH) and cortisol and mediates arousal through the autonomic nervous system. Acute stress results in increased sympathetic activation and increases in norepinephrine (NE) and epinephrine (E). These systems subsequently interact with ovarian hormones and metabolites in a complex manner. Specifically, the activity of the HPA axis influences both hormone levels and sensitivity to ovarian hormones in females. For example, glucocorticoids (GCs) regulate estrogen synthesis via regulation of *CYP19* expression (Simpson et al., 1997) and E2 directly stimulates the CRH

Table 1 Clinical Studies of Fear-Related Circuitry in Women

	Sample			Results				Hormone Levels		
Study	**Overview**	**Age/Hormone Considerations**	***N***	**Measure**	**Unit**	**Outcome**	**Phase**	**E2**	**P4**	**ALLO**
Bryant et al. (2011)	Female patients admitted to hospital after traumatic injury	16–50 years old; cycles < 32 days	138	Clinician Administered PTSD Scale for DSM-IV	Self-report	Women who experienced their trauma during the luteal phase were 3.64 times, and women in the luteal phase at assessment were 4.89 times more likely to have flashbacks in the acute period after trauma	Mid-luteal phase (days 18–24)			
Glover et al. (2013)	Exp. 1: Healthy women Exp. 2: Female traumatized clinical population	Exp. 1: 18–55 years old; regular menstrual cycles Exp. 2: Not pregnant	Exp. 1: 28 Exp. 2: 44	Conditional discrimination paradigm	Paradigm	Exp. 1: Women in the luteal, but not follicular, phase showed significant discrimination and fear inhibition Exp. 2: Women in the high, but not low, estradiol group showed significant discrimination and fear inhibition	Exp. 1: Follicular (days 2–12); Luteal (days 17–25)	Exp. 2: Low mean 6.2 (2.68) pg/mL High mean (SD) 95.4 (106.24) pg/mL (serum)		
Glover et al. (2012)	Female traumatized clinical population	Not pregnant	81	Extinction; fear-potentiated startle; Modified PTSD Symptom Scale	Paradigm; physiology; self-report	In PTSD + the low E2 group had significantly higher levels of fear-potentiated startle during extinction compared to the high E2 group		Low mean 8.00 pg/mL; High mean 92.50 pg/mL (serum)		

Continued

Table 1 Clinical Studies of Fear-Related Circuitry in Women—cont'd

	Sample			Results				Hormone Levels		
Study	**Overview**	**Age/Hormone Considerations**	***N***	**Measure**	**Unit**	**Outcome**	**Phase**	**E2**	**P4**	**ALLO**
Pineles et al. (2016)	Nonmedicated female trauma survivors with and without PTSD	19–46 years old; regular menstrual periods; not using hormonal contraceptives; not pregnant	32	Fear conditioning, Skin conductance; Clinician Administered PTSD Scale for DSM-IV	Paradigm; physiology; self-report	Women with, compared to without, PTSD showed facilitated extinction and impaired retention of extinction learning in the mid-luteal phase. Women without, compared to with, PTSD showed facilitated acquisition and extinction in the early follicular phase but impaired retention of extinction		Early follicular phase mean (SD) 26.52 (14.59) pg/mL Mid-luteal phase mean (SD) 72.22 (22.95) pg/mL (plasma)	Early follicular phase mean (SD) 0.64 (0.93) ng/ml Mid-luteal phase mean (SD) 13.12 (6.65) ng/ml (plasma)	
Wegerer et al. (2014)	Healthy women	18–35 years old; free of hormonal contraceptives for at least 3 months, report regular menstrual cycles, and no diagnosis of premenstrual dysphoric disorder	37	Adapted conditioned-intrusion paradigm; skin conductance; adapted version of the Intrusion Memory Questionnaire	Paradigm; physiology; self-report	Women with lower levels of estradiol displayed larger differential responses during extinction and reported overall larger intrusive memory strength relating to the film clips used in the conditioning procedure		Mean (SD) 6.23 (3.90) pg/mL (saliva)	Mean (SD) 191.17 (89.01) pg/mL (saliva)	

Graham and Milad (2013)	Healthy women	18–30 years old Exp. 1: Naturally cycling women and those using combined monophasic HCs Exp. 5: Naturally cycling women ~5 days after the onset of menstruation (early follicular phase)	Exp. 1: 45 Exp. 5: 31	Fear conditioning; skin conductance	Paradigm; physiology	Exp. 1: Women using hormonal contraceptives exhibited significantly poorer extinction recall compared to naturally cycling women Exp. 5: A single administration of estradiol to naturally cycling women with low E2 enhanced their ability to recall extinction memories	Exp. 1: Naturally cycling: High E2 mean (SEM) 222.6 (22.4) pg/mL Low E2 mean (SEM) 51.9 (8.3) pg/mL [a]Exp. 5 Estradiol-treated mean (SD) 137.98 (113.59) pg/mL Placebo mean (SD) 43.43 (20.72) pg/mL (serum)	Exp. 1: Naturally cycling: High E2 mean 9.61 (2.06) ng/mL Low E2 mean 3.18 (1.19) ng/mL [a]Exp. 5 Estradiol-treated mean (SD) 0.74 (0.3) ng/mL Placebo mean (SD) 1.24 (3.07) ng/mL (serum)
Cheung et al. (2013)	University students		55 (26 men/ 29 women)	Cortisol measurements; adapted intrusion subscale of the impact of event Scale; viewing negative and neutral images	Molecules; self-report; paradigms	Levels of estradiol rather than progesterone were positively related to intrusive memories of negative images	Stress condition (cold water) mean (SD) 2.15 (0.96) pmol/L Control condition (warm water) mean (SD) 2.25 (0.89) pmol/L (saliva)	Stress condition (cold water) mean (SD) 51.52 (30.95) nmol/L Control condition (warm water) mean (SD) 43.79 (21.85) nmol/L (saliva)

Continued

Table 1 Clinical Studies of Fear-Related Circuitry in Women—cont'd

	Sample			Results				Hormone Levels		
Study	**Overview**	**Age/Hormone Considerations**	***N***	**Measure**	**Unit**	**Outcome**	**Phase**	**E2**	**P4**	**ALLO**
Goldstein et al. (2015)	Women with psychoses and healthy controls	Not in menopause	32	Cortisol; medial prefrontal and orbitofrontal cortices; diagnostic interviews; visual stress challenge	Molecules; circuits; self-report; paradigms	In women with psychoses hypercortisolemia was associated with hyporeactivity in the medial prefrontal and orbitofrontal cortices		Psychoses mean (SD) 63.76 (43.61) pg/ml Healthy control mean (SD) 90.96 (64.77) pg/ml (serum)	Psychoses mean (SD) 4.20 (5.82) ng/ml Healthy control mean (SD) 2.02 (2.94) ng/ml (serum)	
Soni et al. (2013)	Healthy women	18–35 years old; predictable menstrual cycle length of between 26 and 34 days; no hormonal contraception use within the past 3 months	45	Online diary to report intrusive memories; viewing a distressing film	Self-report; paradigms	Intrusion frequency occurred more frequently in the early-luteal group compared to mid-follicular and late-luteal groups and was negatively correlated with the E2-to-P4 ratio, but not E2 or P4 alone	Mid follicular: 25–39% of way through cycle, i.e., for a 28 day cycle days 7–11; early-luteal: 57–71% of cycle; and late-luteal: 86–100% of cycle	Mid-follicular mean (SD) 5.63 (2.02) pg/ml Early luteal mean (SD) 6.11 (2.57) pg/ml Late luteal mean (SD) 5.86 (2.52) pg/ml (saliva)	Mid-follicular mean (SD) 47.25 (48.93) pg/ml Early luteal mean (SD) 125.95 (124.24) pg/ml Late luteal mean (SD) 200.18 (183.88) pg/ml (saliva)	
Nielsen et al. (2015)	Female undergraduates	18–34 years old; naturally cycling women and women on combined contraceptive formulations	96	Pupil dilation; viewing a series of negative, positive and neutral images	Physiology; paradigms	Low estradiol levels in women with increased sympathetic arousal facilitated a memory bias for negative events compared to neutral ones		High mean (SD) 3.05 (0.08) pg/mL Low mean (SD) 1.56 (0.08) pg/mL (saliva)	High P4 mean 145.4 (6.9) pg/mL Low P4 mean (SD) 38.6 (6.8) pg/mL (saliva)	

(Petersen et al., 2014)	Healthy controls; naturally cycling women and women using combined oral contraceptive pills	18–40 years old; not pregnant or breastfeeding	91	Default mode network and executive control network	Circuits	Endogenous cycling hormones and exogenous hormones (i.e., oral contraceptives) were associated with decreased functional connectivity in the anterior portion of the default mode network and executive control network	(saliva)			
Brambilla et al. (2003)	Female outpatients with panic disorder and agoraphobia and healthy female volunteers matched for age	19–45 years old	35 (25 patients, 11 controls)	Mini International Neuropsychiatric Interview-Plus (MINI) for DSM-IV	Self-report	Patients with panic disorder displayed higher levels of P4 in the mid-luteal phase, and increased ALLO in the early follicular phase		Panic disorder day 7 mean (SD) 9.2 (4.8) ηg/ml Control day 7 mean (SD) 5.3 (3.6) ηg/ml (plasma)	Panic disorder day 22 mean (SD) 4.0 (1.7) ηg/ml Control day 22 mean (SD) 2.6 (0.8) ηg/ml (plasma)
Strohle et al. (2002)	Patients with panic disorder and control participants		10 patients (7 women and 3 men); 10 healthy age- and sex-matched subjects	Structured Clinical Interview for DSM-IV	Self-report	Patients with panic disorder had greater concentrations of 3α,5α-THP and 3α,5β-THP and significantly lower concentrations of 3β,5α-THP			Panic disorder mean (SD) 2.6 (0.4) nmol/liter Control mean (SD) 2.1 (0.6) nmol/liter (plasma)
Rasmusson et al. (2006)	Premenopausal women with and without PTSD	Lumbar CSF samples collected in follicular phase	9 women with PTSD; 10 controls	Structured Clinical Interview for DSM-IV	Self-report	Decreased central levels of ALLO in women with PTSD			ALLO levels in PTSD women <39% that of controls (CSF)

[a]Femtrace administration selectively enhanced serum estradiol during extinction training, as estradiol was significantly greater in estradiol-treated women at T2, but not at any other point and as such data for this time point are represented.

receptor gene promoter (Vamvakopoulos & Chrousos, 1993). Animal model studies in rodents and nonhuman primates have employed chronic stress models to show that stress exposure disrupts the behavioral and physiological effects of E2 (Keen-Rhinehart et al., 2009; Michopoulos et al., 2011).

Chronic stress exposure can exert an inhibitory effect on the female reproductive system. For example, stress-induced CRH and cortisol release inhibit the secretion of the primary regulator of the female reproductive system, gonadotropin-releasing factor (GnRH) and result in subsequent reduction of LH (Petraglia, Sutton, Vale, & Plotsky, 1987). Thus, heightened cortisol and CRH activity under conditions of chronic stress inhibit the secretion of E2 and P4 thus inducing a hypogonadal state (for a review, see Meczekalski, Katulski, Czyzyk, Podfigurna-Stopa, & Maciejewska-Jeske, 2014). GCs also inhibit E2-stimulated uterine growth (Rabin, Johnson, Brandon, Liapi, & Chrousos, 1990). Understanding the relationship between traumatic stress exposure, chronic stress, and fertility is especially relevant in certain traumatized populations, like female Veterans. Specifically, one study demonstrated 31% of female Veterans reported problems with ovulation and 15.8% reported a lifetime history of infertility for themselves or their partners (Katon et al., 2014). Additionally, female Veterans with mental health disorders had a greater prevalence of reproductive health problems including infertility compared to women without mental health disorders; however, the effects of trauma have not been considered in these studies (Cohen et al., 2012). Interestingly, studies have shown that cognitive behavioral therapy, a treatment option also utilized for PTSD, restores neuroendocrine aberrations that contribute to reproductive dysfunction and infertility in women (Berga & Loucks, 2006; Michopoulos, Mancini, Loucks, & Berga, 2013).

Sex hormones may account for some of the variability in the findings of alterations in HPA function in PTSD. Specifically, PTSD has been associated with low basal cortisol levels (Yehuda, 2002), exacerbated GC negative feedback inhibition of the HPA axis (Yehuda, Halligan, Grossman, Golier, & Wong, 2002), and increases in cortisol response to trauma-related cues (Elzinga, Schmahl, Vermetten, van Dyck, & Bremner, 2003; Meewisse, Reitsma, De Vries, Gersons, & Olff, 2007; Yehuda, Golier, & Kaufman, 2005; Yehuda, Teicher, Trestman, Levengood, & Siever, 1996). However, other studies have shown heightened and no group differences in basal cortisol and cortisol response to a stressor (Bremner et al., 2003; Gola et al., 2012). Some discrepancies in the data could be due to the sample composition of each study (including sex differences). For example, a recent study that included only women showed that a subgroup with severe PTSD displayed a blunted HPA axis response following exposure to an acute social stress (Zaba et al., 2015). In an earlier study conducted with healthy controls, exposure to an acute stressor resulted in salivary cortisol responses that differed across sex and cycle phase (Kirschbaum, Kudielka, Gaab, Schommer, & Hellhammer, 1999). Further, data indicate that decreased basal cortisol levels are found in women, but not men, with PTSD (Freidenberg et al., 2010). Taken together, these somewhat equivocal results represent the complex relationship between chronic and acute stress, sex hormone levels, and trauma history. Disentangling the relative contributions of each factor may elucidate how high levels of traumatic stress exposure interact with gonadal hormones to influence dysregulated fear circuitry and PTSD symptoms. Taken together, these data indicate that while HPA axis circuitry and autonomic regulation are dysregulated in PTSD, it remains unclear how these perturbations interact with ovarian hormones to influence fear-related symptomology and psychophysiology.

1.4 Translational Approaches to Uncovering Sex-Specific Differences in PTSD

While clinical diagnostic criteria for PTSD include hyperarousal, intrusive reminders of the trauma, avoidance of trauma-related cues, negative cognition, and depressed mood, many research studies have adopted a biological approach to the study of mental illness, consistent with Research Domain Criteria (RDoC, Cuthbert, 2015), to study PTSD. This growing multidimensional approach focuses on the underlying neurobiological mechanisms that are shared among mental disorders.

Prior to the development of RDoC, our understanding of PTSD has greatly benefited from the use of translational measures to assess PTSD. Cumulative efforts to classify PTSD-related psychopathology based on "dimensions of observable behavioral and neurobiological measures" have identified promising putative biomarkers for PTSD (Cuthbert, 2015), which include monoaminergic neurotransmitter systems, the HPA axis, metabolic hormonal pathways, inflammatory mechanisms, psychophysiological reactivity, and neural circuits (for a review, see Michopoulos, Norrholm, & Jovanovic, 2015). In order to better understand the interaction between neurobiological influences on disease risk and symptom progression, we must continue to employ multidimensional models that are derived from comprehensive molecular, neurobiological, behavioral, and clinical phenotypes (Michopoulos et al., 2015).

1.4.1 RDoC Constructs

PTSD is a complex, heterogeneous disorder and it has recently been redefined in the *Diagnostic and Statistical Manual of Mental Disorders*, 5th edition (DSM-5) from an anxiety disorder to a trauma- and stressor-related disorder. Further, the RDoC matrix has multiple constructs related to threat responses, specifically (1) acute threat (fear), (2) potential threat (anxiety), and (3) sustained threat. It is important to consider that in women exposed to trauma and those with PTSD, chronic and traumatic stress may present in different forms. This review will focus within the negative valence system and on the following constructs: acute threat (fear) and potential harm (anxiety).

The RDoC matrix (http://www.nimh.nih.gov/research-priorities/rdoc/constructs/rdoc-matrix.shtml) defines fear as "activation of the brain's defensive motivational system to promote behaviors that protect the organism from perceived danger. Normal fear involves a pattern of adaptive responses to conditioned or unconditioned threat stimuli (exteroceptive or interoceptive). Fear can involve internal representations and cognitive processing, and can be modulated by a variety of factors" and anxiety as "activation of a brain system in which harm may potentially occur but is distant, ambiguous, or low/uncertain in probability, characterized by a pattern of responses such as enhanced risk assessment (vigilance). These responses to low imminence threats are qualitatively different than the high imminence threat behaviors that characterize fear."

Notably, there are potential confounds in assessing the range of normal to abnormal fear responses as we move across the "translational bridge" from animal models to clinical studies. Briefly, in order to determine if animal models meet the criteria for modeling a psychiatric disorder, introduced by Willner (1986) and later refined by several groups including Belzung and Lemoine (2011), the model has to meet the following five core criteria: homological validity, pathogenic validity, mechanistic validity, face validity, and predictive validity (Belzung & Lemoine, 2011; Willner, 1986). Homological validity (which includes species and strain validity) requires that the animal model chosen be appropriate considering the targeted human condition. Pathogenic validity requires that the disease process of interest proceed in a similar manner in both the animal model and targeted human system. Mechanistic validity encompasses a requirement that underlying cognitive and biological processes occur in both the model and target systems. Face validity is evident if similar behavioral outcomes are observed in both the animal model and the human target. Lastly, predictive validity requires that the progression of the disease process and any response to treatment is similar across the model and target organism. Animal models of PTSD have achieved some success in terms of pathogenic validity for some of the clusters of the disorder but not the complete disease process; however, complex human behaviors such as intrusive thoughts and nightmares are difficult if not impossible to model in animals.

2. CASE EXAMPLE: PACAP RECEPTOR GENETICS

In 2011, our lab implicated the pituitary adenylate cyclase-activating polypeptide (PACAP)-PAC1 receptor pathway in human psychological stress responses, specifically, in PTSD (Ressler et al., 2011). We found, in heavily traumatized subjects, a sex-specific association of PACAP blood levels with fear physiology, PTSD diagnosis and symptoms in women, but not men (Ressler et al., 2011). Further, we found a sex-specific association of a single nucleotide polymorphism (SNP) residing in a putative ERE within the *ADCYAP1R1* gene (PAC1R) that predicts dysregulated psychophysiological responses, PTSD diagnosis, and symptoms in women (Ressler et al., 2011). PACAP has pleiotropic functions and has recently been identified as a key regulator of the stress response. The resulting neurobiological and behavioral sequelae associated with the "CC" risk genotype demonstrate the utility of employing a dimensional neurobiological approach to studying neuropsychiatric disorders.

PACAP was discovered from hypothalamic tissue through its ability to stimulate adenylyl cyclase from rat anterior pituitary cell cultures (Miyata et al., 1989). While the highest amounts are found in the hypothalamus (Arimura et al., 1991), significant concentrations are also found in other brain regions, including the cerebral cortex, the hippocampus, the thalamus, the striatum, the nucleus accumbens, the substantia nigra, the locus coeruleus, and the pineal gland (Ghatei et al., 1993; Koves, Arimura, Gorcs, & Somogyvari-Vigh, 1991; Palkovits, Somogyvari-Vigh, & Arimura, 1995). PACAP exists as a 38-residue polypeptide (PACAP-38), which is more abundant than the 27-residue peptide, PACAP-27, in most tissues (e.g., hypothalamus, adrenal gland, testes, digestive tract, testes, and peripheral tissues; Arimura et al., 1991). PACAP is likely the ancestral molecule to its superfamily of peptides consisting of nine hormones, which includes glucagon, secretin, and vasoactive intestinal peptide (VIP; Sherwood, Krueckl, & McRory, 2000). It is tightly conserved, and the sequences of human, mouse, rat, and sheep PACAP-38 are identical (Harmar et al., 2012). The PACAP gene is composed of five exons; the sequence of PACAP is encoded by exon 5 (Vaudry et al., 2009). Shorter PACAP mRNA has been found in the mouse, bovine, and human testes (Hurley, Gardiner, Jones, & Bloom, 1995), and another short PACAP transcript is produced in sympathetic neurons (Harakall, Brandenburg, Gilmartin, May, & Braas, 1998).

There are three G protein-coupled PACAP receptor subtypes; one with much greater affinity for PACAP than for VIP (the "PACAP type I receptor," PAC1) and two with high affinity for both PACAP and VIP (the "PACAP type II receptors," VPAC1 and VPAC2). PAC1 mRNA is expressed predominantly in the CNS, most abundantly in the olfactory bulb, thalamus, hypothalamus, dentate gyrus of the hippocampus, and in granule cells of the cerebellum (Hashimoto et al., 1996; Shioda et al., 1997). The receptor is also highly expressed in a number of peripheral tissues, most abundantly in the adrenal medulla (Shivers, Gorcs, Gottschall, & Arimura, 1991).

The PAC1 receptor gene contains 18 exons (Lutz et al., 2006) and undergoes extensive alternative splicing resulting in multiple isoforms of the PAC1 receptor. PACAP-27 and PACAP-38 are greater than 100-fold more potent than VIP as agonists of most isoforms of the PAC1 receptor. VPAC1 and VPAC2 receptors have comparable affinity for PACAP and VIP (Harmar et al., 2012). While the PAC1-null receptor, with no splice deletions or insertions is the most widely expressed in the brain, there are region-specific variants (Blechman & Levkowitz, 2013). These result in differential coupling of adenylyl cycle and phospholipase C to activate signaling pathways that can result in enhanced calcium mobilization, membrane depolarization, action potential frequency, and neurotransmitter synthesis and release (Blechman & Levkowitz, 2013; Spengler et al., 1993).

PACAP functions in development, cell signaling, metabolism, homeostasis, and cell protection (Ghzili et al., 2008; Hashimoto, Shintani, & Baba, 2006; Spengler et al., 1993; Watanabe et al., 2007; Zhong, 1995), and importantly is as a key regulator of the stress response. Evidence suggests that it functions upstream of CRH in stress-related circuits. Specifically, CRH neurons in the paraventricular nucleus (PVN) of the hypothalamus receive substantial innervation by PACAP-containing nerve terminals (Legradi, Hannibal, & Lechan, 1998). In nonstressed rats, intracerebroventricular (icv) injections of PACAP resulted in increased CRH, corticosterone, and anxiety-like behavior, fully recapitulating a stress response (Agarwal, Halvorson, & Legradi, 2005). Reduced anxiety-like phenotypes are observed PACAP and PAC1 receptor null mice (Otto et al., 2001). Further, PACAP may augment behavioral responses to stressors (Hammack et al., 2009; Norrholm, Das, & Legradi, 2005).

Studies suggest that PACAP signaling in the bed nucleus of the stria terminalis (BNST) may be one mechanism by which PACAP is involved with stressor responding. The BNST is a component of the extended amygdala and has been shown to mediate anxiety-like responding following chronic/repeated stress or sustained threats (Walker, Toufexis, & Davis, 2003). The BNST is believed to mediate slower-onset, longer-lasting responses that frequently accompany sustained threats, thus maladaptive BNST activity may be critical for the development of trauma- and stressor-related disorders. For example, there was a 12–14-fold increase in BNST PACAP transcripts in stressed vs nonstressed animals (Hammack et al., 2009). Exposure to repeated variate stress has been shown to increase PACAP and PAC1R expression in the BNST in rodents (Lezak et al., 2014).

The BNST is a sexually dimorphic structure in both rodents (De Vries & Panzica, 2006; Hines, Davis, Coquelin, Goy, & Gorski, 1985; Toufexis, 2007) and humans (Chung, De Vries, & Swaab, 2002) and is associated with regulation of gonadal hormones, neurotransmitters, and autonomic activity (De Vries & Panzica, 2006). Interestingly, direct PACAP infusions into the BNST increased plasma corticosterone levels in male and female rats, which were not blocked by BNST CRH receptor antagonism; however, the increased corticosterone was independent of estradiol. Female rats were ovariectomized rats and implanted with subcutaneous implants of either cholesterol or E2, so while the results indicate the effects of BNST PACAP on corticosterone may not depend on E2 levels, it may be that the effects of E2 may require lower or higher replacement levels, and/or a regimen that allows E2 to cycle (Lezak et al., 2014).

We examined PAC1R gene expression in the BNST in ovariectomized female rats and found that following 21-day implantation of a continuous release estrogen pellet, PAC1R transcripts in the dorsal and ventral BNST were increased (Ressler et al., 2011). Importantly, the PAC1 receptor SNP is within a predicted ERE; however, rs2267735 is positioned within the central variable region of the consensus sequence, so in silico analyses do not currently allow us to predict how the "C" vs "G" allele may differentially alter ERE function and further in vitro analyses are warranted.

Previous evidence suggests a relationship between PACAP signaling and estrogen. Primarily, the amount of PACAP in hypophyseal portal blood has shown to be higher in female than in male rats (Dow, Bennie, & Fink, 1994). PACAP mRNA expression levels within the PVN show significant variations during the rat estrous cycle (Moore, Burger, Dalkin, & Winters, 2005). Hypothalamic and pituitary PACAP are involved in the dynamic control of gonadotropic hormone secretion. In female rats, during the proestrus stage PACAP in the PVN is upregulated in the morning and PACAP mRNA peak precedes secretion of GnRH into the portal circulation, whereas pituitary PACAP is upregulated in the late evening. During a critical period just prior to GnRH release, PACAP level starts to decline allowing a critical amount GnRH to be released overwhelming the PACAP level and inducing the LH surge that evokes ovulation; icv-administered exogenous PACAP just prior to this critical period has been shown to block the LH surge and ovulation (Koves et al., 2014).

Additional clinical data have supported further investigation into the sex-specific differences associated with this risk genotype. Specifically, the risk genotype for PAC1R is associated with exaggerated amygdala activity (Stevens et al., 2014). The risk genotype was associated with increased reactivity of the amygdala and hippocampus to threat stimuli and decreased functional connectivity between the amygdala and hippocampus, indicating that the PACAP system modulates medial temporal lobe function in humans.

Our lab also demonstrated sex-specific differences in psychophysiological measures during a fear conditioning experiment. Specifically, women with high PACAP levels in blood showed enhanced startle responses to both fear cues (reinforced conditioned stimulus, CS+) and safety cues (non-reinforced conditioned stimulus, CS−) and women with the PAC1R receptor (*ADCYAP1R1*) rs2267735 "CC" genotype were significantly less able to discriminate CS+ from CS− signals. In women, but not men, the "CC" genotype predicted significantly more startle in the dark compared to the light. "Dark-enhanced" startle (DES), or the elevated acoustic startle response produced when individuals are exposed to startling stimuli in the context of darkness provides a psychophysiological measure of non-specific anxiety (Grillon & Ameli, 1998) given darkness is commonly recognized as an aversive context in most human beings. In rodents, the BNST has been shown to be the neuroanatomical basis of the analogous "light-enhanced" startle (Walker & Davis, 1997). We found that the same gene polymorphism associated with PTSD risk in adult females is also associated with increased DES in both male and female children prior to puberty, further supporting the activational role of sex hormones (Jovanovic et al., 2013). These data suggest that genetic vulnerability for anxiety is evident in both males and females during child development, but may only be present in females after adolescence due to changes in the estrogen system, given the effects of estrogen on PAC1R regulation.

REFERENCES

Agarwal, A., Halvorson, L. M., & Legradi, G. (2005). Pituitary adenylate cyclase-activating polypeptide (PACAP) mimics neuroendocrine and behavioral manifestations of stress: Evidence for PKA-mediated expression of the corticotropin-releasing hormone (CRH) gene. *Brain Research. Molecular Brain Research*, *138*(1), 45–57.

Arimura, A., Somogyvari-Vigh, A., Miyata, A., Mizuno, K., Coy, D. H., & Kitada, C. (1991). Tissue distribution of PACAP as determined by RIA: Highly abundant in the rat brain and testes. *Endocrinology*, *129*(5), 2787–2789.

Balthazart, J., Foidart, A., & Hendrick, J. C. (1990). The induction by testosterone of aromatase activity in the preoptic area and activation of copulatory behavior. *Physiology & Behavior*, *47*(1), 83–94.

Barbaccia, M. L., Roscetti, G., Bolacchi, F., Concas, A., Mostallino, M. C., Purdy, R. H., et al. (1996). Stress-induced increase in brain neuroactive steroids: Antagonism by abecarnil. *Pharmacology, Biochemistry, and Behavior*, *54*(1), 205–210.

Barbaccia, M. L., Roscetti, G., Trabucchi, M., Mostallino, M. C., Concas, A., Purdy, R. H., et al. (1996). Time-dependent changes in rat brain neuroactive steroid concentrations and GABAA receptor function after acute stress. *Neuroendocrinology*, *63*(2), 166–172.

Belzung, C., & Lemoine, M. (2011). Criteria of validity for animal models of psychiatric disorders: Focus on anxiety disorders and depression. *Biology of Mood & Anxiety Disorders*, *1*(1), 9.

Berga, S. L., & Loucks, T. L. (2006). Use of cognitive behavior therapy for functional hypothalamic amenorrhea. *Annals of the New York Academy of Sciences*, *1092*, 114–129.

Bethea, C. L., Lu, N. Z., Gundlah, C., & Streicher, J. M. (2002). Diverse actions of ovarian steroids in the serotonin neural system. *Frontiers in Neuroendocrinology*, *23*(1), 41–100.

Bitran, D., Shiekh, M., & McLeod, M. (1995). Anxiolytic effect of progesterone is mediated by the neurosteroid allopregnanolone at brain GABAA receptors. *Journal of Neuroendocrinology*, 7(3), 171–177.

Blechman, J., & Levkowitz, G. (2013). Alternative splicing of the pituitary adenylate cyclase-activating polypeptide receptor PAC1: Mechanisms of fine tuning of brain activity. *Frontiers in Endocrinology*, *4*, 55.

Brambilla, F., Biggio, G., Pisu, M. G., Bellodi, L., Perna, G., Bogdanovich-Djukic, V., et al. (2003). Neurosteroid secretion in panic disorder. *Psychiatry Research*, *118*(2), 107–116.

Brann, D. W., Dhandapani, K., Wakade, C., Mahesh, V. B., & Khan, M. M. (2007). Neurotrophic and neuroprotective actions of estrogen: Basic mechanisms and clinical implications. *Steroids*, *72*(5), 381–405.

Bremner, J. D., Vythilingam, M., Vermetten, E., Adil, J., Khan, S., Nazeer, A., et al. (2003). Cortisol response to a cognitive stress challenge in posttraumatic stress disorder (PTSD) related to childhood abuse. *Psychoneuroendocrinology*, *28*(6), 733–750.

Breslau, N., Davis, G. C., Andreski, P., Peterson, E. L., & Schultz, L. R. (1997). Sex differences in posttraumatic stress disorder. *Archives of General Psychiatry*, *54*(11), 1044–1048.

Bryant, R. A., Felmingham, K. L., Silove, D., Creamer, M., O'Donnell, M., & McFarlane, A. C. (2011). The association between menstrual cycle and traumatic memories. *Journal of Affective Disorders*, *131*(1–3), 398–401.

Cheung, J., Chervonsky, L., Felmingham, K. L., & Bryant, R. A. (2013). The role of estrogen in intrusive memories. *Neurobiology of Learning and Memory, 106*, 87–94.
Christensen, L. W., & Gorski, R. A. (1978). Independent masculinization of neuroendocrine systems by intracerebral implants of testosterone or estradiol in the neonatal female rat. *Brain Research, 146*(2), 325–340.
Chung, W. C., De Vries, G. J., & Swaab, D. F. (2002). Sexual differentiation of the bed nucleus of the stria terminalis in humans may extend into adulthood. *The Journal of Neuroscience, 22*(3), 1027–1033.
Cloitre, M., Yonkers, K. A., Pearlstein, T., Altemus, M., Davidson, K. W., Pigott, T. A., et al. (2004). Women and anxiety disorders: Implications for diagnosis and treatment. *CNS Spectrums, 9*(9), 1–16.
Cohen, B. E., Maguen, S., Bertenthal, D., Shi, Y., Jacoby, V., & Seal, K. H. (2012). Reproductive and other health outcomes in Iraq and Afghanistan women veterans using VA health care: Association with mental health diagnoses. *Womens Health Issues, 22*(5), e461–e471.
Cuthbert, B. N. (2015). Research Domain Criteria: Toward future psychiatric nosologies. *Dialogues in Clinical Neuroscience, 17*(1), 89–97.
Dagnault, A., & Richard, D. (1997). Involvement of the medial preoptic area in the anorectic action of estrogens. *The American Journal of Physiology, 272*(1 Pt. 2), R311–R317.
De Vries, G. J., & Panzica, G. C. (2006). Sexual differentiation of central vasopressin and vasotocin systems in vertebrates: Different mechanisms, similar endpoints. *Neuroscience, 138*(3), 947–955.
Dean, C., & Kendell, R. E. (1981). The symptomatology of puerperal illnesses. *The British Journal of Psychiatry, 139*, 128–133.
Douma, S. L., Husband, C., O'Donnell, M. E., Barwin, B. N., & Woodend, A. K. (2005). Estrogen-related mood disorders: Reproductive life cycle factors. *Advances in Nursing Science, 28*(4), 364–375.
Dow, R. C., Bennie, J., & Fink, G. (1994). Pituitary adenylate cyclase-activating peptide-38 (PACAP)-38 is released into hypophysial portal blood in the normal male and female rat. *The Journal of Endocrinology, 142*(1), R1–R4.
Edman, C. D., & MacDonald, P. C. (1978). Effect of obesity on conversion of plasma androstenedione to estrone in ovulatory and anovulator young women. *American Journal of Obstetrics and Gynecology, 130*(4), 456–461.
Elzinga, B. M., Schmahl, C. G., Vermetten, E., van Dyck, R., & Bremner, J. D. (2003). Higher cortisol levels following exposure to traumatic reminders in abuse-related PTSD. *Neuropsychopharmacology, 28*(9), 1656–1665.
Engin, E., & Treit, D. (2007). The anxiolytic-like effects of allopregnanolone vary as a function of intracerebral microinfusion site: The amygdala, medial prefrontal cortex, or hippocampus. *Behavioural Pharmacology, 18*(5–6), 461–470.
Epperson, C. N., Kim, D. R., & Bale, T. L. (2014). Estradiol modulation of monoamine metabolism: One possible mechanism underlying sex differences in risk for depression and dementia. *JAMA Psychiatry, 71*(8), 869–870.
Ertman, N., Andreano, J. M., & Cahill, L. (2011). Progesterone at encoding predicts subsequent emotional memory. *Learning & Memory, 18*(12), 759–763.
Fernandez-Guasti, A., & Picazo, O. (1995). Flumazenil blocks the anxiolytic action of allopregnanolone. *European Journal of Pharmacology, 281*(1), 113–115.
Freidenberg, B. M., Gusmano, R., Hickling, E. J., Blanchard, E. B., Bremner, J. D., & Frye, C. (2010). Women with PTSD have lower basal salivary cortisol levels later in the day than do men with PTSD: A preliminary study. *Physiology & Behavior, 99*(2), 234–236.
Ghatei, M. A., Takahashi, K., Suzuki, Y., Gardiner, J., Jones, P. M., & Bloom, S. R. (1993). Distribution, molecular characterization of pituitary adenylate cyclase-activating

polypeptide and its precursor encoding messenger RNA in human and rat tissues. *The Journal of Endocrinology*, *136*(1), 159–166.

Ghzili, H., Grumolato, L., Thouennon, E., Tanguy, Y., Turquier, V., Vaudry, H., et al. (2008). Role of PACAP in the physiology and pathology of the sympathoadrenal system. *Frontiers in Neuroendocrinology*, *29*(1), 128–141.

Glover, E. M., Jovanovic, T., Mercer, K. B., Kerley, K., Bradley, B., Ressler, K. J., et al. (2012). Estrogen levels are associated with extinction deficits in women with post-traumatic stress disorder. *Biological Psychiatry*, *72*(1), 19–24.

Glover, E. M., Mercer, K. B., Norrholm, S. D., Davis, M., Duncan, E., Bradley, B., et al. (2013). Inhibition of fear is differentially associated with cycling estrogen levels in women. *Journal of Psychiatry & Neuroscience*, *38*(5), 341–348.

Gola, H., Engler, H., Schauer, M., Adenauer, H., Riether, C., Kolassa, S., et al. (2012). Victims of rape show increased cortisol responses to trauma reminders: A study in individuals with war- and torture-related PTSD. *Psychoneuroendocrinology*, *37*(2), 213–220.

Goldstein, J. M., Lancaster, K., Longenecker, J. M., Abbs, B., Holsen, L. M., Cherkerzian, S., et al. (2015). Sex differences, hormones, and fMRI stress response circuitry deficits in psychoses. *Psychiatry Research*, *232*(3), 226–236.

Graham, J. D., & Clarke, C. L. (1997). Physiological action of progesterone in target tissues. *Endocrine Reviews*, *18*(4), 502–519.

Graham, B. M., & Daher, M. (2016). Estradiol and progesterone have opposing roles in the regulation of fear extinction in female rats. *Neuropsychopharmacology*, *41*(3), 774–780. http://dx.doi.org/10.1038/npp.2015.202. Epub 2015 Jul 9.

Graham, B. M., & Milad, M. R. (2013). Blockade of estrogen by hormonal contraceptives impairs fear extinction in female rats and women. *Biological Psychiatry*, *73*(4), 371–378.

Grillon, C., & Ameli, R. (1998). Effects of threat of shock, shock electrode placement and darkness on startle. *International Journal of Psychophysiology*, *28*(3), 223–231.

Grodin, J. M., Siiteri, P. K., & MacDonald, P. C. (1973). Source of estrogen production in postmenopausal women. *The Journal of Clinical Endocrinology and Metabolism*, *36*(2), 207–214.

Gronemeyer, H. (1991). Transcription activation by estrogen and progesterone receptors. *Annual Review of Genetics*, *25*, 89–123.

Gundlah, C., Lu, N. Z., Mirkes, S. J., & Bethea, C. L. (2001). Estrogen receptor beta (ERbeta) mRNA and protein in serotonin neurons of macaques. *Brain Research. Molecular Brain Research*, *91*(1–2), 14–22.

Hammack, S. E., Cheung, J., Rhodes, K. M., Schutz, K. C., Falls, W. A., Braas, K. M., et al. (2009). Chronic stress increases pituitary adenylate cyclase-activating peptide (PACAP) and brain-derived neurotrophic factor (BDNF) mRNA expression in the bed nucleus of the stria terminalis (BNST): Roles for PACAP in anxiety-like behavior. *Psychoneuroendocrinology*, *34*(6), 833–843.

Harakall, S. A., Brandenburg, C. A., Gilmartin, G. A., May, V., & Braas, K. M. (1998). Induction of multiple pituitary adenylate cyclase activating polypeptide (PACAP) transcripts through alternative cleavage and polyadenylation of proPACAP precursor mRNA. *Annals of the New York Academy of Sciences*, *865*, 367–374.

Harmar, A. J., Fahrenkrug, J., Gozes, I., Laburthe, M., May, V., Pisegna, J. R., et al. (2012). Pharmacology and functions of receptors for vasoactive intestinal peptide and pituitary adenylate cyclase-activating polypeptide: IUPHAR review 1. *British Journal of Pharmacology*, *166*(1), 4–17.

Hashimoto, H., Nogi, H., Mori, K., Ohishi, H., Shigemoto, R., Yamamoto, K., et al. (1996). Distribution of the mRNA for a pituitary adenylate cyclase-activating polypeptide receptor in the rat brain: An in situ hybridization study. *The Journal of Comparative Neurology*, *371*(4), 567–577.

Hashimoto, H., Shintani, N., & Baba, A. (2006). New insights into the central PACAPergic system from the phenotypes in PACAP- and PACAP receptor-knockout mice. *Annals of the New York Academy of Sciences*, *1070*, 75–89.

Heldring, N., Pike, A., Andersson, S., Matthews, J., Cheng, G., Hartman, J., et al. (2007). Estrogen receptors: How do they signal and what are their targets. *Physiological Reviews*, *87*(3), 905–931.

Hellstrom, B., & Anderberg, U. M. (2003). Pain perception across the menstrual cycle phases in women with chronic pain. *Perceptual and Motor Skills*, *96*(1), 201–211.

Hines, M., Davis, F. C., Coquelin, A., Goy, R. W., & Gorski, R. A. (1985). Sexually dimorphic regions in the medial preoptic area and the bed nucleus of the stria terminalis of the guinea pig brain: A description and an investigation of their relationship to gonadal steroids in adulthood. *The Journal of Neuroscience*, *5*(1), 40–47.

Holbrook, T. L., Hoyt, D. B., Stein, M. B., & Sieber, W. J. (2002). Gender differences in long-term posttraumatic stress disorder outcomes after major trauma: Women are at higher risk of adverse outcomes than men. *The Journal of Trauma*, *53*(5), 882–888.

Hurley, J. D., Gardiner, J. V., Jones, P. M., & Bloom, S. R. (1995). Cloning and molecular characterization of complementary deoxyribonucleic acid corresponding to a novel form of pituitary adenylate cyclase-activating polypeptide messenger ribonucleic acid in the rat testis. *Endocrinology*, *136*(2), 550–557.

Jovanovic, T., Norrholm, S. D., Davis, J., Mercer, K. B., Almli, L., Nelson, A., et al. (2013). PAC1 receptor (ADCYAP1R1) genotype is associated with dark-enhanced startle in children. *Molecular Psychiatry*, *18*(7), 742–743.

Kang, H., Dalager, N., Mahan, C., & Ishii, E. (2005). The role of sexual assault on the risk of PTSD among Gulf War veterans. *Annals of Epidemiology*, *15*(3), 191–195.

Kato, J., Onouchi, T., & Okinaga, S. (1978). Hypothalamic and hypophysial progesterone receptors: Estrogen-priming effect, differential localization, 5alpha-dihydroprogesterone binding, and nuclear receptors. *Journal of Steroid Biochemistry*, *9*(5), 419–427.

Katon, J., Cypel, Y., Raza, M., Zephyrin, L., Reiber, G., Yano, E. M., et al. (2014). Self-reported infertility among male and female veterans serving during Operation Enduring Freedom/Operation Iraqi Freedom. *Journal of Women's Health (2002)*, *23*(2), 175–183.

Keane, T., Marx, B., & Sloan, D. (2009). *Post-traumatic stress disorder: Definition, prevalence, and risk factors*. New York: Humana.

Keen-Rhinehart, E., Michopoulos, V., Toufexis, D. J., Martin, E. I., Nair, H., Ressler, K. J., et al. (2009). Continuous expression of corticotropin-releasing factor in the central nucleus of the amygdala emulates the dysregulation of the stress and reproductive axes. *Molecular Psychiatry*, *14*(1), 37–50.

Kessler, R. C., Sonnega, A., Bromet, E., Hughes, M., & Nelson, C. B. (1995). Posttraumatic stress disorder in the National Comorbidity Survey. *Archives of General Psychiatry*, *52*(12), 1048–1060.

Kirschbaum, C., Kudielka, B. M., Gaab, J., Schommer, N. C., & Hellhammer, D. H. (1999). Impact of gender, menstrual cycle phase, and oral contraceptives on the activity of the hypothalamus–pituitary–adrenal axis. *Psychosomatic Medicine*, *61*(2), 154–162.

Koopman, P., Gubbay, J., Vivian, N., Goodfellow, P., & Lovell-Badge, R. (1991). Male development of chromosomally female mice transgenic for Sry. *Nature*, *351*(6322), 117–121.

Koves, K., Arimura, A., Gorcs, T. G., & Somogyvari-Vigh, A. (1991). Comparative distribution of immunoreactive pituitary adenylate cyclase activating polypeptide and vasoactive intestinal polypeptide in rat forebrain. *Neuroendocrinology*, *54*(2), 159–169.

Koves, K., Kantor, O., Lakatos, A., Szabo, E., Kirilly, E., Heinzlmann, A., et al. (2014). Advent and recent advances in research on the role of pituitary adenylate cyclase-activating polypeptide (PACAP) in the regulation of gonadotropic hormone secretion of female rats. *Journal of Molecular Neuroscience*, *54*(3), 494–511.

Kramar, E. A., Babayan, A. H., Gall, C. M., & Lynch, G. (2013). Estrogen promotes learning-related plasticity by modifying the synaptic cytoskeleton. *Neuroscience*, *239*, 3–16.

Kushner, P. J., Agard, D. A., Greene, G. L., Scanlan, T. S., Shiau, A. K., Uht, R. M., et al. (2000). Estrogen receptor pathways to AP-1. *The Journal of Steroid Biochemistry and Molecular Biology*, *74*(5), 311–317.

Leavitt, W. W., Chen, T. J., & Allen, T. C. (1977). Regulation of progesterone receptor formation by estrogen action. *Annals of the New York Academy of Sciences*, *286*, 210–225.

Legradi, G., Hannibal, J., & Lechan, R. M. (1998). Pituitary adenylate cyclase-activating polypeptide-nerve terminals densely innervate corticotropin-releasing hormone-neurons in the hypothalamic paraventricular nucleus of the rat. *Neuroscience Letters*, *246*(3), 145–148.

Lezak, K. R., Roelke, E., Harris, O. M., Choi, I., Edwards, S., Gick, N., et al. (2014). Pituitary adenylate cyclase-activating polypeptide (PACAP) in the bed nucleus of the stria terminalis (BNST) increases corticosterone in male and female rats. *Psychoneuroendocrinology*, *45*, 11–20.

Lezak, K. R., Roman, C. W., Braas, K. M., Schutz, K. C., Falls, W. A., Schulkin, J., et al. (2014). Regulation of bed nucleus of the stria terminalis PACAP expression by stress and corticosterone. *Journal of Molecular Neuroscience*, *54*(3), 477–484.

Lutz, E. M., Ronaldson, E., Shaw, P., Johnson, M. S., Holland, P. J., & Mitchell, R. (2006). Characterization of novel splice variants of the PAC1 receptor in human neuroblastoma cells: Consequences for signaling by VIP and PACAP. *Molecular and Cellular Neurosciences*, *31*(2), 193–209.

McCarthy, M. M. (2008). Estradiol and the developing brain. *Physiological Reviews*, *88*(1), 91–124.

Meczekalski, B., Katulski, K., Czyzyk, A., Podfigurna-Stopa, A., & Maciejewska-Jeske, M. (2014). Functional hypothalamic amenorrhea and its influence on women's health. *Journal of Endocrinological Investigation*, *37*(11), 1049–1056.

Meewisse, M., Reitsma, J. B., De Vries, G. J., Gersons, B. P. R., & Olff, M. (2007). Cortisol and post-traumatic stress disorder in adults: Systematic review and meta-analysis. *The British Journal of Psychiatry*, *191*(5), 387–392.

Michopoulos, V., Checchi, M., Sharpe, D., & Wilson, M. E. (2011). Estradiol effects on behavior and serum oxytocin are modified by social status and polymorphisms in the serotonin transporter gene in female rhesus monkeys. *Hormones and Behavior*, *59*(4), 528–535.

Michopoulos, V., Mancini, F., Loucks, T. L., & Berga, S. L. (2013). Neuroendocrine recovery initiated by cognitive behavioral therapy in women with functional hypothalamic amenorrhea: A randomized, controlled trial. *Fertility and Sterility*, *99*(7), 2084–2091. e2081.

Michopoulos, V., Norrholm, S. D., & Jovanovic, T. (2015). Diagnostic biomarkers for post-traumatic stress disorder: Promising horizons from translational neuroscience research. *Biological Psychiatry*, *78*(5), 344–353.

Michopoulos, V., Rothbaum, A. O., Corwin, E., Bradley, B., Ressler, K. J., & Jovanovic, T. (2014). Psychophysiology and posttraumatic stress disorder symptom profile in pregnant African-American women with trauma exposure. *Archives of Women's Mental Health*, *18*, 639–648.

Miyata, A., Arimura, A., Dahl, R. R., Minamino, N., Uehara, A., Jiang, L., et al. (1989). Isolation of a novel 38 residue-hypothalamic polypeptide which stimulates adenylate cyclase in pituitary cells. *Biochemical and Biophysical Research Communications*, *164*(1), 567–574.

Moore, J. P., Jr., Burger, L. L., Dalkin, A. C., & Winters, S. J. (2005). Pituitary adenylate cyclase activating polypeptide messenger RNA in the paraventricular nucleus and anterior pituitary during the rat estrous cycle. *Biology of Reproduction*, *73*(3), 491–499.

Naftolin, F., Ryan, K. J., Davies, I. J., Reddy, V. V., Flores, F., Petro, Z., et al. (1975). The formation of estrogens by central neuroendocrine tissues. *Recent Progress in Hormone Research*, *31*, 295–319.

Nielsen, S. E., Barber, S. J., Chai, A., Clewett, D. V., & Mather, M. (2015). Sympathetic arousal increases a negative memory bias in young women with low sex hormone levels. *Psychoneuroendocrinology*, *62*, 96–106.

Nilsson, S., Makela, S., Treuter, E., Tujague, M., Thomsen, J., Andersson, G., et al. (2001). Mechanisms of estrogen action. *Physiological Reviews*, *81*(4), 1535–1565.

Norrholm, S. D., Das, M., & Legradi, G. (2005). Behavioral effects of local microinfusion of pituitary adenylate cyclase activating polypeptide (PACAP) into the paraventricular nucleus of the hypothalamus (PVN). *Regulatory Peptides*, *128*(1), 33–41.

Norrholm, S. D., & Jovanovic, T. (2010). Tailoring therapeutic strategies for treating post-traumatic stress disorder symptom clusters. *Neuropsychiatric Disease and Treatment*, *6*, 517–532.

Norris, F. H. (1992). Epidemiology of trauma: Frequency and impact of different potentially traumatic events on different demographic groups. *Journal of Consulting and Clinical Psychology*, *60*(3), 409–418.

Otto, C., Martin, M., Wolfer, D. P., Lipp, H. P., Maldonado, R., & Schutz, G. (2001). Altered emotional behavior in PACAP-type-I-receptor-deficient mice. *Brain Research. Molecular Brain Research*, *92*(1–2), 78–84.

Palkovits, M., Somogyvari-Vigh, A., & Arimura, A. (1995). Concentrations of pituitary adenylate cyclase activating polypeptide (PACAP) in human brain nuclei. *Brain Research*, *699*(1), 116–120.

Petersen, N., Kilpatrick, L. A., Goharzad, A., & Cahill, L. (2014). Oral contraceptive pill use and menstrual cycle phase are associated with altered resting state functional connectivity. *NeuroImage*, *90*, 24–32.

Petraglia, F., Sutton, S., Vale, W., & Plotsky, P. (1987). Corticotropin-releasing factor decreases plasma luteinizing hormone levels in female rats by inhibiting gonadotropin-releasing hormone release into hypophysial-portal circulation. *Endocrinology*, *120*(3), 1083–1088.

Phoenix, C. H., Goy, R. W., Gerall, A. A., & Young, W. C. (1959). Organizing action of prenatally administered testosterone propionate on the tissues mediating mating behavior in the female guinea pig. *Endocrinology*, *65*, 369–382.

Pigott, T. A. (2003). Anxiety disorders in women. *The Psychiatric Clinics of North America*, *26*(3), 621–672. vi–vii.

Pineles, S. L., Nillni, Y. I., King, M. W., Patton, S. C., Bauer, M. R., Mostoufi, S. M., et al. (2016). Extinction retention and the menstrual cycle: Different associations for women with posttraumatic stress disorder. *Journal of Abnormal Psychology*, *125*(3), 349–355.

Rabin, D. S., Johnson, E. O., Brandon, D. D., Liapi, C., & Chrousos, G. P. (1990). Glucocorticoids inhibit estradiol-mediated uterine growth: Possible role of the uterine estradiol receptor. *Biology of Reproduction*, *42*(1), 74–80.

Rapkin, A. J., Mikacich, J. A., Moatakef-Imani, B., & Rasgon, N. (2002). The clinical nature and formal diagnosis of premenstrual, postpartum, and perimenopausal affective disorders. *Current Psychiatry Reports*, *4*(6), 419–428.

Rasmusson, A. M., Pinna, G., Paliwal, P., Weisman, D., Gottschalk, C., Charney, D., et al. (2006). Decreased cerebrospinal fluid allopregnanolone levels in women with post-traumatic stress disorder. *Biological Psychiatry*, *60*(7), 704–713.

Resko, J. A., & Roselli, C. E. (1992). Brain steroid synthesis and metabolism. *Neuroprotocols*, *1*(1), 27–34.

Resnick, H. S., Kilpatrick, D. G., Dansky, B. S., Saunders, B. E., & Best, C. L. (1993). Prevalence of civilian trauma and posttraumatic stress disorder in a representative national sample of women. *Journal of Consulting and Clinical Psychology*, *61*(6), 984–991.

Ressler, K. J., Mercer, K. B., Bradley, B., Jovanovic, T., Mahan, A., Kerley, K., et al. (2011). Post-traumatic stress disorder is associated with PACAP and the PAC1 receptor. *Nature, 470*(7335), 492–497.

Romeo, E., Strohle, A., Spalletta, G., di Michele, F., Hermann, B., Holsboer, F., et al. (1998). Effects of antidepressant treatment on neuroactive steroids in major depression. *The American Journal of Psychiatry, 155*(7), 910–913.

Rubinow, D. R., Schmidt, P. J., & Roca, C. A. (1998). Estrogen-serotonin interactions: Implications for affective regulation. *Biological Psychiatry, 44*(9), 839–850.

Santollo, J., Torregrossa, A. M., & Eckel, L. A. (2011). Estradiol acts in the medial preoptic area, arcuate nucleus, and dorsal raphe nucleus to reduce food intake in ovariectomized rats. *Hormones and Behavior, 60*(1), 86–93.

Seal, K. H., Bertenthal, D., Miner, C. R., Sen, S., & Marmar, C. (2007). Bringing the war back home: Mental health disorders among 103,788 US veterans returning from Iraq and Afghanistan seen at Department of Veterans Affairs facilities. *Archives of Internal Medicine, 167*(5), 476–482.

Seedat, S., Stein, D. J., & Carey, P. D. (2005). Post-traumatic stress disorder in women: Epidemiological and treatment issues. *CNS Drugs, 19*(5), 411–427.

Sherwin, B. B. (1994). Estrogenic effects on memory in women. *Annals of the New York Academy of Sciences, 743*, 213–230. discussion 230–211.

Sherwood, N. M., Krueckl, S. L., & McRory, J. E. (2000). The origin and function of the pituitary adenylate cyclase-activating polypeptide (PACAP)/glucagon superfamily. *Endocrine Reviews, 21*(6), 619–670.

Shioda, S., Shuto, Y., Somogyvari-Vigh, A., Legradi, G., Onda, H., Coy, D. H., et al. (1997). Localization and gene expression of the receptor for pituitary adenylate cyclase-activating polypeptide in the rat brain. *Neuroscience Research, 28*(4), 345–354.

Shivers, B. D., Gorcs, T. J., Gottschall, P. E., & Arimura, A. (1991). Two high affinity binding sites for pituitary adenylate cyclase-activating polypeptide have different tissue distributions. *Endocrinology, 128*(6), 3055–3065.

Sigmon, S. T., Dorhofer, D. M., Rohan, K. J., Hotovy, L. A., Boulard, N. E., & Fink, C. M. (2000). Psychophysiological, somatic, and affective changes across the menstrual cycle in women with panic disorder. *Journal of Consulting and Clinical Psychology, 68*(3), 425–431.

Simpson, E. R. (2003). Sources of estrogen and their importance. *The Journal of Steroid Biochemistry and Molecular Biology, 86*(3–5), 225–230.

Simpson, E. R. (2004). Aromatase: Biologic relevance of tissue-specific expression. *Seminars in Reproductive Medicine, 22*(1), 11–23.

Simpson, E. R., Mahendroo, M. S., Means, G. D., Kilgore, M. W., Hinshelwood, M. M., Graham-Lorence, S., et al. (1994). Aromatase cytochrome P450, the enzyme responsible for estrogen biosynthesis. *Endocrine Reviews, 15*(3), 342–355.

Simpson, E. R., Michael, M. D., Agarwal, V. R., Hinshelwood, M. M., Bulun, S. E., & Zhao, Y. (1997). Cytochromes P450 11: Expression of the CYP19 (aromatase) gene: An unusual case of alternative promoter usage. *The FASEB Journal, 11*(1), 29–36.

Soni, M., Curran, V. H., & Kamboj, S. K. (2013). Identification of a narrow post-ovulatory window of vulnerability to distressing involuntary memories in healthy women. *Neurobiology of Learning and Memory, 104*, 32–38.

Spengler, D., Waeber, C., Pantaloni, C., Holsboer, F., Bockaert, J., Seeburg, P. H., et al. (1993). Differential signal transduction by five splice variants of the PACAP receptor. *Nature, 365*(6442), 170–175.

Stevens, J. S., Almli, L. M., Fani, N., Gutman, D. A., Bradley, B., Norrholm, S. D., et al. (2014). PACAP receptor gene polymorphism impacts fear responses in the amygdala and hippocampus. *Proceedings of the National Academy of Sciences of the United States of America, 111*(8), 3158–3163.

Street, A. E., Vogt, D., & Dutra, L. (2009). A new generation of women veterans: Stressors faced by women deployed to Iraq and Afghanistan. *Clinical Psychology Review, 29*(8), 685–694.
Strohle, A., Romeo, E., di Michele, F., Pasini, A., Yassouridis, A., Holsboer, F., et al. (2002). GABA(A) receptor-modulating neuroactive steroid composition in patients with panic disorder before and during paroxetine treatment. *The American Journal of Psychiatry, 159*(1), 145–147.
Suris, A., Lind, L., Kashner, T. M., Borman, P. D., & Petty, F. (2004). Sexual assault in women veterans: An examination of PTSD risk, health care utilization, and cost of care. *Psychosomatic Medicine, 66*(5), 749–756.
Toufexis, D. (2007). Region- and sex-specific modulation of anxiety behaviours in the rat. *Journal of Neuroendocrinology, 19*(6), 461–473.
Unger, E. K., Burke, K. J., Jr., Yang, C. F., Bender, K. J., Fuller, P. M., & Shah, N. M. (2015). Medial amygdalar aromatase neurons regulate aggression in both sexes. *Cell Reports, 10*(4), 453–462.
Vamvakopoulos, N. C., & Chrousos, G. P. (1993). Evidence of direct estrogenic regulation of human corticotropin-releasing hormone gene expression. Potential implications for the sexual dimophism of the stress response and immune/inflammatory reaction. *The Journal of Clinical Investigation, 92*(4), 1896–1902.
Vaudry, D., Falluel-Morel, A., Bourgault, S., Basille, M., Burel, D., Wurtz, O., et al. (2009). Pituitary adenylate cyclase-activating polypeptide and its receptors: 20 years after the discovery. *Pharmacological Reviews, 61*(3), 283–357.
Walker, D. L., & Davis, M. (1997). Double dissociation between the involvement of the bed nucleus of the stria terminalis and the central nucleus of the amygdala in startle increases produced by conditioned versus unconditioned fear. *The Journal of Neuroscience, 17*(23), 9375–9383.
Walker, D. L., Toufexis, D. J., & Davis, M. (2003). Role of the bed nucleus of the stria terminalis versus the amygdala in fear, stress, and anxiety. *European Journal of Pharmacology, 463*(1–3), 199–216.
Watanabe, J., Nakamachi, T., Matsuno, R., Hayashi, D., Nakamura, M., Kikuyama, S., et al. (2007). Localization, characterization and function of pituitary adenylate cyclase-activating polypeptide during brain development. *Peptides, 28*(9), 1713–1719.
Wegerer, M., Kerschbaum, H., Blechert, J., & Wilhelm, F. H. (2014). Low levels of estradiol are associated with elevated conditioned responding during fear extinction and with intrusive memories in daily life. *Neurobiology of Learning and Memory, 116*, 145–154.
Whalen, R. E., & Olsen, K. L. (1981). Role of aromatization in sexual differentiation: Effects of prenatal ATD treatment and neonatal castration. *Hormones and Behavior, 15*(2), 107–122.
Willner, P. (1986). Validation criteria for animal models of human mental disorders: Learned helplessness as a paradigm case. *Progress in Neuro-Psychopharmacology & Biological Psychiatry, 10*(6), 677–690.
Yehuda, R. (2002). Current status of cortisol findings in post-traumatic stress disorder. *The Psychiatric Clinics of North America, 25*(2), 341–368. vii.
Yehuda, R., Golier, J. A., & Kaufman, S. (2005). Circadian rhythm of salivary cortisol in Holocaust survivors with and without PTSD. *The American Journal of Psychiatry, 162*(5), 998–1000.
Yehuda, R., Halligan, S. L., Grossman, R., Golier, J. A., & Wong, C. (2002). The cortisol and glucocorticoid receptor response to low dose dexamethasone administration in aging combat veterans and holocaust survivors with and without posttraumatic stress disorder. *Biological Psychiatry, 52*(5), 393–403.

Yehuda, R., Teicher, M. H., Trestman, R. L., Levengood, R. A., & Siever, L. J. (1996). Cortisol regulation in posttraumatic stress disorder and major depression: A chronobiological analysis. *Biological Psychiatry*, *40*(2), 79–88.

Yonkers, K. A. (1997). Anxiety symptoms and anxiety disorders: How are they related to premenstrual disorders? *The Journal of Clinical Psychiatry*, *58*(3), 62–67.

Young, E. A., Altemus, M., Parkison, V., & Shastry, S. (2001). Effects of estrogen antagonists and agonists on the ACTH response to restraint stress in female rats. *Neuropsychopharmacology*, *25*(6), 881–891.

Zaba, M., Kirmeier, T., Ionescu, I. A., Wollweber, B., Buell, D. R., Gall-Kleebach, D. J., et al. (2015). Identification and characterization of HPA-axis reactivity endophenotypes in a cohort of female PTSD patients. *Psychoneuroendocrinology*, *55C*, 102–115.

Zhong, Y. (1995). Mediation of PACAP-like neuropeptide transmission by coactivation of Ras/Raf and cAMP signal transduction pathways in Drosophila. *Nature*, *375*(6532), 588–592.

Zinzow, H. M., Grubaugh, A. L., Monnier, J., Suffoletta-Maierle, S., & Frueh, B. C. (2007). Trauma among female veterans: A critical review. *Trauma Violence Abuse*, *8*(4), 384–400.

CHAPTER FOUR

The Big Role of Small RNAs in Anxiety and Stress-Related Disorders

S. Malan-Müller[1], S.M.J. Hemmings
Faculty of Medicine and Health Sciences, Department of Psychiatry, Stellenbosch University, Cape Town, South Africa
[1]Corresponding author: e-mail address: smalan@sun.ac.za

Contents

Abstract

In the study of complex, heterogeneous disorders, such as anxiety and stress-related disorders, epigenetic factors provide an additional level of heritable complexity. MicroRNAs (miRNAs) are a class of small, noncoding RNAs that function as epigenetic modulators of gene expression by binding to target messenger RNAs (mRNAs) and subsequently blocking translation or accelerating their degradation. In light of their abundance in the central nervous system (CNS) and their involvement in synaptic plasticity and neuronal differentiation, miRNAs represent an exciting frontier to be explored in the etiology and treatment of anxiety and stress-related disorders. This chapter will present a thorough review of miRNAs, their functions, and mRNA targets in the CNS, focusing on their role in anxiety and stress-related disorders as described by studies performed in animals and human subjects.

Vitamins and Hormones, Volume 103
ISSN 0083-6729
http://dx.doi.org/10.1016/bs.vh.2016.08.001

1. INTRODUCTION

According to the Diagnostic and Statistical Manual of Mental Disorders, version five (DSM-V), the category of anxiety disorders includes disorders that share features of excessive fear and anxiety, including Separation Anxiety Disorder, Specific Phobia, Social Anxiety Disorder (Social Phobia), Panic Disorder (PD), Panic Attack Specifier, Agoraphobia, Generalized Anxiety Disorder, and a few others (DSM-5, APA 2013). Disorders such as posttraumatic stress disorder (PTSD), acute stress disorder, adjustment disorders, reactive attachment disorder, and disinhibited social engagement disorder all fall under the category of trauma and stressor-related disorders. For these disorders, exposure to a traumatic or stressful event is listed explicitly as a diagnostic criterion. The placement of the trauma and stressor-related disorders chapter in the DSM-V indicates the close relationship between these diagnoses and that of disorders in the surrounding chapters on anxiety disorders, obsessive–compulsive and related disorders, as well as dissociative disorders (DSM-5, APA 2013).

The aforementioned psychiatric disorders are debilitating and significantly impair normal functioning and quality of life, in addition to placing a heavy health and economic burden on society. Despite these facts, there are a limited range of available pharmacotherapies to treat these disorders, with the majority of treatments suboptimal with regard to efficacy and tolerability (Holmes, Heilig, Rupniak, Steckler, & Griebel, 2003; Kasper et al., 2010; Kessler, Chiu, Demler, Merikangas, & Walters, 2005). One of the hurdles in the treatment of these multifactorial, complex, and heterogeneous disorders is the fact that the genetic architecture of these disorders is not yet fully elucidated.

Genome-wide association studies and smaller case–control association studies have yielded numerous statistically significant associations between genetic variants and the anxiety and stress-related disorders; however, most of these variants have small effect sizes and only explain a small amount of the heritability of these complex disorders (Manolio et al., 2009). The search for the missing heritability (heritability influenced by unknown factors, since single genetic variations cannot account for the heritability of diseases or behaviors) has shifted attention to rare variants, copy number variants, copy neutral variants, gene–gene and gene–environment interactions, and epigenetic modifications (Manolio et al., 2009).

A multitude of preclinical studies have reported activity-dependent changes in the expression of certain candidate genes in several brain regions in animal models of complex psychiatric disorders (reviewed in Malan-Müller, 2013). Changes in gene expression profiles were shown to mediate fear conditioning and extinction processes in an animal model of PTSD (Yamamoto et al., 2010); investigating differential gene expression patterns in specific brain regions may therefore broaden our understanding of the processes involved with these complex disorders. Through the quantification of gene expression, the biological pathways involved in complex disorders can be identified; however, the exact mechanisms contributing to the observed changes in expression profiles should also be explored. These could be attributed to altered transcription of the gene itself, changes in mRNA turnover in the cell (Akbarian & Huang, 2009), or posttranscriptional modifications, such as epigenetics.

2. EPIGENETICS

The term *epigenetics* means "outside conventional genetics" and it is the study of stable changes in gene expression that are not attributable to changes in DNA sequence (Bjornsson, Fallin, & Feinberg, 2004). These epigenetic changes are heritable and potentially reversible (Jaenisch & Bird, 2003) and offer an additional layer of transcriptional control that may facilitate the delineation of the interaction between genetic predisposition, environmental factors, and subsequent changes in neural functioning (Bjornsson et al., 2004). Epigenetic modifications may thus explain the long-lasting effects of trauma exposure and interindividual variation in response to trauma (Yehuda & Bierer, 2009).

Epigenetic mechanisms include DNA methylation, posttranscriptional modifications of histone proteins (acetylation, methylation, phosphorylation, ubiquitination, and sumoylation), and noncoding RNA-mediated alterations (such as microRNAs (miRNAs) and small interfering RNAs (siRNAs)) (Sontheimer & Carthew, 2005). The study of epigenetics provides an exciting frontier of investigation in psychiatric disorders, as it provides a direct link between environmental risk factors and genetic changes, mostly alterations in gene and protein expression, thus forming a basis for interindividual differences in gene function and vulnerability to specific disorders. This is likely to be the case in some of the anxiety and stress-related disorders and might explain the origin of glucocorticoid (GC)-related alterations associated with these disorders (Yehuda & LeDoux, 2007). Unraveling the role of the epigenome and its influence on vulnerability to develop anxiety disorders is vital to understanding and effectively treating these disorders (Chertkow-Deutsher, Cohen, Klein, & Ben-Shachar, 2010; Kajantie, 2006; Seckl, 1994; Stankiewicz, Swiergiel, & Lisowski, 2013). Early-life trauma has been found to be associated with psychopathology in adult life (Brown & Harris, 1993; Brown, Harris, & Eales, 1993; Oomen et al., 2010, 2011). Recent research has shown that epigenetic regulation may be involved in shaping the effect of the early environment on the development of certain processes, such as neural plasticity and fear extinction. This was illustrated in an animal study where maternal care resulted in stable epigenetic markings that contributed toward the anxiety-like phenotype in adult rats (Murgatroyd et al., 2009; Szyf, Weaver, Champagne, Diorio, & Meaney, 2005; Weaver, Meaney, & Szyf, 2006; Weaver et al., 2004, 2005). These results have recently been extrapolated to human subjects (Franklin et al., 2010; McGowan et al., 2009; Yehuda et al., 2015).

This review will focus on miRNAs and their involvement in anxiety and stress-related disorders. Relevant findings in both animal and human studies as well as the role of miRNAs in pharmacotherapies for anxiety disorders will be discussed.

3. MicroRNAs

miRNAs form part of the class of noncoding RNAs; these small RNAs are epigenetic modulators of gene expression and have more recently gained popularity in the study of psychiatric disorders (Hunsberger, Austin, Chen, & Manji, 2009). The first miRNA, lin-4, was discovered in 1993 in *Caenorhabditis elegans* through genetic screening for deficiencies in the temporal control of postembryonic development (Lee, Feinbaum, & Ambros, 1993). The role of miRNAs as a novel regulator of gene regulation was finally realized in 2001 (Lagos-Quintana, Rauhut, Lendeckel, & Tuschl, 2001; Lau, Lim, Weinstein, & Bartel, 2001; Lee & Ambros, 2001).

miRNAs are single-stranded RNA species approximately 22 nucleotides (nt) long that form part of a large class of small, noncoding RNAs. miRBase is the major online repository for published miRNA sequences and associated annotations; the most recent version of the database, release 21, included 28,645 entries of hairpin precursor miRNAs (pre-miRNAs) that express 35,828 mature miRNA products in 223 species (Kozomara &

Griffiths-Jones, 2011, 2014). miRNAs are one of the most abundant classes of regulators in the genome (Espinosa-Parrilla & Muiños-Gimeno, 2011), with half of all miRNAs expressed from nonprotein-coding transcripts and the other half from intronic regions of protein-coding genes (Saini, Griffiths-Jones, & Enright, 2007). miRNAs show strong evolutionary conservation and are involved in vital central nervous system (CNS) processes, such as stress responses (Babenko, Golubov, Ilnytskyy, Kovalchuk, & Metz, 2012; He et al., 2007; Rinaldi et al., 2010; Schmidt et al., 2013), learning and memory (Fiore & Schratt, 2007; Lin et al., 2011), and synaptic plasticity (Lee et al., 2012; Lugli, Larson, Demars, & Smalheiser, 2012; Schratt et al., 2006). Certain miRNAs are ubiquitously expressed (e.g., let-7b, miR-17-5p, and miR-21) (Tang et al., 2007), while the expression of others is cell type specific or developmental stage dependent (Brown & Naldini, 2009), for example, brain and spinal cord-specific miR-34a (Tang et al., 2007) and miR-409-3p specific for brain development in mice (Krichevsky, King, Donahue, Khrapko, & Kosik, 2003).

The production of mature miRNAs is a complex process. In short, miRNA genes are transcribed in the nucleus by RNA polymerase II and the transcripts undergo splicing and polyadenylation to form the primary transcript miRNAs (pri-miRNAs). The pri-miRNAs are cleaved by ribonuclease III (Drosha) enzymes and the DiGeorge syndrome critical region gene 8 protein (DGCR8) in the nucleus, which results in a pre-miRNA of approximately 70–100 nt long that is actively transported to the cytoplasm by exportin 5. In the cytoplasm, the pre-miR is cleaved by Dicer (an RNaseIII enzyme) and the transactivation-responsive RNA-binding protein (TRBP) to produce double-stranded miRNAs of approximately 22 nt long. A helicase subsequently unwinds the dsRNA and one of the strands is degraded, while the other, the guide strand, functions as the mature miRNA. The mature miRNA is incorporated into a miRNA-induced silencing complex (miRISC), which target mRNAs based on sequence complementarity, mostly in the 3′untranslated regions (UTRs) (Zhang et al., 2007). The seed region consists of nucleotides two to seven of the miRNA sequence and is the most critical region for target recognition (Brennecke, Stark, Russell, & Cohen, 2005). When perfect complementarity exists between the miRNA and target mRNA, the target mRNA will be degraded. In the absence of perfect complementarity, the target will be deadenylated which results in decapping and subsequent exonucleolytic digestion or translational repression of the target mRNA (through a different mechanism at each translational step, namely initiation, elongation, and termination) (de Kloet, Fitzsimons, Datson, Meijer, & Vreugdenhil, 2009; Kuss & Chen, 2008) (Fig. 1).

Not all mRNA targets are, however, directly targeted by miRNAs through binding to the 3′UTR of the mRNA. Indirect targets lack structural affinity for the miRNA but form part of an miRNA-mediated regulatory pathway and the expression of these targets is indirectly affected by another target of the miRNA (Reyes-Herrera, Ficarra, Acquaviva, & Macii, 2011).

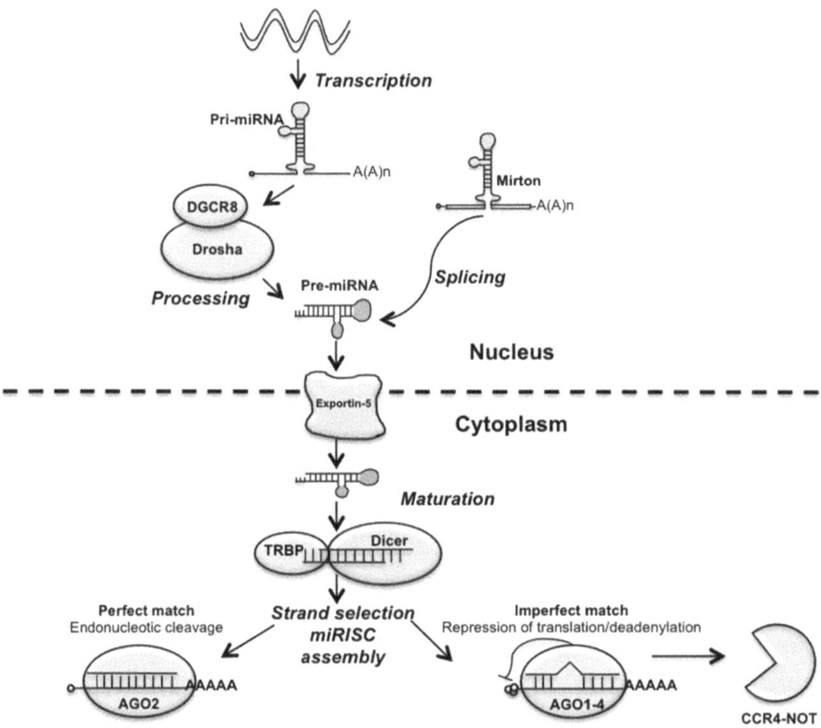

Fig. 1 Schematic representation of the production of mature miRNAs. miRNAs are encoded in the genome, their genes usually transcribed by RNA polymerase II. The transcripts undergo splicing and polyadenylation. The pri-miRNA is processed in the nucleus by the Drosha RNaseIII enzyme and the DGCR8 protein, producing the pre-miRNA. The pre-miRNA is exported to the cytoplasm by exportin-5 where Dicer and the TRBP cleave the pre-miRNA to yield a miRNA duplex (about 22-bp long). One strand is selected to function as a mature miRNA, and the other strand is usually degraded. Mature miRNAs are then incorporated in a miRNA-induced silencing complex (miRISC) that recognizes and binds to the 3′UTR of the target mRNA, which results in either endonucleotic cleavage or the repression of translation/deadenylation (AGO-argonaute) (Kuss & Chen, 2008). *Adapted From Malan-Muller, S., Hemmings, S. M., & Seedat, S. (2013). Big effects of small RNAs: A review of microRNAs in anxiety.* Molecular Neurobiology, 47, *726–739.*

Indirect targets subsequently often have a delayed response in expression changes over time compared to direct targets (as described for miR-124) (Parker & Wen, 2009). Thus, miRNAs clearly do not function as mere switches that turn genes on and off, but form part of an organized regulatory network that fine-tunes gene expression levels (Sun & Tsao, 2008).

Approximately 20–30% of all genes are regulated by at least one miRNA (Bartel, 2004; Krek et al., 2005; Lewis, Shih, Jones-Rhoades, Bartel, & Burge, 2003; Volonté, Apolloni, & Parisi, 2015). In addition, a single miRNA can target hundreds of genes and one gene can also be targeted by more than one miRNA (Wang, Li, & Hu, 2011). The typical target site of miRNAs is the 3′UTR of mRNAs; however, target sites within the coding region have also been reported (Duursma, Kedde, Schrier, Le Sage, & Agami, 2008; Kloosterman, Wienholds, Ketting, & Plasterk, 2005). Further complicating miRNA-mediated gene expression regulation is the fact that miRNAs are prone to tissue-specific RNA editing, a posttranscriptional mechanism whereby certain RNA molecules are changed to contain bases not originally encoded in the genome (via nucleotide insertion, deletion, or modification). This could result in altered properties of miRNAs and alternative mRNA: miRNA interactions (Erson & Petty, 2008). Furthermore, variations in mRNA target sites or in the miRNA itself could result in altered gene expression patterns and ultimately contribute to disease susceptibility (Muiños-Gimeno et al., 2009). *PolymiRTS* (Polymorphism in miRNAs and their Target Sites) is a database of naturally occurring DNA sequence variants in miRNA seed regions and miRNA target sites. This database predicts or identifies potential downstream effects of the variants on gene expression and higher-order phenotypes (Bhattacharya, Ziebarth, & Cui, 2014; Ziebarth, Bhattacharya, Chen, & Cui, 2012). miRSNPs represent a novel class of functional polymorphisms, which could play an important role in the etiology of complex disorders (Bertino, Banerjee, & Mishra, 2007).

It is therefore clear that, in order to understand miRNAs and their functions, the identification of their mRNA targets is crucial. This step has proven to be computationally challenging. Although great advances have been made in the field of miRNA target prediction, the false-positive rates of target-predicting software are relatively high (between 24% and 70%) (Baek et al., 2008; Easow, Teleman, & Cohen, 2007; Selbach et al., 2008). Experimental strategies therefore have to be implemented to validate predicted targets in an attempt to identify genuine miRNA targets and to determine miRNA function (Thomson, Bracken, & Goodall, 2011). Methods to identify and validate miRNA targets are a fascinating topic that is, unfortunately, beyond the scope of this review. The reader is, however, referred to reviews by Thomson et al. (2011), Schratt et al. (2006), Pasquinelli (2012), and Karres, Hilgers, Carrera, Treisman, and Cohen (2007) for further detail on in vitro and in vivo experimental strategies for miRNA target identification.

miRNAs have been implicated in the regulation of various neurobiological processes, such as neurogenesis, neurite outgrowth, synaptogenesis, and synaptic and neural plasticity (Zhou et al., 2009), especially through regulating mRNA targets that encode proteins involved in the plasticity of neural networks, maintenance of neuronal function, and specific neurodevelopmental and neurodegenerative diseases (Rogaev, 2005). miRNA expression profiles can also be altered by various factors, such as stress, GCs, and mood stabilizers (Hunsberger et al., 2009), implicating these small RNAs as role players in the etiology of anxiety and stress-related disorders. The following section will discuss evidence from animal studies to illustrate the involvement of miRNAs in anxiety and stress responses. Table 1 summarizes the miRNAs that have been described in animal models of anxiety.

4. INSIGHTS INTO miRNAs THROUGH THE USE OF ANIMAL MODELS

Animal models provide the unique opportunity to perform tissue-specific, genetic, epigenetic, and even transgenerational research on behavioral phenotypes. The high level of genomic conservation between murine models and humans makes them good models for several areas of medical research (Cheng et al., 2014; Gibbs et al., 2004) and allows for some degree of extrapolation to humans.

Neural activity in various brain regions is sensitive to acute and repeated stress (Anisman & Zacharko, 1992); short-term changes occur in neural transmission and gene regulation (Alfonso et al., 2006; Gao et al., 2006; Xu, He, Richardson, & Li, 2004), whereas long-term changes occur in structural modification (Cook & Wellman, 2004; Donohue et al., 2006; Magarinos & McEwen, 1995). It is therefore possible that miRNAs are involved in these processes. Indeed, a marked increase in expression levels of various miRNAs was noted in the frontal cortex of CD1 mice following acute stress, while minor changes in miRNA expression were observed after repeated restraint stress (RRS) (Rinaldi et al., 2010). The authors postulated that acute stress rapidly regulated miRNA expression, but that these effects were transient. Increased levels of let-7a, miR-9, and miR-26-a/b following acute restraint stress were

Table 1 MicroRNAs Investigated in Animal Studies or in vitro Cell Lines of Anxiety and Stress-Related Disorders, and Pharmacotherapies for These Disorders

miRNA	Involvement with Anxiety Disorders	Species	References
Let-7a-1	Upregulated expression in the frontal cortex following acute stress	*Mus musculus*	Rinaldi et al. (2010)
	Downregulated in amygdala after acute and chronic stress	*Rattus norvegicus*	Meerson et al. (2010)
Let-7b	Increased expression in the hippocampus due to treatment with lithium and sodium valproate	*Rattus norvegicus*	Zhou et al. (2009)
Let-7c	Decreased expression in the hippocampus due to treatment with lithium and sodium valproate	*Rattus norvegicus*	Zhou et al. (2009)
miR-1	Upregulated in amygdala under chronic stress and downregulated in the hippocampus under acute stress	*Rattus norvegicus*	Meerson et al. (2010)
miR-9	Involved in neural lineage differentiation in ESCs	*Mus musculus* and in vitro cell line	Krichevsky, Sonntag, Isacson, and Kosik (2006)
	Upregulated expression in the frontal cortex following acute stress	*Mus musculus*	Rinaldi et al. (2010)
	Upregulated expression in the medial prefrontal cortex following maternal separation	*Rattus norvegicus*	Uchida et al. (2010)
miR-9	Downregulated in striatum following maternal deprivation, downregulated in nucleus accumbens following chronic unpredictable stress with subsequent increased expression of *DRD2* Significantly associated with depressive phenotypes	*Rattus norvegicus*	Zhang et al. (2015)

Continued

Table 1 MicroRNAs Investigated in Animal Studies or in vitro Cell Lines of Anxiety and Stress-Related Disorders, and Pharmacotherapies for These Disorders—cont'd

miRNA	Involvement with Anxiety Disorders	Species	References
miR-9-1	Pre-miRNA upregulated expression in the medial prefrontal cortex following maternal separation	*Rattus norvegicus*	Uchida et al. (2010)
	Downregulated in CA1 region of hippocampus under acute or chronic stress		Meerson et al. (2010)
miR-9-3	Pre-miRNA upregulated expression in the medial prefrontal cortex following maternal separation	*Rattus norvegicus*	Uchida et al. (2010)
miR-16	Upregulated in serotonergic raphe nuclei following treatment with SSRI fluoxetine, this resulted in reduced SERT expression	*Mus musculus*	Baudry, Mouillet-Richard, Schneider, Launay, and Kellermann (2010)
miR-17-5p	Upregulated in the hippocampus CA1 region under chronic stress	*Rattus norvegicus*	Meerson et al. (2010)
	Controls neuronal development and differentiation	In vitro cell line	Hebert et al. (2009)
miR-17-5p	Significantly increased in stressed rats compared to controls, possibly controls transcription of *gp130* and *STAT3*	*Rattus norvegicus*	Bradesi et al. (2015)
miR-18	Downregulated in hypothalamus and subsequent normalized GR protein levels detected in PVN of YKS-pretreated mice following stress exposure	*Mus musculus*	Shimizu, Tanaka, Takeda, Tohyama, and Miyata (2015)

Table 1 MicroRNAs Investigated in Animal Studies or in vitro Cell Lines of Anxiety and Stress-Related Disorders, and Pharmacotherapies for These Disorders—cont'd

miRNA	Involvement with Anxiety Disorders	Species	References
miR-18a	Possible repressor of the glucocorticoid receptor gene in the hypothalamic paraventricular nucleus regulating stress responses	*Rattus norvegicus*	Uchida et al. (2008) and Vreugdenhil et al. (2009)
miR-19b	Stress-responsive miRNA, differentially expressed in serum and amygdala 14 days following stress exposure; may regulate genes associated with delayed and exaggerated fear	*Rattus norvegicus*	Balakathiresan et al. (2014)
miR-21	Involved in the control of glial cell differentiation	In vitro cell line	Chan, Krichevsky, and Kosik (2005)
miR-24a	Decreased expression in the hippocampus due to treatment with lithium and sodium valproate	*Rattus norvegicus*	Zhou et al. (2009)
miR-26a/b	Upregulated expression in the frontal cortex following acute stress	*Mus musculus*	Rinaldi et al. (2010)
miR-29a	Upregulated expression in the medial prefrontal cortex following maternal separation	*Rattus norvegicus*	Uchida et al. (2010)
miR-30c	Decreased expression in the hippocampus due to treatment with lithium and sodium valproate	*Rattus norvegicus*	Zhou et al. (2009)
miR-31	Correlated with anxiety and alcohol-related behaviors	*Mus musculus*	Parsons et al. (2012)
miR-34a	Correlation between differential expression of this miRNA and behavioral measures for exploration on the elevated plus-maze task	*Mus musculus*	Parsons et al. (2008)

Continued

Table 1 MicroRNAs Investigated in Animal Studies or in vitro Cell Lines of Anxiety and Stress-Related Disorders, and Pharmacotherapies for These Disorders—cont'd

miRNA	Involvement with Anxiety Disorders	Species	References
	Decreased expression in the hippocampus due to treatment with lithium and sodium valproate	*Rattus norvegicus* and in vitro cell line	Zhou et al. (2009)
miR-34c	Correlation between differential expression of this miRNA and behavioral measures for anxiety in mice.	*Mus musculus*	Parsons et al. (2008)
	Upregulated following exposure to stress; this upregulation reduced anxiety symptoms possibly by targeting *CRFR1*		Haramati et al. (2011)
miR-124	Upregulated in response to physical environmental enrichment and subsequent improvements in cognition and neurogenesis in the dentate gyrus	*Rattus norvegicus*	Brenes et al. (2015)
miR-124	Upregulated expression in the medial prefrontal cortex following maternal separation	*Rattus norvegicus*	Uchida et al. (2010)
miR-124	Upregulated following environmental enrichment which resulted in modulation of glucocorticoid receptor expression and subsequently reduced anxiety	*Mus booduga*	Ragu, Marimuthu, and Rajan (2013)
	Upregulated in response to physical environmental enrichment, which resulted in improvements in cognition and neurogenesis in dentate gyrus	*Rattus norvegicus*	Brenes et al. (2015)

Table 1 MicroRNAs Investigated in Animal Studies or in vitro Cell Lines of Anxiety and Stress-Related Disorders, and Pharmacotherapies for These Disorders—cont'd

miRNA	Involvement with Anxiety Disorders	Species	References
miR-124-1	Pre-miRNA upregulated expression in the medial prefrontal cortex following maternal separation	*Rattus norvegicus*	Uchida et al. (2010)
	Downregulated in the hippocampus under acute stress.		Meerson et al. (2010)
	Controls neuronal development and differentiation		Hebert et al. (2009)
miR-124a	Involved in neural lineage differentiation in ESCs	*Mus musculus* and in vitro cell line	Krichevsky et al. (2006); Lim et al. (2005); and Makeyev, Zhang, Carrasco, and Maniatis (2007)
	Downregulates glucocorticoid receptor	*Rattus norvegicus*	Vreugdenhil et al. (2009)
miR-128a	Decreased expression in the hippocampus due to treatment with lithium and sodium valproate	*Rattus norvegicus*	Zhou et al. (2009)
miR-128b	Regulates formation of fear extinction memory in the infralimbic prefrontal cortex	*Mus musculus*	Lin et al. (2011)
miR-132	One of the most highly inducible CREB targets plays a role in neurite outgrowth and neuronal plasticity	In vitro neural cell line	Vo et al. (2005)
	Upregulated expression of miRNA and pre-miRNA in the medial prefrontal cortex following maternal separation	*Rattus norvegicus*	Uchida et al. (2010)

Continued

Table 1 MicroRNAs Investigated in Animal Studies or in vitro Cell Lines of Anxiety and Stress-Related Disorders, and Pharmacotherapies for These Disorders—cont'd

miRNA	Involvement with Anxiety Disorders	Species	References
miR-132	Upregulated in response to physical environmental enrichment and subsequent improvements in cognition and neurogenesis in the dentate gyrus	*Rattus norvegicus*	Brenes et al. (2015)
miR-134	Modulates synaptic plasticity in hippocampus	*Rattus norvegicus*	Hansen et al. (2007)
	Upregulated expression in the central amygdala and hippocampus after acute stress.	*Rattus norvegicus*	Meerson et al. (2010)
	Downregulated expression in the central amygdala and hippocampus after chronic stress	*Rattus norvegicus*	Meerson et al. (2010)
miR-142-5p	Stress-responsive miRNA, differentially expressed in serum and amygdala 14 days following stress exposure; may regulate genes associated with delayed and exaggerated fear	*Rattus norvegicus*	Balakathiresan et al. (2014)
miR-144	Decreased expression in the hippocampus due to treatment with lithium and sodium valproate	*Rattus norvegicus*	Zhou et al. (2009)
miR-146b	Downregulated in humanized *BDNF* Val66Met knock-in mice with Met/Met genotype	*Mus musculus*	Hsu, Xu, Mukai, Karayiorgou, and Gogos (2015)
miR-183	Upregulated expression in the central amygdala following acute stress	*Rattus norvegicus*	Meerson et al. (2010)

Table 1 MicroRNAs Investigated in Animal Studies or in vitro Cell Lines of Anxiety and Stress-Related Disorders, and Pharmacotherapies for These Disorders—cont'd

miRNA	Involvement with Anxiety Disorders	Species	References
	Upregulated following exposure to the elevated plus maize in mice subjected to environmental enrichment, but downregulated following exposure to the elevated plus maize in mice that received standard care	*Mus booduga*	Ragu et al. (2013)
miR-186	Upregulated in the cerebellum of animals subjected to daily restrained stress compared to the naïve controls	*Rattus norvegicus*	Babenko et al. (2012)
	Significantly upregulated intestinal levels of miR-186 and activities of BChE and AChE-R 1-week postexposure to predator scent stress	*Mus musculus*	Nadorp and Soreq (2014)
miR-188	Upregulated in hippocampus during LTP	*Rattus norvegicus*	Lee et al. (2012)
miR-200b-3p	Upregulated in sperm of F1 progeny following MSUS	*Mus musculus*	Gapp et al. (2014)
miR-208	Upregulated in CA1 region of hippocampus under acute or chronic stress	*Rattus norvegicus*	Meerson et al., 2010
miR-212	Pre-miRNA upregulated expression in the medial prefrontal cortex following maternal separation	*Rattus norvegicus*	Uchida et al. (2010)
miR-221	Decreased expression in the hippocampus due to treatment with lithium and sodium valproate	*Rattus norvegicus*	Zhou et al. (2009)

Continued

Table 1 MicroRNAs Investigated in Animal Studies or in vitro Cell Lines of Anxiety and Stress-Related Disorders, and Pharmacotherapies for These Disorders—cont'd

miRNA	Involvement with Anxiety Disorders	Species	References
miR-223-3p	Stress-responsive miRNA, differentially expressed in serum and amygdala 14 days following stress exposure; may regulate genes associated with delayed and exaggerated fear	*Rattus norvegicus*	Balakathiresan et al. (2014)
miR-273	Plays a role in neuronal differentiation	*C. elegans*	Chang, Johnston, Frokjaer-Jensen, Lockery, and Hobert (2004); and Johnston and Hobert (2003)
miR-322*	Stress-responsive miRNA, differentially expressed in serum and amygdala 14 days following stress exposure	*Rattus norvegicus*	Balakathiresan et al. (2014)
miR-323	Correlation between differential expression of this miRNA and behavioral measures for anxiety in mice	*Mus musculus*	Parsons et al. (2008)
miR-324	Stress-responsive miRNA, differentially expressed in serum and amygdala 14 days following stress exposure	*Rattus norvegicus*	Balakathiresan et al. (2014)
miR-326	Increased expression in nucleus accumbens and decreased expression in the striatum, following chronic unpredictable stress. Increased expression in striatum following maternal deprivation. Associated with anxiety- and depression-like symptoms. Escitalopram normalized microRNA-326 expression levels in nucleus accumbens and striatum	*Rattus norvegicus*	Zhang et al. (2015)

Table 1 MicroRNAs Investigated in Animal Studies or in vitro Cell Lines of Anxiety and Stress-Related Disorders, and Pharmacotherapies for These Disorders—cont'd

miRNA	Involvement with Anxiety Disorders	Species	References
miR-375-3p and -5p	Upregulated in sperm and hippocampus of F1 and hippocampus of F2 progeny following MSUS	*Mus musculus*	Gapp et al. (2014)
miR-376	Upregulated in CA1 region of hippocampus under acute or chronic stress	*Rattus norvegicus*	Meerson et al. (2010)
miR-378	Association between miRNA and behavioral measures (exploration, learning, and memory) for anxiety in mice	*Mus musculus*	Parsons et al. (2008)
miR-381	Upregulated in the cerebellum of animals subjected to daily restrained stress compared to the naïve controls	*Rattus norvegicus*	Babenko et al. (2012)
miR-421-3p	Stress-responsive miRNA, differentially expressed in serum and amygdala 14 days following stress exposure; may regulate genes associated with delayed and exaggerated fear	*Rattus norvegicus*	Balakathiresan et al. (2014)
miR-451	Association between miRNA and behavioral measures (exploration, learning, and memory) for anxiety in mice	*Mus musculus*	Parsons et al. (2008)
miR-463*	Stress-responsive miRNA, differentially expressed in serum and amygdala 14 days following stress exposure	*Rattus norvegicus*	Balakathiresan et al. (2014)
miR-466-5p	Upregulated in sperm and hippocampus of F1 and hippocampus of F2 progeny following MSUS	*Mus musculus*	Gapp et al. (2014)

Continued

Table 1 MicroRNAs Investigated in Animal Studies or in vitro Cell Lines of Anxiety and Stress-Related Disorders, and Pharmacotherapies for These Disorders—cont'd

miRNA	Involvement with Anxiety Disorders	Species	References
miR-667-3p	Facilitates downregulation of *Comt* mRNA in the brain	*Mus musculus*	Segall et al. (2015)
miR-672-5p	Upregulated in sperm of F1 and hippocampus of F2 progeny following MSUS	*Mus musculus*	Gapp et al. (2014)
miR-674*	Stress-responsive miRNA, differentially expressed in serum and amygdala 14 days following stress exposure	*Rattus norvegicus*	Balakathiresan et al. (2014)
miR-709	Downregulated in the cerebellum of animals subjected to daily restrained stress compared to the naïve controls	*Rattus norvegicus*	Babenko et al. (2012)
miR-1928	Stress-responsive miRNA, differentially expressed in serum and amygdala 14 days following stress exposure; may regulate genes associated with delayed and exaggerated fear	*Rattus norvegicus*	Balakathiresan et al. (2014)
miR-1971	Downregulation of this miRNA in PFC is associated with therapeutic action of fluoxetine in shocked mice	*Mus musculus*	Schmidt et al. (2013)

AChE-R, acetylcholinesterase rare soluble form; *BDNF*, brain-derived neurotrophic factor; *BChE*, butyrylcholinesterase; *Comt*, catechol-O-methyltransferase; *CREB*, cAMP response element-binding protein; *CRFR1*, corticotrophin-releasing factor receptor type 1; *ESCs*, embryonic stem cells; *gp130*, interleukin 6 signal transducer; *LTP*, long-term potentiation; *MSUS*, unpredictable maternal stress; *PFC*, prefrontal cortex; *PVN*, paraventricular nucleus; *STAT3*, signal transducer and activator of transcription 3; *YKS*, Yokukansan.

Adapted from Malan-Muller, S., Hemmings, S. M., & Seedat, S. (2013). Big effects of small RNAs: A review of microRNAs in anxiety. Molecular Neurobiology, 47, 726–739.

verified using Northern blot analysis. These changes were found to be region-specific, present only in the frontal cortex but not in the hippocampus, suggesting that miRNAs in the frontal cortex are involved in the translation of stressful events into alterations in protein expression (Rinaldi et al., 2010).

Also using an animal model of RRS in F344 rats, Uchida et al. (2008) investigated neuroendocrine and biochemical responses to RRS and miRNAs that could mediate these responses. They found reduced levels of glucocorticoid receptor (GR) protein expression in the paraventricular nucleus (PVN) compared to control SD rats. Data from cultured neuronal cells indicated that miR-18a inhibited translation of GR mRNA, and in addition, they also detected increased expression levels of miR-18a in the PVN of RRS F344 rats compared to control SD rats. Later in vitro experiments confirmed the results for miR-18a and also established a similar role for miR-124 (Vreugdenhil et al., 2009). These studies indicated the usefulness of F344 rats as a model to study vulnerability to repeated stress and illustrated that the translational repression of GR by elevated levels of miR-18a could result in increased susceptibility to stress-related disorders (Uchida et al., 2008). Additional studies also suggested that miR-18 may regulate early postnatal hippocampal neurogenesis through its repression of GR expression (Molitoris, McColl, & Distelhorst, 2011; Vreugdenhil et al., 2009). Furthermore, the presence of a regulatory feedback loop between miRNA-18 and the estrogen-receptor α (Castellano et al., 2009) implicates the long-lasting changes in miR-18 expression in the observed sex differences in the levels of adult neurogenesis following early-life stress (Oomen et al., 2010, 2011; Schouten, Aschrafi, Bielefeld, Doxakis, & Fitzsimons, 2013).

Studies have indicated that regional gene expression and epigenomic responses to chronic stress in the brain could also be mediated by region-specific miRNA-mediated reprogramming (Babenko et al., 2012). The authors found that expression of miR-186 and miR-381 was increased, whereas expression of miR-709 was decreased in the cerebellum of stressed animals compared to nonstressed controls. Interestingly, downregulation of miR-709 was evident even after 2 weeks after recovery from stress, indicating that downregulation of miR-709 in the cerebellum is possibly involved in long-term adaptation to chronic stress. Differences in expression profiles of miR-709 were, however, noted between the cerebellum, hippocampus, and PFC, where expression of miR-709 in the hippocampus and PFC was unchanged following 2 weeks of chronic stress, and miR-709 expression was upregulated in the PFC after 2 weeks of recovery. These findings emphasize the fact that stress can affect regional expression profiles of certain miRNAs (Babenko et al., 2012) and this might be a mechanism that mediates region-specific gene expression.

Differential miRNA expression was investigated in the hippocampus of four common inbred mouse strains (A/J, BALB/cJ, C57BL/6J, and

DBA/2J) prone to anxiety-like behavior. This investigation resulted in the identification of 11 differentially expressed miRNAs that correlated with different anxiety measures (Parsons et al., 2008). The expression of miR-34a, miR-323, miR-378, and miR-451 correlated with lower levels of anxiety (increased explorative behavior), whereas expression of miR-34c and miR-323 correlated with higher levels of anxiety (reduced explorative behavior) as measured by the elevated plus-maze task. Expression of miR-34c, miR-323, miR-378, and miR-451 correlated with learning and memory tests (Parsons et al., 2008). This was one of the first studies to demonstrate the involvement of miRNAs in anxiety, learning, and memory.

Early-life adversity is a strong risk factor for the development of anxiety and stress-related disorders later in life (Brown & Harris, 1993; Brown et al., 1993). In animal models, maternal separation is mostly used to mimic early-life trauma in humans. In rats subjected to maternal separation a significant increase in several miRNAs as well as RE1 silencing transcription factor (REST) target gene mRNAs was detected in the medial prefrontal cortex (mPFC). Expression levels of pre-miR-132, -124-1, -9-1, -9-3, -212, and -29a as well as the mature miR-132, -124, -9, and -29a were significantly augmented in maternally separated rats compared to control rats (Uchida et al., 2010). Interestingly, several miRNAs, including miR-132, -124-1, -9-1, -9-3, -212, and -29a, contain a repressor element-1 (RE-1) site within 50 kb of their promoter regions (Otto et al., 2007). Differential expression of mRNAs and miRNAs that contain RE-1 might be in response to alterations in RE-1-mediated gene transcription in the mPFC of maternally separated rats due to altered REST4 expression. Indeed, Uchida and colleagues were able to detect increased expression levels of genes and miRNAs possibly regulated by REST4, such as glutamate receptor subunit (*GLUR2*), calcium/calmodulin-dependent protein kinase II (*CAMKIIα*), and adenylate cyclase 5 (ADCY5) and precursors for miR-132, -124, and -212. These findings implicate a REST4-mediated gene network and specific miRNAs within the mPFC that might be involved in susceptibility to developing mood and anxiety disorders in adulthood following exposure to early-life stress (Uchida et al., 2010).

In an animal model that incorporated early-life stress followed by chronic unpredictable stress, researchers found that miR-9 downregulation could be associated with increased susceptibility to features of anxiety and depression in the context of early-life stress (Zhang et al., 2015). The mechanism whereby miR-9 exerts this effect is by directly targeting dopamine receptor

D2 (*DRD2*) and subsequently repressing the translation of the mRNA, which results in decreased DRD2 protein expression. Chronic, unpredictable stress also increased the expression of miR-326 in the nucleus accumbens and downregulated it in the striatum, while increased miR-326 expression was only induced in the striatum following maternal deprivation. Dysregulated miR-326 expression in the nucleus accumbens was positively correlated with depressive-like phenotypes. These results highlight the distinct role of miR-326 in different brain areas in stress-induced depression-like phenotypes. In addition, miR-326 expression levels normalized after a 4-week treatment of escitalopram, suggesting that miR-326 in the nucleus accumbens and striatum is associated with the molecular mechanism associated with escitalopram treatment (Zhang et al., 2015).

One of the hypotheses is that miRNAs mediate regulation of the stress response by alternative splicing (Meerson et al., 2010). By investigating miRNA expression profiles in the hippocampal CA1 region and the central amygdala in acute and chronically stressed rats, distinct miRNA expression profiles were identified in these two brain regions in response to stress. Upregulation of miR-134 and miR-183 was evident in the amygdala in response to acute stress, while miR-134 was downregulated in the hippocampus and amygdala following chronic stress. These two miRNAs share several common predicted mRNA targets, some of which are known mediators of neuronal stress reactions, such as the serine/arginine-rich splicing factor 2 (SC35). SC35 is augmented in response to stress, which facilitates alternative splicing of acetylcholinesterase (AChE) from the synapse-associated isoform (AChE-S) to the rare soluble form (AChE-R). The authors confirmed the miR-183-mediated suppression of SC35 in cultured cells (Meerson et al., 2010). This demonstrates that stress can alter the expression of miR-183 and miR-134, and through regulating splice factors and their targets, these miRNAs are able to modify alternative splicing as well as cholinergic neurotransmission under stress conditions, providing a link between the molecular and physiological responses of different brain regions to psychological stress (Meerson et al., 2010).

Another study produced similar findings in an Indian field mouse, *Mus booduga*, where they detected increased expression levels of miR-183 following exposure to the elevated plus maize in mice subjected to environmental enrichment, but not in mice that received standard care (Ragu et al., 2013). Upregulated levels of miR-183 subsequently suppressed *SC35* expression, which shifts splicing of AChE from synaptic (AChE-S) to the read-through (AChE-R) form. Under standard housing conditions, however, miR-183 was downregulated, which induced the splicing of AChE from AChE-S to AChE-R. The upregulated AChE-R form could terminate acetylcholine transmission, which is reflected in the level of anxiety-like behavior. In addition, miR-124 was found to be upregulated in response to environmental enrichment, resulting in modulation of GR expression and subsequently reduced anxiety (Ragu et al., 2013). Animal studies have shown that an enriched environmental condition has the ability to reduce anxiety (Benaroya-Milshtein et al., 2004; Prior & Sachser, 1995). Interestingly, another study showed that miR-124 is upregulated in response to physical environmental enrichment, which resulted in subsequent improvements in cognition and neurogenesis in the dentate gyrus of rats (Brenes et al., 2015). These studies suggest that environmental enrichment regulates several miRNAs, which facilitates a reduction in anxiety-like behavior through the intricate regulation of their target mRNAs (Brenes et al., 2015; Ragu et al., 2013).

Nadorp and Soreq (2014) determined that miR-186 targets not only the stress-inducible soluble "read-through" variant (AChE-R) but also the major peripheral cholinesterase, butyrylcholinesterase (BChE). They investigated the in vivo relevance of this dual interaction by exposing mice to predator scent stress and investigating the subsequent changes in miR-186, AChE-R, and BChE levels. Intestinal levels of miR-186 and the activities of BChE and AChE-R were significantly upregulated 1 week postexposure. Their results revealed that overlapping miRNA regulation is a surveillance mechanism controlling cholinergic neurotransmission in health and disease, and illuminated the previously unknown involvement of miR-186 and BChE in psychological stress responses.

Long-term potentiation (LTP) is a model of synaptic plasticity. LTP is similar to the plasticity that underlies learning and memory (Bliss & Collingridge, 1993; Hölscher, 1999), which are the two cognitive processes that are impaired in PTSD (Friedman, 1997; Vermetten & Bremner, 2002). The role of miRNAs in processes such as LTP has also been established; LTP was shown to induce differential expression of several miRNAs in the rat hippocampus, with the largest upregulation detected for miRNA-188. miR-188 in hippocampal neurons could therefore be involved in the fine-tuning of synaptic plasticity (Lee et al., 2012).

In order to investigate the functional role of miRNAs in stress response regulation, Haramati et al. (2011) followed an interesting approach and inactivated the Dicer gene (a key enzyme in miRNA synthesis pathway) in the central amygdala, which subsequently inactivated miRNA processing in this brain region (Haramati et al., 2011). Mice deficient in Dicer and mature miRNAs in their amygdala exhibited a sharp increase in anxiety-like behavior (Haramati et al., 2011). They also showed that exposure to acute stress resulted in differential expression of numerous miRNAs in the amygdala of wild-type mice. miR-34c, a prominent stress-induced miRNA, was also shown to be strongly upregulated by exposure to stress; this upregulation subsequently resulted in reduced anxiety symptoms in normal mice. One of the targets of miR-34c is corticotrophin-releasing factor receptor type 1 (*CRFR1*), and the authors were able to show that one of the effects of miR-34c in the amygdala is to target an evolutionarily conserved region in the 3′UTR of *CRFR1* mRNA. They postulated that miR-34c facilitated the downregulation of stress-related proteins like CRFR1 and promoted the stress recovery process in these mice (Haramati et al., 2011).

Bradesi et al. (2015) demonstrated that chronic water avoidance stress in rats (a characterized model of visceral hypersensitivity and anxiety) (Bradesi et al., 2005) results in altered expression levels of spinal genes and miRNAs involved in neuroinflammation. They identified 39 miRNAs that were differentially regulated by stress, which formed part of two main networks relevant to inflammatory processes. Of interest was the significant increase in miR-17-5p in stressed rats compared to controls; increased miR-17-5p subsequently resulted in changes in the expression of predicted targets, interleukin 6 signal transducer (*gp130*) and signal transducer and activator of transcription 3 (*STAT3*) (involved in intracellular signaling cascades in response to gp130 activation). Furthermore, the authors confirmed a modulatory role of spinal miR-17-5p in the modulation of visceral sensitivity in vivo (Bradesi et al., 2015). This study provided novel insights into the possible mechanisms that mediate the effect of chronic stress on neuroinflammation in the spinal cord.

Animal models provide the opportunity to investigate the transgenerational effects of stress. Researchers were able to illustrate that sperm RNAs could mediate the transgenerational inheritance of the effects of early trauma (unpredictable maternal stress [MSUS]) in mice (Gapp et al., 2014). Furthermore, traumatic stress in early life was also shown to have the ability to alter mouse miRNA expression as well as behavioral and metabolic responses in the progeny. A large set of miRNAs were differentially expressed in the sperm of animals subjected to MSUS; miR-375-3p, miR-375-5p, miR-200b-3p, miR-672-5p, and miR-466-5p were verified to be upregulated in F1 MSUS sperm. Differential expression of these miRNAs was also detected in the hypothalamus, hippocampus, and serum; however, in some instances the difference in expression between controls and MSUS was tissue specific. Expression of all the aforementioned miRNAs, except for miR-200b-3p, was comparable between F1 sperm and F2 hippocampal samples. Expression of sperm miR-375-3p and miR-466-5p was comparable to that of F1 hippocampal samples. Furthermore, by injecting sperm RNAs from traumatized males into fertilized wild-type oocytes, researchers were able to reproduce the behavioral and metabolic modifications in the resulting offspring (Gapp et al., 2014). These findings highlight the importance of miRNAs in germ cells and their sensitivity to early traumatic stress and demonstrate their role in the cross-generational consequences of early-life stress.

Catechol-O-methyltransferase (COMT) is a critical enzyme in catechol catabolism and is involved in a several essential processes such as stress response (Desbonnet et al., 2012; McLean et al., 2011), mood (Tost, Alam, & Meyer-Lindenberg, 2010), and cognition (Buckert, Kudielka, Reuter, & Fiebach, 2012). An insertion of a transposon, B2 short interspersed element (SINE) in the 3′UTR of *Comt*, has been detected in mouse laboratory strains and has been associated with decreased stress responses, anxiety, and pain phenotypes mediated by increased mRNA and protein amounts (Kember et al., 2010; Li et al., 2010; Segall et al., 2010). The B2 SINE insertion possibly shortens the 3′UTR of the mRNA and it was therefore hypothesized that it might result in an alteration in miRNA binding sites (Segall et al., 2015). Comparative analysis and in silico prediction of miRNAs that could target the transcript 3′ to the B2 element of *Comt* were performed and mmu-miR-3470a, mmu-miR-3470b, and mmu-miR-667 were selected as candidates for experimental screening. Transfection with each miRNA resulted in the downregulation of expression and enzymatic activity of the ancestral *COMT* transcript. In vivo experiments showed that mmu-miR-667-3p is significantly associated with downregulation of *Comt* mRNA in the brain; furthermore, lentiviral injections of mmu-miR-3470a, mmu-miR-3470b, and mmu-miR-667 increased hypersensitivity in the mouse formalin model, consistent with reduced COMT activity (Segall et al., 2015).

As mentioned earlier, SNPs account for a significant portion of the variation in gene expression, although the mechanisms whereby *cis*-acting variants affect gene expression are well documented, the exact way in which *trans*-acting variants modulate gene expression and disease susceptibility is largely unknown (Cookson, Liang, Abecasis, Moffatt, & Lathrop, 2009).

The Val66Met (rs6265) variant in brain-derived neurotrophic factor (*BDNF*) has been associated with several psychiatric disorders including anxiety disorders, depression, and schizophrenia (Neves-Pereira et al., 2005; Tocchetto et al., 2011; Verhagen et al., 2010). Hsu et al. (2015) wanted to determine the downstream effects of this *BDNF* SNP on miRNA expression profiles and used humanized *BDNF* Val66Met knock-in mice and investigated global miRNA expression profiles in the hippocampus. They showed that this variant resulted in the dysregulation of numerous miRNAs. Of particular interest was the downregulation of miR-146b, and subsequent increased mRNA levels of some of its target genes, Per1 and Npas4 as well as Irak1 protein levels detected in vitro, with similar changes evident in the hippocampus of the humanized Met/Met mice. miR-146b and its target genes are particularly enriched in neuronal genes and could therefore account for the biological effects of the Val66Met variant (Hsu et al., 2015). These findings underscore the potential *trans* effects of common variants on miRNA-mediated gene expression, which could form a fundamental part of the genetic underpinnings of complex disorders.

It is important to note that, when attempting to extrapolate results from animal studies to humans, surrogate human tissues are required. Researchers therefore also investigate circulating miRNAs that can be detected in several fluids, to determine to which degree the miRNA profiles in these fluids correlate with the levels of miRNAs in the brain. Studies showed that circulating miRNAs can be detected in fluids such as serum and urine and many studies are therefore interested to use circulating miRNAs as novel biomarkers of disease. Balakathiresan et al. (2014) examined the expression of miRNAs in serum and the amygdala in an animal model of PTSD. miRNAs detected to be significant differentially expressed in serum 14 days after stress exposure (immobilization stress for 2 h per day along with tail shocks for three consecutive days) were validated for similar regulation in amygdala of the same animals. Several stress-responsive miRNAs, namely miR-142-5p, miR-19b, miR-1928, miR-223-3p, miR-322*, miR-324, miR-421-3p, miR-463*, and miR-674*, were differentially expressed and may have potential as biomarkers for PTSD. Further validations revealed that five miRNAs, including miR-142-5p, miR-19b, miR-1928, miR-223, and miR-421-3p, may regulate genes associated with delayed and exaggerated fear (Balakathiresan et al., 2014).

The authors conclude that this is possibly the first study that identified serum miRNAs that may have potential for the diagnosis of PTSD. This rather sweeping statement should however be made with much caution due to the fact that several factors could contribute to the differential regulation of miRNAs and the same miRNA could even be differentially expressed in different diseases, such as depression, PTSD, and obsessive–compulsive disorder (OCD), which would therefore not enable clear diagnosis of a specific disorder. For instance, increased neuroinflammation has been shown to underlie a plethora of psychiatric disorders and a vast array of miRNAs play a role in inflammatory processes (Iborra et al., 2013; Manca et al., 2011; Murata et al., 2013), and therefore, similar miRNAs might be regulated in a similar fashion in these psychiatric disorders, making a clear diagnosis challenging. In addition, there could be other differentially expressed miRNAs in serum that correlate with expression profiles in other brain regions affected by PTSD, which could also be important and should be investigated. Refer to Table 1 for a summary of miRNAs that have been described in animal models of anxiety.

Although miRNAs and their 3′UTR target regions are highly conserved between species, it is important to realize that, in order to extrapolate any miRNA findings from animal studies, they first have to be investigated in human subjects. Furthermore, animal studies have the advantage of tissue-specific miRNA detection; therefore, brain-specific miRNAs should be investigated in surrogate tissues in humans to determine their suitability as true biological markers of anxiety and stress-related disorders. The following section will discuss miRNAs that have been implicated in anxiety and stress-related disorders in humans.

5. INVESTIGATING miRNAs IN ANXIETY DISORDERS

The abundance of miRNAs in the brain, together with their functions in neuronal differentiation and synaptic plasticity, implicates these tiny regulators in the etiology of psychiatric disorders. miRNA biogenesis has been shown to occur locally in the vicinity of synapses; postsynaptic densities are especially rich in Drosha and DGCR8 (the microprocessor components) and their pri-miRNA substrate (Lugli et al., 2012). This provides miRNAs with the ideal position to rapidly respond to synaptic activity and regulate translation. Findings by Lugli, Larson, Martone, Jones, and Smalheiser (2005) demonstrated that Dicer and Argonaute 1 are highly enriched in postsynaptic densities and that their levels are modulated by neuronal activity (Lugli et al., 2005). The role of miRNAs has been reported in several psychiatric disorders, including schizophrenia (postmortem brain samples) (Beveridge, Gardiner, Carroll, Tooney, & Cairns, 2010; Beveridge et al.,

Table 2 MicroRNAs Investigated in Individuals With Anxiety and Stress-Related Disorders

miRNA	Involvement With Anxiety Disorders	References
miR-22	Associated with PD—targets *RGS2*, *BDNF*, *HTR2C*, and *MAOA*	Muiños-Gimeno et al. (2011)
miR-128	Association of an allelic variant in the target site for miR-128 in *NTRK3* (ss102661458) with PD—reduction of *NTRK3* repression	Muiños-Gimeno et al. (2009)
miR-138-2	Associated with age at onset in PD and PD subphenotypes (PD with and without agoraphobia)—targets *GABRA6*, *CCKBR*, and *POMC*	Muiños-Gimeno et al. (2011)
miR-144-5p	Lower plasma expression levels in patients with depression. These levels increased following treatment (mindfulness group therapy)	Wang et al. (2015)
miR-148a	rs73531 (tags miR-148a) is associated with age at onset in PD. miR-148a associated with PD subphenotypes (PD with and without agoraphobia)—targets *CCKBR*	Muiños-Gimeno et al. (2011)
miR-204	Association of an allelic variant in the 3′UTR of *ALAD* with social phobia	Donner et al. (2008)
miRNA-206	Significant gene–gene interaction between rs16882131 variant in *MIR206* gene and rs6265 in *BDNF* that contributed to BD-I susceptibility and treatment response	Wang et al. (2014)
miR-211	Association of an allelic variant in the 3′UTR of *ALAD* with social phobia	Donner et al. (2008)
miR-330-3p	Association of an allelic variant (rs41305272) in the target site for miR-330-3p in *MAP2K5* with PD and depression	
miR-339	Associated with PD, targets *ADORA2A*, *BDNF*, and *CRHR2*	Muiños-Gimeno et al. (2011)
miR-485-3p	Significantly associated with hoarding subtype of OCD	Muiños-Gimeno et al. (2009)
miR-488	Associated with PD subphenotypes (PD with and without agoraphobia), targets *POMC*	Muiños-Gimeno et al. (2011)
miR-491	Associated with PD subphenotypes (PD with and without agoraphobia)	Muiños-Gimeno et al. (2011)

Continued

Table 2 MicroRNAs Investigated in Individuals With Anxiety and Stress-Related Disorders—cont'd

miRNA	Involvement With Anxiety Disorders	References
miR-509	Association of an allelic variant in the target site for miR-509 in *NTRK3* (ss102661458) with PD—reduction of *NTRK3* repression	Muiños-Gimeno et al. (2009)
miR-608	rs17228616 in *AChE* resulted in a weakened interaction of miR-608 with *AChE*, resulting in increased levels of *AChE* and a decrease in *Cdc42* and *Il6* in postmortem human cortices homozygous for the minor rs17228616 allele	Hanin et al. (2014)
miR-668	Minor *A*-allele of *WFS1* variant rs1046322 (associated with high levels of aggression, impulsivity, anxiety, and depression) generates a mismatch in target site of the seed region, which results in increased *WFS1* expression	Kovacs-Nagy et al. (2013)
miR-765	Association of an allelic variant in the target site for miR-765 in *NTRK3* (*ss102661460*) with PD reduction of *NTRK3* repression	Muiños-Gimeno et al. (2009)
miR-4717-5p	Regulates human *RGS2* and could contribute to genetic risk toward anxiety-related traits	Hommers et al. (2015)

AChE, acetylcholinesterase; *ADORA2A*, adenosine receptor A2a; *ALAD*, aminolevulinate dehydratase gene; *BD*, bipolar disorder; *BDNF*, brain-derived neurotrophic factor; *CCKBR*, cholecystokinin B receptor; *Cdc42*, cell division cycle 42; *CRHR2*, corticotropin-releasing hormone receptor 2; *GABRA6*, gamma-aminobutyric acid A receptor, alpha 6; *HTR2C*, serotonin 5-HT2C receptor; *Il6*, interleukin-6; *MAOA*, monoamine oxidase; *MAP2K5*, mitogen-activated protein kinase kinase 5; *NTRK3*, neurotrophin-3 growth factor receptor; OCD, obsessive–compulsive disorder; *PD*, panic disorder; *POMC*, pro-opiomelanocortin preproprotein; *RGS2*, regulator of G-protein signaling 2; *UTR*, untranslated region; *WFS1*, wolfram syndrome 1.

2008; Perkins et al., 2007), Rett syndrome (Abu-Elneel et al., 2008), autism spectrum disorders (Abu-Elneel et al., 2008; Talebizadeh, Butler, & Theodoro, 2008), and substance abuse disorders (Chandrasekar & Dreyer, 2009). However, the number of studies that investigate miRNAs in anxiety disorders in human subjects is still limited. Table 2 summarizes the miRNAs that have been investigated in individuals with anxiety and stress-related disorders and pharmacotherapies for these disorders.

5.1 miRSNPs and Variants Within miRNA Genes

Genetic variants within miRNA binding sites are known as miRSNPs. They represent an exciting new class of functional SNP (Mishra, Mishra,

Banerjee, & Bertino, 2008). Several studies have investigated miRSNPs in complex psychiatric disorders. Variants in wolfram syndrome 1 gene (*WFS1*) have previously been associated with depression, bipolar depression, suicidality, and mood disorders (Kawamoto et al., 2004; Koido et al., 2005; Sequeira et al., 2003). Individuals homozygous for the rs1046322 minor *A*-allele showed higher levels of aggression, impulsivity, anxiety, and depression (Kovacs-Nagy et al., 2013). Researchers discovered that rs1046322 is a miRSNP and in silico studies in conjunction with functional luciferase analyses showed that the minor *A*-allele generates a mismatch in the target site of the seed region of miR-668, which ultimately results in the increased expression of *WFS1* (Kovacs-Nagy et al., 2013). Another study that investigated miRSNPs found that they could facilitate expanded downstream effects that could alter important biological pathways. miR-608 targets *AChE* mRNA and Hanin et al. (2014) found that rs17228616 in *AChE* resulted in a weakened interaction of miR-608 with *AChE*. This weakened interaction mediated miR-608 suppression of other targets, including cell division cycle 42 (*Cdc42*) and interleukin-6 (*Il6*) and increased levels of *AChE*, and decreases in *Cdc42* and *Il6* were noted in postmortem human cortices homozygous for the minor rs17228616 allele. Furthermore, individuals heterozygous and homozygous for the minor allele had reduced cortisol levels and elevated blood pressure, which was found to predict risk of anxiety and hypertension (Hanin et al., 2014).

Another candidate gene for PD and OCD is neurotrophin-3 growth factor receptor (*NTRK3*). Resequencing of the 3′UTRs of two different *NTRK3* isoforms in PD and OCD patients revealed that the *C*-allele of rs28521337, in the truncated isoform of *NTRK3* (positioned within a miR-485-3p functional target site), was associated with the hoarding phenotype of OCD. In addition, two new rare variants were identified, rs72481816 (within a functional target sites for miR-509 and miR-128) and rs72481814 (in a functional target site for miR-765), in the 3′UTR of *NTRK3* of a PD patient (Muiños-Gimeno et al., 2009). miR-128 is enriched in the brain and is involved in neuronal differentiation and synaptic processing (Betel, Wilson, Gabow, Marks, & Sander, 2008). These two variants significantly altered the miRNA-mediated regulation of *NTRK3*, which resulted in the recovery of *NTRK3* expression. These data underscore the role of miRNAs in posttranscriptional regulation, where even allele-specific miRNA regulation is possible, as shown for *NTRK3* in anxiety disorders (Muiños-Gimeno et al., 2009).

Another 3′UTR SNP, rs817782, in the aminolevulinate dehydratase gene (*ALAD*) was found to be associated with social phobia (Donner et al., 2008). The *A*-allele was predicted to generate putative target sites for miR-204 and miR-211 within the *ALAD* 3′UTR (Donner et al., 2008). Elevated expression levels of *ALAD* have been detected in the hippocampus and periaqueductal gray of six inbred, anxious mouse strains. These two brain regions are part of the fear network that was shown to be abnormally sensitive in patients with PD. Future studies should try to established a direct link between this functional SNP, the putative miRNA target sites, and PD (Hovatta et al., 2005).

A variant in mitogen-activated protein kinase kinase 5 (*MAP2K5*), rs41305272, was found to be associated with PD (with and without agoraphobia) and depression in a large European–American and African–American case–control sample of patients who also suffered from comorbid substance or alcohol abuse (Jensen, Kranzler, Stein, & Gelernter, 2014). This variant was found to be a miRSNP that disrupts a binding site of hsa-miR-330-3p, therefore affecting translational repression of *MAP2K5*. Furthermore, additional predicted mRNA targets of miR-330-3p were enriched in pathways relevant to psychiatric disorders. These results suggest that miR-330-3p and *MAP2K5* could be important contributors to mood and anxiety-related traits (Jensen et al., 2014). These results, however, require validation and functional verification.

It is however not only variants in the mRNA target regions that can have an effect on gene and protein expression but also variants within the miRNAs themselves or the miRNA host genes. In a PD association study that investigated 712 SNPs covering 325 miRNA regions, Muiños-Gimeno et al. (2011) found significantly lower SNP coverage in miRNA regions compared to the rest of the genome. In accordance with the reported negative selection at miRNAs and 3′UTR miRNA target sites, researchers noted that mature miRNA sequences did not contain any SNPs (Chen & Rajewsky, 2006). A resequencing study that focused on 117 miRNAs in four different human reference populations confirmed this lower SNP density (Quach et al., 2009). It is therefore probable that variants in miRNA regions are deleterious with possible severe phenotypic implications. Resequencing of miRNAs and their 3′UTR mRNA target regions in patients with anxiety disorders could reveal the role of miRNA-facilitated regulation in anxiety disorders (Quach et al., 2009).

Muiños-Gimeno et al. (2011) were also interested in the functional role of miRNAs in PD and extensively investigated those 712 SNPs and miRNA regions in a Spanish cohort of PD patients. They found two SNPs, rs6502892 and rs11763020, significantly associated with PD; these SNPs also tagged miR-22 and miR-339, respectively. Interestingly, miRNA-22 regulates four important PD candidate genes, including serotonin 5-HT2C receptor (*HTR2C*), *BDNF*, monoamine oxidase A (*MAO-A*), and the regulator of G-protein signaling 2 gene (*RGS2*). Some of the predicted targets for miR-339 include adenosine receptor A2a (*ADORA2A*), *BDNF*, corticotropin-releasing hormone receptor 2 (*CRHR2*), and sodium-dependent noradrenaline transporter (*SLC6A2*). They even found SNPs associated with PD subphenotypes (PD with and without agoraphobia); these SNPs tagged miR-491, miR-488, miR-148a, and miR-138-2. Functional studies revealed that miR-488, miR-148a, and miR-138-2 suppressed (between 30% and 60%) the expression of a number of PD candidate genes, such as cholecystokinin B receptor (*CCKBR*), gamma-aminobutyric acid A receptor, alpha 6 (*GABRA6*), and pro-opiomelanocortin preproprotein (*POMC*), respectively (Muiños-Gimeno et al., 2011). Functional analyses of miR-22 and miR-488 in neuroblastoma cells suggested that these miRNAs and their genes could be involved in physiological pathways related to anxiety. Furthermore, rs73531 that tagged miR-148a was associated with age at onset (AAO), where *GG* homozygotes had an average AAO of 23 years, whereas the average AAO for *AG* heterozygotes and *AA* homozygotes was 30 years (Muiños-Gimeno et al., 2011).

Hommers et al. (2015) investigated whether variations in miRNAs that regulate *RGS2* expression might be associated with PD with comorbid agoraphobia (PD/AG) and anxiety-related traits. *RGS2* was of interest as variations in this gene were previously associated with anxiety-related phenotypes and PD with and without comorbid agoraphobia (Hohoff et al., 2015; Leygraf et al., 2006). Four bioinformatic algorithms predicted several miRNAs that were able to regulate *RGS2* expression and these were experimentally validated using luciferase reporter gene assays. Disruption of the seed sequences of these miRNAs resulted in rescued luciferase activity repression, indicating that the miRNA is not regulating the mRNA target gene to the same extent as previously. Hsa-miR-4717-5p exhibited the most pronounced effect on *RGS2* expression and also regulated two other candidate genes of anxiety disorders (*IKBKE* and *CNR1*). Furthermore, two SNPs (rs150925, rs161427) within a 1000 bp upstream of the host gene of hsa-miR-4717-5p, *MIR4717*, had a minor allele frequency greater than 0.05. Both SNPs were in high linkage disequilibrium with each other, and the major *G*-alleles showed a trend for association with PD with comorbid agoraphobia. It was therefore proposed that hsa-miR-4717-5p regulates human *RGS2* and thereby contributes to the genetic risk toward anxiety-related traits, pointing toward novel miRNA-regulated gene networks involved in anxiety disorders. *MIR4717* should also be investigated further as a potential candidate gene for anxiety-related traits (Hommers et al., 2015).

Similarly, another study wanted to determine whether a variant in the miRNA-206 gene (*MIR206*) confers susceptibility to bipolar disorder type I (BD-I) and affects treatment response to mood stabilizers in a Han Chinese cohort ($n = 568$) (Wang et al., 2014). *MIR206* is located on chromosome 6p, a linkage region that has been implicated in BD (Abou et al., 2007; Schulze et al., 2004). Furthermore, miRNA-206 plays a possible role in the regulation of *BDNF* expression and *BDNF* has previously been implicated in the etiology of BD and the efficacy of mood stabilizers (Post, 2007). A significant gene–gene interaction was observed between the rs16882131 variant in *MIR206* and rs6265 in *BDNF* that contributed to BD-I susceptibility and treatment response. Furthermore, they also detected a significant interaction between *MIR206* and *BDNF* on treatment score: significantly lower mean treatment scores were evident in individuals with at least one *MIR206* *T*-allele and *BDNF* A/A genotype, compared to those with *MIR206* *C/C* and *BDNF A/A* or *A/G* and *MIR206 C/C* and *BDNF G/G* genotypes (Wang et al., 2014). Although the researchers did not establish whether this interaction resulted in altered expression levels of *BDNF*, it provides preliminary results suggesting that the interaction between these two genes could affect BD-I susceptibility and treatment response.

5.2 Differential Expression of miRNAs

In addition to investigating miRSNPs and variants within miRNA genes, investigations into the expression levels of miRNAs can shed light on the etiology of anxiety and stress-related disorders. Wang et al. (2015) recently showed that expression levels of miR-144-5p in plasma were inversely associated with depression scores, and that levels of miR-144-5p were significantly lower in depressed patients than in healthy controls. Furthermore, following treatment, plasma levels of miRNA-144-5p significantly increased in depression/anxiety patients and were significantly higher than levels measured at baseline. These findings seem contradictory to results from animal studies which found this miRNA to be downregulated in response to lithium and sodium valproate. However, the animal study used hippocampal tissue, whereas the human study used plasma. Furthermore, the treatment in the human study was not pharmacological; they employed mindfulness group therapy. These findings suggest that miRNA-144-5p may be involved in the pathologic processes of depression, but that its expression might be tissue specific and respond differently to different types of treatment.

Privitera et al. (2015) recently developed an integrative database that provides an online resource containing scientific knowledge, including genetic and epigenetic data about OCD in order to improve the understanding of the molecular etiology of this disorder. The OCD database (OCDB) is a manually curated database which integrates the relationships between candidate genes in OCD, miRNAs involved in the pathophysiology of OCD, and drugs used in its treatments. This database provides researchers with a handy resource to plan, perform, and validate molecular research in OCD. Future studies could extrapolate this to include additional anxiety and stress-related disorders.

5.3 The Role of miRNAs in Pharmacotherapies for Anxiety Disorders

Selective SSRI antidepressants are a widely used for the effective treatment of psychiatric disorders, including anxiety and stress-related disorders (Martenyi, Brown, Zhang, Koke, & Prakash, 2002). The serotonin transporter (SERT) regulates serotonin levels in the brain by ensuring reuptake of serotonin at the synaptic cleft and is also a target of selective SSRI antidepressants (Torres, Gainetdinov, & Caron, 2003). Malfunctioning of serotonergic neurotransmission has been implicated in anxiety, depression, OCD, and suicidal behavior (van Praag, 1996a, 1996b). Researchers discovered that *SERT* is a target of miR-16 (Baudry et al., 2010), and following chronic treatment with the SSRI fluoxetine (Prozac), increased levels of miR-16 were detected in serotonergic raphe nuclei of mice, resulting in reduced SERT expression (Baudry et al., 2010). The regulation of miR-16 may therefore be one of the molecular mechanisms whereby SSRIs elicit their limiting effect on SERT, thereby facilitating an anxiolytic response. Their role in the regulation of several essential stress response pathways underscores the importance of miRNAs in the etiology of anxiety and stress-related disorders and renders further investigations into their potential for drug targets for anxiety disorders necessary.

One of the first studies to confirm a role for miRNAs and their effectors as targets of pharmacotherapeutic drugs was conducted by Zhou et al. (2009). Valproate (VPA) was shown to be a useful adjunct in treatment-refractory anxiety disorder patients and patients with comorbid bipolar disorder; it has also been found to enhance exposure-based cognitive therapy for anxiety disorders and PTSD (Kuriyama, Honma, Soshi, Fujii, & Kim, 2011). Following chronic treatment with lithium and VPA, fluctuating levels of several hippocampal miRNAs were detected, including let-7b, let-7c, miR-24a, miR-30c, miR-34a, miR-128a, miR-144, and miR-221. Furthermore, the predicted targets of these miRNAs are involved in neurite outgrowth, neurogenesis, and signaling of extracellular signal-regulated kinase (Zhou et al., 2009).

Treatment with mood stabilizers such as lithium and VPA was shown to increase the expression of genes encoding dipeptidyl-peptidase 10, metabotropic glutamate receptor 7 (GRM7), and thyroid hormone receptor β in vivo (Zhou et al., 2009), several of which have previously been identified as candidate genes for bipolar disorder. The authors therefore investigated the effects of lithium and VPA on the expression of miRNAs and their effectors in primary cultures (Zhou et al., 2009). Following treatment with lithium or VPA, decreased levels of miR-34a and elevated levels of *GRM7* (a predicted target of miR-34a) were detected in primary cultures. Furthermore, *GRM7* levels were effectively decreased by a miR-34a precursor and increased following treatment with a miR-34a inhibitor. These findings indicate that the levels of *GRM7* are regulated by endogenous miR-34a, which may be one of the therapeutic effects elicited by lithium and VPA on *GRM7* (Zhou et al., 2009).

Another study that investigated the role of miRNAs in anxiolytic drugs investigated miRNA profiles in the PFC of fluoxetine-treated C57BL/6N mice 74 days after they were subjected to a single electric footshock or a mock-treatment. The antidepressant, fluoxetine, has been shown to alleviate PTSD symptoms in animals and humans. Schmidt and colleagues (2013) noted that fluoxetine's therapeutic action in shocked mice was associated with a significant decrease in mmu-miR-1971 expression in the PFC. It is therefore possible that one of the molecular mechanisms of fluoxetine in PTSD is through regulation of mmu-miR-1971 in the PFC (Schmidt et al., 2013). Additional research on the mRNA targets of mmu-miR-1971 in the PFC would shed more light on these findings.

The Kampo medicine Yokukansan (YKS) (Tsumura, TJ-54) is a traditional Japanese medicine used to treat psychological symptoms of impaired stress responses such as depression and anxiety (Hiratsuka et al., 2010). Recent studies have shown that miR-18 and/or miR-124a can act as negative regulators of GR in the brain (Vreugdenhil et al., 2009). Researchers were therefore interested to evaluate the effect of pretreatment with YKS on the levels of miR-18, miR-124a, and GR in PVN of the hypothalamus of mice exposed to water stress (Shimizu et al., 2015). YKS pretreatment was able to normalize elevated corticosterone plasma levels in stress-exposed mice. Furthermore, following stress exposure, a downregulation of GR mRNA levels were detected in the brain. In addition, decreased expression levels of miR-18 in the hypothalamus and normalized GR protein levels were detected in the PVN of YKS-pretreated mice following stress exposure. These findings therefore indicate that YKS normalizes GR protein levels through regulating miRNA-18 expression in the hypothalamus, thereby normalizing HPA axis activity following stress exposure (Shimizu et al., 2015).

6. CONCLUSION

In an attempt to unravel the complexities of anxiety and stress-related disorders, the study of epigenetic factors, such as miRNAs, can offer novel and interesting perspectives. The abundance of miRNAs and their mRNA targets in the CNS, and their involvement in numerous CNS functions suggest that these small RNAs could play a role in neurobiological and psychiatric conditions. This chapter summarized evidence from various sources, including animal studies, human studies, in vitro cell culture, and computational methods, that illustrate the involvement of miRNAs in the etiology of anxiety and stress-related disorders. Preliminary evidence even suggests a role for miRNAs as therapeutic targets in anxiety and mood disorders, and preclinical models demonstrated that altered levels of miRNAs have the ability to elicit anxiolytic responses (O'Connor, Dinan, & Cryan, 2011).

Future studies should aim to improve in silico tools to enable more accurate predictions of miRNA:mRNA interaction. Using high-throughput sequencing technologies that generate miRNA expression profiles for particular brain regions or single cells and surrogate tissues, followed by prediction analyses and target verification, could enable a better understanding of these complex disorders and identification of reliable biomarkers for anxiety disorders or predictors of treatment outcome. Additional investigations into how polymorphisms in miRNA genes might influence the structure and transcription of pre-miRNAs and the processing and expression of mature miRNAs are warranted (Espinosa-Parrilla & Muiños-Gimeno, 2011). Identification of biological pathways involved in anxiety and stress-related disorders, and the roles that miRNAs play within these complex networks, is essential. In addition, the interaction between miRNAs and another class of noncoding RNA, such as long noncoding RNAs (lncRNAs), should also be investigated. Literature suggests the potential of lncRNAs to be processed to smaller RNAs (Jalali, Jayaraj, & Scaria, 2012) and miRNAs can also potentially target lncRNAs and be involved in a novel layer of regulatory interactions between noncoding RNA classes. Convergence of data from several sources, including genetics, epigenetics, preclinical models, clinical samples, neurophysiology, and neuroimaging, could provide a more comprehensive understanding of anxiety and stress-related disorders, to facilitate more effective treatment strategies for these disorders.

ACKNOWLEDGMENTS

This work is based upon research supported by the South African Research Chairs Initiative of the Department of Science and Technology and National Research Foundation, South African Medical Research Council (self-initiated research grant), and Harry Crossley Foundation funding.

REFERENCES

Abou, J. R., Fuerst, R., Kaneva, R., Orozco, D. G., Rivas, F., Mayoral, F., et al. (2007). The first genomewide interaction and locus-heterogeneity linkage scan in bipolar affective disorder: Strong evidence of epistatic effects between loci on chromosomes 2q and 6q. *The American Journal of Human Genetics*, *81*, 974–986.

Abu-Elneel, K., Liu, T., Gazzaniga, F. S., Nishimura, Y., Wall, D. P., Geschwind, D. H., et al. (2008). Heterogeneous dysregulation of microRNAs across the autism spectrum. *Neurogenetics*, *9*, 153–161.

Akbarian, S., & Huang, H. S. (2009). Epigenetic regulation in human brain-focus on histone lysine methylation. *Biological Psychiatry*, *65*, 198–203.

Alfonso, J., Frick, L. R., Silberman, D. M., Palumbo, M. L., Genaro, A. M., & Frasch, A. C. (2006). Regulation of hippocampal gene expression is conserved in two species subjected to different stressors and antidepressant treatments. *Biological Psychiatry*, *59*, 244–251.

Anisman, H., & Zacharko, R. M. (1992). Depression as a consequence of inadequate neurochemical adaptation in response to stressors. *The British Journal of Psychiatry. Supplement*, *15*, 36–43.

Babenko, O., Golubov, A., Ilnytskyy, Y., Kovalchuk, I., & Metz, G. A. (2012). Genomic and epigenomic responses to chronic stress involve miRNA-mediated programming. *PLoS One*, 7, e29441.

Baek, D., Villen, J., Shin, C., Camargo, F. D., Gygi, S. P., & Bartel, D. P. (2008). The impact of microRNAs on protein output. *Nature*, *455*, 64–71.

Balakathiresan, N. S., Chandran, R., Bhomia, M., Jia, M., Li, H., & Maheshwari, R. K. (2014). Serum and amygdala microRNA signatures of posttraumatic stress: Fear correlation and biomarker potential. *Journal of Psychiatric Research*, *57*, 65–73.

Bartel, D. P. (2004). MicroRNAs: Genomics, biogenesis, mechanism, and function. *Cell*, *116*, 281–297.

Baudry, A., Mouillet-Richard, S., Schneider, B., Launay, J. M., & Kellermann, O. (2010). miR-16 targets the serotonin transporter: A new facet for adaptive responses to antidepressants. *Science*, *329*, 1537–1541.

Benaroya-Milshtein, N., Hollander, N., Apter, A., Kukulansky, T., Raz, N., Wilf, A., et al. (2004). Environmental enrichment in mice decreases anxiety, attenuates stress responses and enhances natural killer cell activity. *The European Journal of Neuroscience, 20*, 1341–1347.

Bertino, J. R., Banerjee, D., & Mishra, P. J. (2007). Pharmacogenomics of microRNA: A miRSNP towards individualized therapy. *Pharmacogenomics, 8*, 1625–1627.

Betel, D., Wilson, M., Gabow, A., Marks, D. S., & Sander, C. (2008). The microRNA.org resource: Targets and expression. *Nucleic Acids Research, 36*, D149–D153.

Beveridge, N. J., Gardiner, E., Carroll, A. P., Tooney, P. A., & Cairns, M. J. (2010). Schizophrenia is associated with an increase in cortical microRNA biogenesis. *Molecular Psychiatry, 15*, 1176–1189.

Beveridge, N. J., Tooney, P. A., Carroll, A. P., Gardiner, E., Bowden, N., Scott, R. J., et al. (2008). Dysregulation of miRNA 181b in the temporal cortex in schizophrenia. *Human Molecular Genetics, 17*, 1156–1168.

Bhattacharya, A., Ziebarth, J. D., & Cui, Y. (2014). PolymiRTS database 3.0: Linking polymorphisms in microRNAs and their target sites with human diseases and biological pathways. *Nucleic Acids Research, 42*, D86–D91.

Bjornsson, H. T., Fallin, M. D., & Feinberg, A. P. (2004). An integrated epigenetic and genetic approach to common human disease. *Trends in Genetics, 20*, 350–358.

Bliss, T. V., & Collingridge, G. L. (1993). A synaptic model of memory: Long-term potentiation in the hippocampus. *Nature, 361*, 31–39.

Bradesi, S., Karagiannides, I., Bakirtzi, K., Joshi, S. M., Koukos, G., Iliopoulos, D., et al. (2015). Identification of spinal cord microRNA and gene signatures in a model of chronic stress-induced visceral hyperalgesia in rat. *PLoS One, 10*, e0130938.

Bradesi, S., Schwetz, I., Ennes, H. S., Lamy, C. M., Ohning, G., Fanselow, M., et al. (2005). Repeated exposure to water avoidance stress in rats: A new model for sustained visceral hyperalgesia. American Journal of Physiology. *Gastrointestinal and Liver Physiology, 289*, G42–G53.

Brenes, J. C., Lackinger, M., Hoglinger, G. U., Schratt, G., Schwarting, R. K., & Wohr, M. (2015). Differential effects of social and physical environmental enrichment on brain plasticity, cognition, and ultrasonic communication in rats. *The Journal of Comparative Neurology, 524*, 1586–1607.

Brennecke, J., Stark, A., Russell, R. B., & Cohen, S. M. (2005). Principles of microRNA-target recognition. *PLoS Biology, 3*, e85.

Brown, G. W., & Harris, T. O. (1993). Aetiology of anxiety and depressive disorders in an inner-city population. 1. Early adversity. *Psychological Medicine, 23*, 143–154.

Brown, G. W., Harris, T. O., & Eales, M. J. (1993). Aetiology of anxiety and depressive disorders in an inner-city population. 2. Comorbidity and adversity. *Psychological Medicine, 23*, 155–165.

Brown, B. D., & Naldini, L. (2009). Exploiting and antagonizing microRNA regulation for therapeutic and experimental applications. *Nature Reviews Genetics, 10*, 578–585.

Buckert, M., Kudielka, B. M., Reuter, M., & Fiebach, C. J. (2012). The COMT Val158Met polymorphism modulates working memory performance under acute stress. *Psychoneuroendocrinology, 37*, 1810–1821.

Castellano, L., Giamas, G., Jacob, J., Coombes, R. C., Lucchesi, W., Thiruchelvam, P., et al. (2009). The estrogen receptor-alpha-induced microRNA signature regulates itself and its transcriptional response. *Proceedings of the National academy of Sciences of the United States of America, 106*, 15732–15737.

Chan, J. A., Krichevsky, A. M., & Kosik, K. S. (2005). MicroRNA-21 is an antiapoptotic factor in human glioblastoma cells. *Cancer Research, 65*, 6029–6033.

Chandrasekar, V., & Dreyer, J. L. (2009). microRNAs miR-124, let-7d and miR-181a regulate cocaine-induced plasticity. *Molecular and Cellular Neurosciences, 42*, 350–362.

Chang, S., Johnston, R. J., Jr., Frokjaer-Jensen, C., Lockery, S., & Hobert, O. (2004). MicroRNAs act sequentially and asymmetrically to control chemosensory laterality in the nematode. *Nature*, *430*, 785–789.

Chen, K., & Rajewsky, N. (2006). Natural selection on human microRNA binding sites inferred from SNP data. *Nature Genetics*, *38*, 1452–1456.

Cheng, Y., Ma, Z., Kim, B. H., Wu, W., Cayting, P., Boyle, A. P., et al. (2014). Principles of regulatory information conservation between mouse and human. *Nature*, *515*, 371–375.

Chertkow-Deutsher, Y., Cohen, H., Klein, E., & Ben-Shachar, D. (2010). DNA methylation in vulnerability to post-traumatic stress in rats: Evidence for the role of the post-synaptic density protein Dlgap2. *The International Journal of Neuropharmacology*, *13*, 347–359.

Cook, S. C., & Wellman, C. L. (2004). Chronic stress alters dendritic morphology in rat medial prefrontal cortex. *Journal of Neurobiology*, *60*, 236–248.

Cookson, W., Liang, L., Abecasis, G., Moffatt, M., & Lathrop, M. (2009). Mapping complex disease traits with global gene expression. *Nature Reviews. Genetics*, *10*, 184–194.

de Kloet, E. R., Fitzsimons, C. P., Datson, N. A., Meijer, O. C., & Vreugdenhil, E. (2009). Glucocorticoid signaling and stress-related limbic susceptibility pathway: About receptors, transcription machinery and microRNA. *Brain Research*, *1293*, 129–141.

Desbonnet, L., Tighe, O., Karayiorgou, M., Gogos, J. A., Waddington, J. L., & O'Tuathaigh, C. M. (2012). Physiological and behavioural responsivity to stress and anxiogenic stimuli in COMT-deficient mice. *Behavioural Brain Research*, *228*, 351–358.

Donner, J., Pirkola, S., Silander, K., Kananen, L., Terwilliger, J. D., Lonnqvist, J., et al. (2008). An association analysis of murine anxiety genes in humans implicates novel candidate genes for anxiety disorders. *Biological Psychiatry*, *64*, 672–680.

Donohue, H. S., Gabbott, P. L., Davies, H. A., Rodriguez, J. J., Cordero, M. I., Sandi, C., et al. (2006). Chronic restraint stress induces changes in synapse morphology in stratum lacunosum-moleculare CA1 rat hippocampus: A stereological and three-dimensional ultrastructural study. *Neuroscience*, *140*, 597–606.

Duursma, A. M., Kedde, M., Schrier, M., Le Sage, C., & Agami, R. (2008). miR-148 targets human DNMT3b protein coding region. *RNA-A Publication of the RNA Society*, *14*, 872–877.

Easow, G., Teleman, A. A., & Cohen, S. M. (2007). Isolation of microRNA targets by miRNP immunopurification. *RNA*, *13*, 1198–1204.

Erson, A. E., & Petty, E. M. (2008). MicroRNAs in development and disease. *Clinical Genetics*, *74*, 296–306.

Espinosa-Parrilla, Y., & Muiños-Gimeno, M. (2011). MicroRNA-mediated regulation and the genetic susceptibility to anxiety disorders. In V. Kalinin (Ed.), *Anxiety disorders* (pp. 281–306): Slavka Krautzeka, Croatia: InTech, ISBN: 978-953-307-592-1.

Fiore, R., & Schratt, G. (2007). MicroRNAs in vertebrate synapse development. *ScientificWorldJournal*, *7*, 167–177.

Franklin, T. B., Russig, H., Weiss, I. C., Graff, J., Linder, N., Michalon, A., et al. (2010). Epigenetic transmission of the impact of early stress across generations. *Biological Psychiatry*, *68*, 408–415.

Friedman, M. J. (1997). Posttraumatic stress disorder. *The Journal of Clinical Psychiatry*, *58*(Suppl. 9), 33–36.

Gao, Y., Bezchlibnyk, Y. B., Sun, X., Wang, J. F., McEwen, B. S., & Young, L. T. (2006). Effects of restraint stress on the expression of proteins involved in synaptic vesicle exocytosis in the hippocampus. *Neuroscience*, *141*, 1139–1148.

Gapp, K., Jawaid, A., Sarkies, P., Bohacek, J., Pelczar, P., Prados, J., et al. (2014). Implication of sperm RNAs in transgenerational inheritance of the effects of early trauma in mice. *Nature Neuroscience*, *17*, 667–669.

Gibbs, R. A., Weinstock, G. M., Metzker, M. L., Muzny, D. M., Sodergren, E. J., Scherer, S., et al. (2004). Genome sequence of the Brown Norway rat yields insights into mammalian evolution. *Nature, 428*, 493–521.

Hanin, G., Shenhar-Tsarfaty, S., Yayon, N., Yau, Y. H., Bennett, E. R., Sklan, E. H., et al. (2014). Competing targets of microRNA-608 affect anxiety and hypertension. *Human Molecular Genetics, 23*, 4569–4580.

Hansen, T., Olsen, L., Lindow, M., Jakobsen, K. D., Ullum, H., Jonsson, E., et al. (2007). Brain expressed microRNAs implicated in schizophrenia etiology. *PLoS One, 2*, e873.

Haramati, S., Navon, I., Issler, O., Ezra-Nevo, G., Gil, S., Zwang, R., et al. (2011). MicroRNA as repressors of stress-induced anxiety: The case of amygdalar miR-34. *The Journal of Neuroscience, 31*, 14191–14203.

He, L., He, X., Lim, L. P., de Stanchina, E., Xuan, Z., Liang, Y., et al. (2007). A microRNA component of the p53 tumour suppressor network. *Nature, 447*, 1130–1134.

Hebert, S. S., Horre, K., Nicolai, L., Bergmans, B., Papadopoulou, A. S., Delacourte, A., et al. (2009). MicroRNA regulation of Alzheimer's amyloid precursor protein expression. *Neurobiology of Disease, 33*, 422–428.

Hiratsuka, T., Matsuzaki, S., Miyata, S., Kinoshita, M., Kakehi, K., Nishida, S., et al. (2010). Yokukansan inhibits neuronal death during ER stress by regulating the unfolded protein response. *PLoS One, 5*, e13280.

Hohoff, C., Weber, H., Richter, J., Domschke, K., Zwanzger, P. M., Ohrmann, P., et al. (2015). RGS2 genetic variation: Association analysis with panic disorder and dimensional as well as intermediate phenotypes of anxiety. *American Journal of Medical Genetics. Part B, Neuropsychiatric Genetics, 168B*, 211–222.

Holmes, A., Heilig, M., Rupniak, N. M., Steckler, T., & Griebel, G. (2003). Neuropeptide systems as novel therapeutic targets for depression and anxiety disorders. *Trends in Pharmacological Sciences, 24*, 580–588.

Hölscher, C. (1999). Synaptic plasticity and learning and memory: LTP and beyond. *Journal of Neuroscience Research, 58*, 62–75.

Hommers, L., Raab, A., Bohl, A., Weber, H., Scholz, C. J., Erhardt, A., et al. (2015). MicroRNA hsa-miR-4717-5p regulates RGS2 and may be a risk factor for anxiety-related traits. *American Journal of Medical Genetics. Part B, Neuropsychiatric Genetics, 168B*, 296–306.

Hovatta, I., Tennant, R. S., Helton, R., Marr, R. A., Singer, O., Redwine, J. M., et al. (2005). Glyoxalase 1 and glutathione reductase 1 regulate anxiety in mice. *Nature, 438*, 662–666.

Hsu, P. K., Xu, B., Mukai, J., Karayiorgou, M., & Gogos, J. A. (2015). The BDNF Val66Met variant affects gene expression through miR-146b. *Neurobiology of Disease, 77*, 228–237.

Hunsberger, J. G., Austin, D. R., Chen, G., & Manji, H. K. (2009). MicroRNAs in mental health: From biological underpinnings to potential therapies. *Neuromolecular Medicine, 11*, 173–182.

Iborra, M., Bernuzzi, F., Correale, C., Vetrano, S., Fiorino, G., Beltran, B., et al. (2013). Identification of serum and tissue micro-RNA expression profiles in different stages of inflammatory bowel disease. *Clinical and Experimental Immunology, 173*, 250–258.

Jaenisch, R., & Bird, A. (2003). Epigenetic regulation of gene expression: How the genome integrates intrinsic and environmental signals. *Nature Genetics, 33*, 245–254.

Jalali, S., Jayaraj, G. G., & Scaria, V. (2012). Integrative transcriptome analysis suggest processing of a subset of long non-coding RNAs to small RNAs. *Biology Direct, 7*, 25.

Jensen, K. P., Kranzler, H. R., Stein, M. B., & Gelernter, J. (2014). The effects of a MAP2K5 microRNA target site SNP on risk for anxiety and depressive disorders. *American Journal of Medical Genetics. Part B, Neuropsychiatric Genetics, 165B*, 175–183.

Johnston, R. J., & Hobert, O. (2003). A microRNA controlling left/right neuronal asymmetry in *Caenorhabditis elegans*. *Nature, 426*, 845–849.

Kajantie, E. (2006). Fetal origins of stress-related adult disease. *Annals of the New York Academy of Sciences, 1083*, 11–27.

Karres, J. S., Hilgers, V., Carrera, I., Treisman, J., & Cohen, S. M. (2007). The conserved microRNA miR-8 tunes atrophin levels to prevent neurodegeneration in Drosophila. *Cell*, *131*, 136–145.
Kasper, S., Hajak, G., Wulff, K., Hoogendijk, W. J., Montejo, A. L., Smeraldi, E., et al. (2010). Efficacy of the novel antidepressant agomelatine on the circadian rest-activity cycle and depressive and anxiety symptoms in patients with major depressive disorder: A randomized, double-blind comparison with sertraline. *The Journal of Clinical Psychiatry*, *71*, 109–120.
Kawamoto, T., Horikawa, Y., Tanaka, T., Kabe, N., Takeda, J., & Mikuni, M. (2004). Genetic variations in the WFS1 gene in Japanese with type 2 diabetes and bipolar disorder. *Molecular Genetics and Metabolism*, *82*, 238–245.
Kember, R. L., Fernandes, C., Tunbridge, E. M., Liu, L., Paya-Cano, J. L., Parsons, M. J., et al. (2010). A B2 SINE insertion in the Comt1 gene (Comt1(B2i)) results in an overexpressing, behavior modifying allele present in classical inbred mouse strains. *Genes, Brain, and Behavior*, *9*, 925–932.
Kessler, R. C., Chiu, W. T., Demler, O., Merikangas, K. R., & Walters, E. E. (2005). Prevalence, severity, and comorbidity of 12-month DSM-IV disorders in the National Comorbidity Survey Replication. *Archives of General Psychiatry*, *62*, 617–627.
Kloosterman, W. P., Wienholds, E., Ketting, R. F., & Plasterk, R. H. (2005). Substrate requirements for let-7 function in the developing zebrafish embryo. *Nucleic Acids Research*, *32*, 6284–6291.
Koido, K., Koks, S., Nikopensius, T., Maron, E., Altmae, S., Heinaste, E., et al. (2005). Polymorphisms in wolframin (WFS1) gene are possibly related to increased risk for mood disorders. *The International Journal of Neuropsychopharmacology*, *8*, 235–244.
Kovacs-Nagy, R., Elek, Z., Szekely, A., Nanasi, T., Sasvari-Szekely, M., & Ronai, Z. (2013). Association of aggression with a novel microRNA binding site polymorphism in the wolframin gene. *American Journal of Medical Genetics. Part B, Neuropsychiatric Genetics*, *162B*, 404–412.
Kozomara, A., & Griffiths-Jones, S. (2011). miRBase: Integrating microRNA annotation and deep-sequencing data. *Nucleic Acids Research*, *39*, D152–D157.
Kozomara, A., & Griffiths-Jones, S. (2014). miRBase: Annotating high confidence microRNAs using deep sequencing data. *Nucleic Acids Research*, *42*, D68–D73.
Krek, A., Grun, D., Poy, M. N., Wolf, R., Rosenberg, L., Epstein, E. J., et al. (2005). Combinatorial microRNA target predictions. *Nature Genetics*, *37*, 495–500.
Krichevsky, A. M., King, K. S., Donahue, C. P., Khrapko, K., & Kosik, K. S. (2003). A microRNA array reveals extensive regulation of microRNAs during brain development. *RNA-A Publication of the RNA Society*, *9*, 1274–1281.
Krichevsky, A. M., Sonntag, K. C., Isacson, O., & Kosik, K. S. (2006). Specific MicroRNAs modulate embryonic stem cell-derived neurogenesis. *Stem Cells (Dayton, Ohio)*, *24*(4), 857–864. http://dx.doi.org/10.1634/stemcells.2005-0441.
Kuriyama, K., Honma, M., Soshi, T., Fujii, T., & Kim, Y. (2011). Effect of d-cycloserine and valproic acid on the extinction of reinstated fear-conditioned responses and habituation of fear conditioning in healthy humans: A randomized controlled trial. *Psychopharmacology (Berl)*, *218*, 589–597.
Kuss, A. W., & Chen, W. (2008). MicroRNAs in brain function and disease. *Current Neurology and Neuroscience Reports*, *8*, 190–197.
Lagos-Quintana, M., Rauhut, R., Lendeckel, W., & Tuschl, T. (2001). Identification of novel genes coding for small expressed RNAs. *Science*, *294*, 853–858.
Lau, N. C., Lim, L. P., Weinstein, E. G., & Bartel, D. P. (2001). An abundant class of tiny RNAs with probable regulatory roles in *Caenorhabditis elegans*. *Science*, *294*, 858–862.
Lee, R. C., & Ambros, V. (2001). An extensive class of small RNAs in *Caenorhabditis elegans*. *Science*, *294*, 862–864.

Lee, R. C., Feinbaum, R. L., & Ambros, V. (1993). The *C. elegans* heterochronic gene lin-4 encodes small RNAs with antisense complementarity to lin-14. *Cell*, *75*, 843–854.

Lee, K., Kim, J. H., Kwon, O. B., An, K., Ryu, J., Cho, K., et al. (2012). An activity-regulated microRNA, miR-188, controls dendritic plasticity and synaptic transmission by downregulating neuropilin-2. *The Journal of Neuroscience*, *32*, 5678–5687.

Lewis, B. P., Shih, I. H., Jones-Rhoades, M. W., Bartel, D. P., & Burge, C. B. (2003). Prediction of mammalian microRNA targets. *Cell*, *115*, 787–798.

Leygraf, A., Hohoff, C., Freitag, C., Willis-Owen, S. A., Krakowitzky, P., Fritze, J., et al. (2006). Rgs 2 gene polymorphisms as modulators of anxiety in humans? *Journal of Neural Transmission (Vienna, Austria: 1996)*, *113*, 1921–1925.

Li, Z., Mulligan, M. K., Wang, X., Miles, M. F., Lu, L., & Williams, R. W. (2010). A transposon in Comt generates mRNA variants and causes widespread expression and behavioral differences among mice. *PLoS One*, *5*, e12181.

Lim, L. P., Lau, N. C., Garrett-Engele, P., Grimson, A., Schelter, J. M., Castle, J., et al. (2005). Microarray analysis shows that some microRNAs downregulate large numbers of target mRNAs. *Nature*, *433*, 769–773.

Lin, Q., Wei, W., Coelho, C. M., Li, X., Baker-Andresen, D., Dudley, K., et al. (2011). The brain-specific microRNA miR-128b regulates the formation of fear-extinction memory. *Nature Neuroscience*, *14*, 1115–1117.

Lugli, G., Larson, J., Demars, M. P., & Smalheiser, N. R. (2012). Primary microRNA precursor transcripts are localized at post-synaptic densities in adult mouse forebrain. *Journal of Neurochemistry*, *123*, 459–466.

Lugli, G., Larson, J., Martone, M. E., Jones, Y., & Smalheiser, N. R. (2005). Dicer and eIF2c are enriched at postsynaptic densities in adult mouse brain and are modified by neuronal activity in a calpain-dependent manner. *Journal of Neurochemistry*, *94*, 896–905.

Magarinos, A. M., & McEwen, B. S. (1995). Stress-induced atrophy of apical dendrites of hippocampal CA3c neurons: Involvement of glucocorticoid secretion and excitatory amino acid receptors. *Neuroscience*, *69*, 89–98.

Makeyev, E. V., Zhang, J., Carrasco, M. A., & Maniatis, T. (2007). The microRNA miR-124 promotes neuronal differentiation by triggering brain-specific alternative pre-mRNA splicing. *Molecular Cell*, *27*, 435–448.

Malan-Müller, S. (2013). *Molecular mechanisms of D-cycloserine in a fear extinction posttraumatic stress disorders (PTSD) animal model,* (Doctoral dissertation). http://scholar.sun.ac.za/handle/10019.1/88457

Manca, S., Magrelli, A., Cialfi, S., Lefort, K., Ambra, R., Alimandi, M., et al. (2011). Oxidative stress activation of miR-125b is part of the molecular switch for Hailey-Hailey disease manifestation. *Experimental Dermatology*, *20*, 932–937.

Manolio, T. A., Collins, F. S., Cox, N. J., Goldstein, D. B., Hindorff, L. A., Hunter, D. J., et al. (2009). Finding the missing heritability of complex diseases. *Nature*, *461*, 747–753.

Martenyi, F., Brown, E. B., Zhang, H., Koke, S. C., & Prakash, A. (2002). Fluoxetine v. placebo in prevention of relapse in post-traumatic stress disorder. *The British Journal of Psychiatry*, *181*, 315–320.

McGowan, P. O., Sasaki, A., D'Alessio, A. C., Dymov, S., Labonte, B., Szyf, M., et al. (2009). Epigenetic regulation of the glucocorticoid receptor in human brain associates with childhood abuse. *Nature Neuroscience*, *12*, 342–348.

McLean, S. A., Diatchenko, L., Lee, Y. M., Swor, R. A., Domeier, R. M., Jones, J. S., et al. (2011). Catechol O-methyltransferase haplotype predicts immediate musculoskeletal neck pain and psychological symptoms after motor vehicle collision. *The Journal of Pain*, *12*, 101–107.

Meerson, A., Cacheaux, L., Goosens, K. A., Sapolsky, R. M., Soreq, H., & Kaufer, D. (2010). Changes in brain MicroRNAs contribute to cholinergic stress reactions. *Journal of Molecular Neuroscience*, *40*, 47–55.

Mishra, P. J., Mishra, P. J., Banerjee, D., & Bertino, J. R. (2008). MiRSNPs or MiR-polymorphisms, new players in microRNA mediated regulation of the cell: Introducing microRNA pharmacogenomics. *Cell Cycle*, 7, 853–858.

Molitoris, J. K., McColl, K. S., & Distelhorst, C. W. (2011). Glucocorticoid-mediated repression of the oncogenic microRNA cluster miR-17~92 contributes to the induction of Bim and initiation of apoptosis. *Molecular Endocrinology*, *25*, 409–420.

Muiños-Gimeno, M., Espinosa-Parrilla, Y., Guidi, M., Kagerbauer, B., Sipila, T., Maron, E., et al. (2011). Human microRNAs miR-22, miR-138-2, miR-148a, and miR-488 are associated with panic disorder and regulate several anxiety candidate genes and related pathways. *Biological Psychiatry*, *69*, 526–533.

Muiños-Gimeno, M., Guidi, M., Kagerbauer, B., Martin-Santos, R., Navines, R., Alonso, P., et al. (2009). Allele variants in functional microRNA target sites of the neurotrophin-3 receptor gene (NTRK3) as susceptibility factors for anxiety disorders. *Human Mutation*, *30*, 1062–1071.

Murata, K., Furu, M., Yoshitomi, H., Ishikawa, M., Shibuya, H., Hashimoto, M., et al. (2013). Comprehensive microRNA analysis identifies miR-24 and miR-125a-5p as plasma biomarkers for rheumatoid arthritis. *PLoS One*, *8*, e69118.

Murgatroyd, C., Patchev, A. V., Wu, Y., Micale, V., Bockmuhl, Y., Fischer, D., et al. (2009). Dynamic DNA methylation programs persistent adverse effects of early-life stress. *Nature Neuroscience*, *12*, 1559–1566.

Nadorp, B., & Soreq, H. (2014). Predicted overlapping microRNA regulators of acetylcholine packaging and degradation in neuroinflammation-related disorders. *Frontiers in Molecular Neuroscience*, 7, 9.

Neves-Pereira, M., Cheung, J. K., Pasdar, A., Zhang, F., Breen, G., Yates, P., et al. (2005). BDNF gene is a risk factor for schizophrenia in a Scottish population. *Molecular Psychiatry*, *10*, 208–212.

O'Connor, R. M., Dinan, T. G., & Cryan, J. F. (2011). Little things on which happiness depends: MicroRNAs as novel therapeutic targets for the treatment of anxiety and depression. *Molecular Psychiatry*, *17*, 359–376.

Oomen, C. A., Soeters, H., Audureau, N., Vermunt, L., van Hasselt, F. N., Manders, E. M., et al. (2010). Severe early life stress hampers spatial learning and neurogenesis, but improves hippocampal synaptic plasticity and emotional learning under high-stress conditions in adulthood. *The Journal of Neuroscience*, *30*, 6635–6645.

Oomen, C. A., Soeters, H., Audureau, N., Vermunt, L., van Hasselt, F. N., Manders, E. M., et al. (2011). Early maternal deprivation affects dentate gyrus structure and emotional learning in adult female rats. *Psychopharmacology*, *214*, 249–260.

Otto, S. J., McCorkle, S. R., Hover, J., Conaco, C., Han, J. J., Impey, S., et al. (2007). A new binding motif for the transcriptional repressor REST uncovers large gene networks devoted to neuronal functions. *The Journal of Neuroscience*, *27*, 6729–6739.

Parker, B. J., & Wen, J. (2009). Predicting microRNA targets in time-series microarray experiments via functional data analysis. *BMC Bioinformatics*, *10*(Suppl. 1), S32.

Parsons, M. J., Grimm, C., Paya-Cano, J. L., Fernandes, C., Liu, L., Philip, V. M., et al. (2012). Genetic variation in hippocampal microRNA expression differences in C57BL/6J X DBA/2 J (BXD) recombinant inbred mouse strains. *BMC Genomics*, *13*, 476.

Parsons, M. J., Grimm, C. H., Paya-Cano, J. L., Sugden, K., Nietfeld, W., Lehrach, H., et al. (2008). Using hippocampal microRNA expression differences between mouse inbred strains to characterise miRNA function. *Mammalian Genome*, *19*, 552–560.

Pasquinelli, A. E. (2012). MicroRNAs and their targets: Recognition, regulation and an emerging reciprocal relationship. *Nature Reviews. Genetics*, *13*, 271–282.

Perkins, D. O., Jeffries, C. D., Jarskog, L. F., Thomson, J. M., Woods, K., Newman, M. A., et al. (2007). microRNA expression in the prefrontal cortex of individuals with schizophrenia and schizoaffective disorder. *Genome Biology*, *8*, R27.

Post, R. M. (2007). Role of BDNF in bipolar and unipolar disorder: Clinical and theoretical implications. *Journal of Psychiatric Research*, *41*, 979–990.
Prior, H., & Sachser, N. (1995). Effects of enriched housing environment on the behaviour of young male and female mice in four exploratory tasks. *Journal of Experimental Animal Science*, *37*, 57–68.
Privitera, A. P., Distefano, R., Wefer, H. A., Ferro, A., Pulvirenti, A., & Giugno, R. (2015). OCDB: A database collecting genes, miRNAs and drugs for obsessive-compulsive disorder. *Database (Oxford)*, bav069. http://dx.doi.org/10.1093/database/bav069.
Quach, H., Barreiro, L. B., Laval, G., Zidane, N., Patin, E., Kidd, K. K., et al. (2009). Signatures of purifying and local positive selection in human miRNAs. *The American Journal of Human Genetics*, *84*, 316–327.
Ragu, V. D., Marimuthu, G., & Rajan, K. E. (2013). Environmental enrichment upregulates micro-RNA-183 and alters acetylcholinesterase splice variants to reduce anxiety-like behavior in the little Indian field mouse (Mus booduga). *Journal of Neuroscience Research*, *91*, 426–435.
Reyes-Herrera, P. H., Ficarra, E., Acquaviva, A., & Macii, E. (2011). miREE: miRNA recognition elements ensemble. *BMC Bioinformatics*, *12*, 454.
Rinaldi, A., Vincenti, S., De, V. F., Bozzoni, I., Oliverio, A., Presutti, C., et al. (2010). Stress induces region specific alterations in microRNAs expression in mice. *Behavioural Brain Research*, *208*, 265–269.
Rogaev, E. I. (2005). Small RNAs in human brain development and disorders. *Biochemistry*, *70*, 1404–1407.
Saini, H. K., Griffiths-Jones, S., & Enright, A. J. (2007). Genomic analysis of human microRNA transcripts. *Proceedings of the National academy of Sciences of the United States of America*, *104*, 17719–17724.
Schmidt, U., Herrmann, L., Hagl, K., Novak, B., Huber, C., Holsboer, F., et al. (2013). Therapeutic action of fluoxetine is associated with a reduction in prefrontal cortical miR-1971 expression levels in a mouse model of posttraumatic stress disorder. *Frontiers in Psychiatry*, *4*, 66.
Schouten, M., Aschrafi, A., Bielefeld, P., Doxakis, E., & Fitzsimons, C. P. (2013). microRNAs and the regulation of neuronal plasticity under stress conditions. *Neuroscience*, *241*, 188–205.
Schratt, G. M., Tuebing, F., Nigh, E. A., Kane, C. G., Sabatini, M. E., Kiebler, M., et al. (2006). A brain-specific microRNA regulates dendritic spine development. *Nature*, *439*, 283–289.
Schulze, T. G., Buervenich, S., Badner, J. A., Steele, C. J., tera-Wadleigh, S. D., Dick, D., et al. (2004). Loci on chromosomes 6q and 6p interact to increase susceptibility to bipolar affective disorder in the national institute of mental health genetics initiative pedigrees. *Biological Psychiatry*, *56*, 18–23.
Seckl, J. R. (1994). Glucocorticoids and small babies. *The Quarterly Journal of Medicine*, *87*, 259–262.
Segall, S. K., Nackley, A. G., Diatchenko, L., Lariviere, W. R., Lu, X., Marron, J. S., et al. (2010). Comt1 genotype and expression predicts anxiety and nociceptive sensitivity in inbred strains of mice. *Genes, Brain, and Behavior*, *9*, 933–946.
Segall, S. K., Shabalina, S. A., Meloto, C. B., Wen, X., Cunningham, D., Tarantino, L. M., et al. (2015). Molecular genetic mechanisms of allelic specific regulation of murine Comt expression. *Pain*, *156*, 1965–1977.
Selbach, M., Schwanhausser, B., Thierfelder, N., Fang, Z., Khanin, R., & Rajewsky, N. (2008). Widespread changes in protein synthesis induced by microRNAs. *Nature*, *455*, 58–63.
Sequeira, A., Kim, C., Seguin, M., Lesage, A., Chawky, N., Desautels, A., et al. (2003). Wolfram syndrome and suicide: Evidence for a role of WFS1 in suicidal and impulsive behavior. *American Journal of Medical Genetics. Part B, Neuropsychiatric Genetics*, *119B*, 108–113.

Shimizu, S., Tanaka, T., Takeda, T., Tohyama, M., & Miyata, S. (2015). The kampo medicine yokukansan decreases microRNA-18 expression and recovers glucocorticoid receptors protein expression in the hypothalamus of stressed mice. *BioMed Research International, 2015*, 797280.

Sontheimer, E. J., & Carthew, R. W. (2005). Silence from within: Endogenous siRNAs and miRNAs. *Cell, 122*, 9–12.

Stankiewicz, A. M., Swiergiel, A. H., & Lisowski, P. (2013). Epigenetics of stress adaptations in the brain. *Brain Research Bulletin, 98*, 76–92.

Sun, B. K., & Tsao, H. (2008). Small RNAs in development and disease. *Journal of the American Academy of Dermatology, 59*, 725–737.

Szyf, M., Weaver, I. C., Champagne, F. A., Diorio, J., & Meaney, M. J. (2005). Maternal programming of steroid receptor expression and phenotype through DNA methylation in the rat. *Frontiers in Neuroendocrinology, 26*, 139–162.

Talebizadeh, Z., Butler, M. G., & Theodoro, M. F. (2008). Feasibility and relevance of examining lymphoblastoid cell lines to study role of microRNAs in autism. *Autism Research, 1*, 240–250.

Tang, X. Q., Gal, J., Zhuang, X., Wang, W. X., Zhu, H. N., & Tang, G. L. (2007). A simple array platform for microRNA analysis and its application in mouse tissues. *RNA, 13*, 1803–1822.

Thomson, D. W., Bracken, C. P., & Goodall, G. J. (2011). Experimental strategies for microRNA target identification. *Nucleic Acids Research, 39*, 6845–6853.

Tocchetto, A., Salum, G. A., Blaya, C., Teche, S., Isolan, L., Bortoluzzi, A., et al. (2011). Evidence of association between Val66Met polymorphism at BDNF gene and anxiety disorders in a community sample of children and adolescents. *Neuroscience Letters, 502*, 197–200.

Torres, G. E., Gainetdinov, R. R., & Caron, M. G. (2003). Plasma membrane monoamine transporters: Structure, regulation and function. *Nature Reviews. Neuroscience, 4*, 13–25.

Tost, H., Alam, T., & Meyer-Lindenberg, A. (2010). Dopamine and psychosis: Theory, pathomechanisms and intermediate phenotypes. *Neuroscience and Biobehavioral Reviews, 34*, 689–700.

Uchida, S., Hara, K., Kobayashi, A., Funato, H., Hobara, T., Otsuki, K., et al. (2010). Early life stress enhances behavioral vulnerability to stress through the activation of REST4-mediated gene transcription in the medial prefrontal cortex of rodents. *The Journal of Neuroscience, 30*, 15007–15018.

Uchida, S., Nishida, A., Hara, K., Kamemoto, T., Suetsugi, M., Fujimoto, M., et al. (2008). Characterization of the vulnerability to repeated stress in Fischer 344 rats: Possible involvement of microRNA-mediated down-regulation of the glucocorticoid receptor. *The European Journal of Neuroscience, 27*, 2250–2261.

van Praag, H. M. (1996a). Faulty cortisol/serotonin interplay. Psychopathological and biological characterisation of a new, hypothetical depression subtype (SeCA depression). *Psychiatry Research, 65*, 143–157.

van Praag, H. (1996b). Serotonin-related, anxiety/aggression-driven, stressor-precipitated depression. A psycho-biological hypothesis. *European Psychiatry, 11*, 57–67.

Verhagen, M., van der, M. A., van Deurzen, P. A., Janzing, J. G., rias-Vasquez, A., Buitelaar, J. K., et al. (2010). Meta-analysis of the BDNF Val66Met polymorphism in major depressive disorder: Effects of gender and ethnicity. *Molecular Psychiatry, 15*, 260–271.

Vermetten, E., & Bremner, J. D. (2002). Circuits and systems in stress. II. Applications to neurobiology and treatment in posttraumatic stress disorder. *Depression and Anxiety, 16*, 14–38.

Vo, N., Klein, M. E., Varlamova, O., Keller, D. M., Yamamoto, T., Goodman, R. H., et al. (2005). A cAMP-response element binding protein-induced microRNA regulates neuronal morphogenesis. *Proceedings of the National academy of Sciences of the United States of America, 102*, 16426–16431.

Volonté, C., Apolloni, S., & Parisi, C. (2015). MicroRNAs: Newcomers into the ALS picture. *CNS & Neurological Disorders: Drug Targets*, *14*, 24–32.

Vreugdenhil, E., Verissimo, C. S., Mariman, R., Kamphorst, J. T., Barbosa, J. S., Zweers, T., et al. (2009). MicroRNA 18 and 124a down-regulate the glucocorticoid receptor: Implications for glucocorticoid responsiveness in the brain. *Endocrinology*, *150*, 2220–2228.

Wang, Y., Li, X., & Hu, H. (2011). Transcriptional regulation of co-expressed microRNA target genes. *Genomics*, *98*, 445–452.

Wang, X., Sundquist, K., Hedelius, A., Palmer, K., Memon, A. A., & Sundquist, J. (2015). Circulating microRNA-144-5p is associated with depressive disorders. *Clinical Epigenetics*, 7, 69.

Wang, Z., Zhang, C., Huang, J., Yuan, C., Hong, W., Chen, J., et al. (2014). MiRNA-206 and BDNF genes interacted in bipolar I disorder. *Journal of Affective Disorders*, *162*, 116–119.

Weaver, I. C., Cervoni, N., Champagne, F. A., D'Alessio, A. C., Sharma, S., Seckl, J. R., et al. (2004). Epigenetic programming by maternal behavior. *Nature Neuroscience*, 7, 847–854.

Weaver, I. C., Champagne, F. A., Brown, S. E., Dymov, S., Sharma, S., Meaney, M. J., et al. (2005). Reversal of maternal programming of stress responses in adult offspring through methyl supplementation: Altering epigenetic marking later in life. *The Journal of Neuroscience*, *25*, 11045–11054.

Weaver, I. C., Meaney, M. J., & Szyf, M. (2006). Maternal care effects on the hippocampal transcriptome and anxiety-mediated behaviors in the offspring that are reversible in adulthood. *Proceedings of the National academy of Sciences of the United States of America*, *103*, 3480–3485.

Xu, H., He, J., Richardson, J. S., & Li, X. M. (2004). The response of synaptophysin and microtubule-associated protein 1 to restraint stress in rat hippocampus and its modulation by venlafaxine. *Journal of Neurochemistry*, *91*, 1380–1388.

Yamamoto, S., Morinobu, S., Iwamoto, Y., Ueda, Y., Takei, S., Fujita, Y., et al. (2010). Alterations in the hippocampal glycinergic system in an animal model of posttraumatic stress disorder. *Journal of Psychiatric Research*, *44*, 1069–1074.

Yehuda, R., & Bierer, L. M. (2009). The relevance of epigenetics to PTSD: Implications for the DSM-V. *Journal of Traumatic Stress*, *22*, 427–434.

Yehuda, R., Daskalakis, N. P., Bierer, L. M., Bader, H. N., Klengel, T., Holsboer, F., et al. (2015). Holocaust exposure induced intergenerational effects on FKBP5 methylation. *Biological Psychiatry*, *80*, 372–380.

Yehuda, R., & LeDoux, J. (2007). Response variation following trauma: A translational neuroscience approach to understanding PTSD. *Neuron*, *56*, 19–32.

Zhang, L., Ding, L., Cheung, T. H., Dong, M. Q., Chen, J., Sewell, A. K., et al. (2007). Systematic identification of *C. elegans* miRISC proteins, miRNAs, and mRNA targets by their interactions with GW182 proteins AIN-1 and AIN-2. *Molecular Cell*, *28*, 598–613.

Zhang, Y., Wang, Y., Wang, L., Bai, M., Zhang, X., & Zhu, X. (2015). Dopamine receptor D2 and associated microRNAs are involved in stress susceptibility and resistance to escitalopram treatment. *The International Journal of Neuropsychopharmacology*, *18*, pyv025. http://dx.doi.org/10.1093/ijnp/pyv025.

Zhou, R., Yuan, P., Wang, Y., Hunsberger, J. G., Elkahloun, A., Wei, Y., et al. (2009). Evidence for selective microRNAs and their effectors as common long-term targets for the actions of mood stabilizers. *Neuropsychopharmacology*, *34*, 1395–1405.

Ziebarth, J. D., Bhattacharya, A., Chen, A., & Cui, Y. (2012). PolymiRTS database 2.0: Linking polymorphisms in microRNA target sites with human diseases and complex traits. *Nucleic Acids Research*, *40*, D216–D221.

CHAPTER FIVE

Natriuretic Peptides in Anxiety and Panic Disorder

T. Meyer*,†,1, C. Herrmann-Lingen*,†
*University of Göttingen Medical Centre, Göttingen, Germany
†German Centre for Cardiovascular Research, University of Göttingen, Göttingen, Germany
[1]Corresponding author: e-mail address: thomas.meyer@med.uni-goettingen.de

Contents

Abstract

Natriuretic peptides exert pleiotropic effects on the cardiovascular system, including natriuresis, diuresis, vasodilation, and lusitropy, by signaling through membrane-bound guanylyl cyclases. In addition to their use as diagnostic and prognostic markers for heart failure, accumulating behavioral evidence suggests that these hormones also modulate anxiety symptoms and panic attacks. This review summarizes our current knowledge of the role of natriuretic peptides in animal and human anxiety and highlights some novel aspects from recent clinical studies on this topic.

1. NEUROENDOCRINE FACTORS IN ANXIETY AND FEAR-RELATED DISORDERS

Anxiety and fear-related disorders, such as generalized anxiety disorder, panic disorder, agoraphobia, and specific and social phobia, are clinically well-studied conditions characterized by an unpleasant state of inner tension and the expectation of future threat. The nosological entities subsumed under the rubric anxiety and fear-related disorders are usually accompanied by a variety of somatic symptoms, such as sweating, autonomic dysfunction, altered heart rate, abdominal distress, and nausea. Whereas diagnostic

Vitamins and Hormones, Volume 103
ISSN 0083-6729
http://dx.doi.org/10.1016/bs.vh.2016.08.002

categories for anxiety disorders are clinically well established, much less is known about the underlying pathophysiological pathways that trigger or maintain anxiety symptoms. Neurovegetative manifestations of anxiety have long been attributed an important role in aversive conditioning, and there is compelling evidence obtained from both clinical and laboratory observations supporting the participation of neuroendocrine pathways in the negative emotional state that defines anxiety (Dunsmoor & Paz, 2015). It is generally accepted that anxiety symptoms seem to be, at least in part, aggravated by interoceptive stimuli, although numerous endogenous factors may buffer patients from the behavioral and physiological manifestations associated with panic attacks and clinically relevant episodes of anxiety. Studying the physiological mechanisms behind the neuroendocrine networks may help to better understand the behavioral dysfunction and emotion generation that play a fundamental role in the vulnerability to anxiety.

From a clinical perspective, it is desirable to assess the effects of neuroendocrine factors on anxiety in defined subsets of patients, for example, those who are at risk of experiencing cardiovascular events. The current review summarizes the existing literature with regard to the role of natriuretic peptides, a class of hormones with a broad range of neuromodulatory functions, in the onset of anxiety symptoms and highlights the use of experimental and clinical studies on anxiety in various settings and clinical samples.

2. MOLECULAR MECHANISMS INVOLVED IN NATRIURETIC PEPTIDE SYNTHESIS

Natriuretic peptides are endogenous molecules that were first shown to act in an endocrine or paracrine fashion to regulate extracellular fluid volume and blood pressure by inhibiting sodium and water reabsorption in the kidney. The first clue of an endocrine link between the heart and kidneys came from the pioneering work of de Bold and colleagues who described a potent diuretic and natriuretic response to intravenous injection of atrial myocardial extracts in rats whereas no such effect was observed after injection of supernatants from ventricular myocardial homogenates (de Bold, Borenstein, Veress, & Sonnenberg, 1981). The authors concluded that an unknown factor most probably present in membrane-bound storage granules in atrial cardiomyocytes caused the increased NaCl excretion by inhibiting its reabsorption in the medullary collecting duct (Sonnenberg, Cupples, de Bold, & Veress, 1982).

The discovery of a polypeptide factor blocking renal tubular NaCl reabsorption in crude extracts from the atria led to the characterization of the A-type (atrial) natriuretic factor, more recently after identification of its chemical structure termed atrial natriuretic peptide (ANP) (Flynn, Davies, Kennedy, de Bold, & de Bold, 1985; Flynn, de Bold, & de Bold, 1983; Matsuo & Kangawa, 1984; Oikawa et al., 1984). ANP was the first and two other natriuretic peptides derived from different precursors have been identified so far: B-type (BNP) and C-type (CNP) natriuretic peptides. While ANP is predominantly released from preformed granules in atrial cardiomyocytes and BNP (also known as brain natriuretic peptide because it was first isolated from brain tissue) is produced in both atrial and ventricular cardiomyocytes, CNP expression is largely restricted to neurons and vascular endothelial cells in the brain. Expression of ANP has been found not only in the ordinary myocardium but also in cells of the cardiac conduction system (Hansson, 2002). In addition, ANP immunostaining was demonstrated in the nodose ganglion and the superior cervical ganglion of rats, suggesting that peripheral ANP may affect the autonomic regulation of the heart via sympathoinhibitory and/or vagoexcitatory actions (Hansson, 2002).

All three natriuretic peptides are expressed as preprohormones which undergo proteolytical processing to the mature peptides (Nemer et al., 1984; Ogawa et al., 1994). In humans, the genes encoding the precursors of ANP and BNP, termed *NPPA* (natriuretic peptide precursor A) and *NPPB*, are located in tandem orientation on chromosome 1, while the *NPPC* gene is on chromosome 2 (Ogawa et al., 1994; Yang-Feng, Floyd-Smith, Nemer, Drouin, & Francke, 1985). Upon mechanical stretch of the heart muscle by pressure overload or stimulation by vasoactive factors such as endothelin, angiotensin II, vasopressin, the sympathetic nervous system, hypoxia, cold and exercise, the transcription factor GATA4 binds to the promoter of the ANP-encoding gene to induce *NPPA* gene activation (Thuerauf, Hanford, & Glembotski, 1994). The mRNA transcript is then translated to 151-preproANP, which is cleaved by removal of a 25-amino acid signaling peptide to yield 126-amino acid proANP (Fig. 1A). Finally, proANP is proteolytically cleaved by the membrane-bound serine protease corin to the biologically inactive NT-proANP and the carboxy-terminal active ANP1-28 (Yan, Wu, Morser, & Wu, 2000). Similar cleavage steps have been reported in the posttranslational processing of proBNP and proCNP, yielding the bioactive BNP1-32 and CNP1-22 fragments (Fig. 1B and C). The proteolytic processing site generating BNP is identical to the consensus site of another serine endoprotease, furin (Steinhelper, 1993). All three carboxy-terminally cleaved natriuretic peptides share a

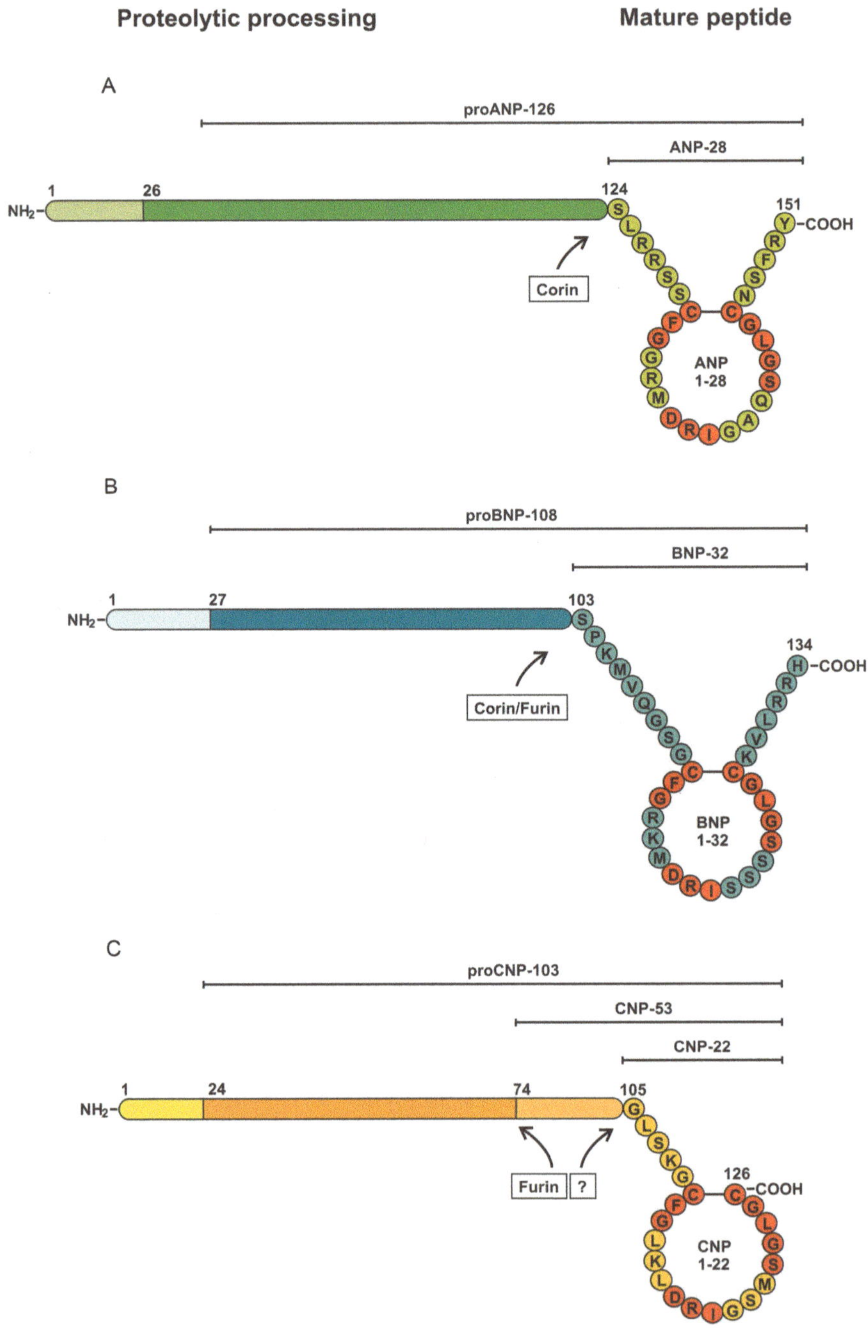

Fig. 1 Proteolytic processing of the preprohormones of natriuretic peptides including the sequences of their mature, biologically active peptides. The models show the protease cleavage sites in the precursor molecules of the three major human endogenous natriuretic peptides ANP (A), BNP (B), and CNP (C). *Arrows* indicate processing sites by corin or furin.

similar ring structure with two pairs of cysteine residues forming intramolecular disulfide bonds in conserved positions that are required for receptor binding. The biologically active natriuretic hormones are less stable in circulation and have shorter half-lives than the higher molecular mass forms of NT-proANP and NT-proBNP. Hence, the amino-terminal fragments, and not the bioactive carboxy-terminal peptides, of the prohormones have been preferentially used as diagnostic biomarkers in cardiac conditions such as heart failure.

3. EXPRESSION OF NATRIURETIC PEPTIDES AND THEIR RECEPTORS IN THE BRAIN

Natriuretic peptides exert their intracellular actions by binding to single-membrane-spanning guanylyl cyclase receptors (Fig. 2). Ligand

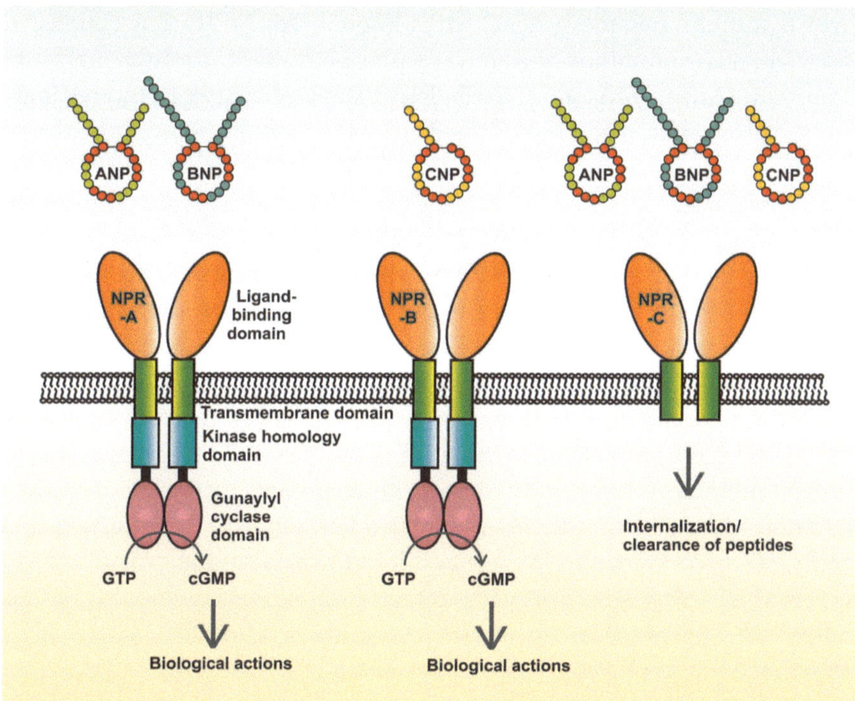

Fig. 2 Binding specificity of the three natriuretic peptide receptors for their endogenous ligands. Hormone binding to NPR-A and NPR-B induces the guanylyl cyclase activity located at the carboxy-termini of the receptor dimer and subsequently increases the intracellular concentration of cyclic guanosine monophosphate (cGMP). In contrast, binding of all three natriuretic peptides to NPR-C leads to their internalization and biological inactivity.

binding to the receptor induces the intracellular guanylyl cyclase activity and results in the conversion of guanosine-5′-triphosphate to cyclic guanosine monophosphate (cGMP) in the cytoplasm, which then stimulates cGMP-dependent protein kinase to induce smooth muscle relaxation, lowering of intracellular calcium, and phosphorylation of a number of target proteins (Winquist et al., 1984). Three subtypes of natriuretic peptide receptors have been described so far: natriuretic peptide receptor-A (NPR-A, also termed guanylyl cyclase-A), natriuretic peptide receptor-B (NPR-B or guanylyl cyclase-B), and natriuretic peptide receptor-C (NPR-C or clearance receptor). Upon binding of ANP or BNP to its extracellular ligand-binding domain (Fig. 3A), NPR-A produces the second messenger cGMP, while CNP binds to the NPR-B receptor resulting also in increased intracellular cGMP levels (Koller et al., 1991).

The NPR-C receptor contains only a short intracellular domain lacking guanylyl cyclase activity and acts predominantly as a clearance receptor for all three natriuretic peptides (Fig. 2). It is known that NPR-C is coupled to adenylyl cyclase inhibition and activates phospholipase C through inhibitory G proteins (Anand-Srivastava, 2005; Matsukawa et al., 1999). Interestingly, the CNP-specific receptor NPR-B is widely expressed in the brain, suggesting that CNP has neuromodulatory functions rather than acting as a peripheral hormone (Kaneko et al., 1993). Using in situ hybridization histochemistry, expression of NPR-B mRNA was detected throughout the limbic cortex and neocortex, olfactory bulb, hippocampus and amygdala, where CNP most likely acts as the primary biologically active natriuretic peptide (Herman, Dolgas, Rucker, & Langub, 1996). Similarly, immunohistochemical staining of the brain of the cynomolgus monkey, *Macaca fascicularis*, revealed a wide distribution of NPR-A in neurons of specific brainstem nuclei (Abdelalim, Osman, Takada, Torii, & Tooyama, 2007).

Although CNP is the most abundantly expressed natriuretic peptide in the brain, the two other family members, ANP and BNP, including their cognate receptors, are also located in distinct regions of the central nervous system. *ANP* gene transcripts were detected in different regions in the central nervous system, often in colocalization with *CNP* mRNA, such as in the hypothalamus, brainstem, cerebellum, and cerebral cortex (Brown & Czarnecki, 1990; Gardner, Vlasuk, Baxter, Fiddes, & Lewicki, 1987; Kawata et al., 1985; Langub, Watson, & Herman, 1995; McKenzie et al., 1994; McKenzie, Juan, Thomas, Berman, & Klein, 2001; Ryan & Gundlach, 1995; Saavedra & Kurihara, 1991; Standaert, Needleman, & Saper, 1986; Tanaka, Misono, & Inagami, 1984).

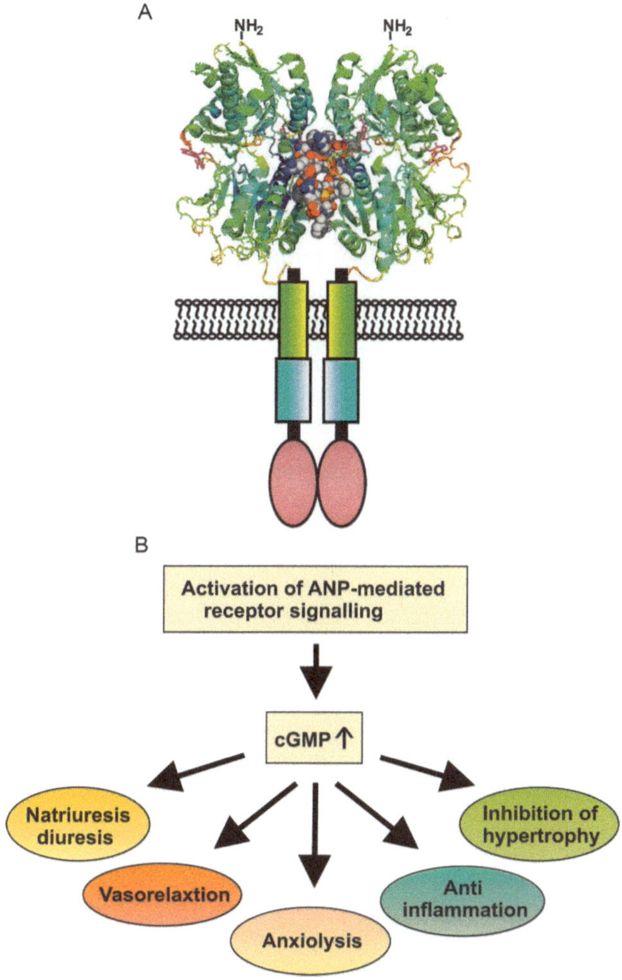

Fig. 3 (A) Crystal structure of the extracellular ligand-binding domain of the atrial natriuretic (ANP) receptor dimer complexed with rat ANP7-27. The receptor structure is shown as a ribbon model with carbohydrate structures depicted in *magenta*, while the receptor-bound hormone is depicted in a space-filling model with nitrogen atoms marked in *blue*, oxygen atoms in *red*, and sulfur atoms in *yellow*. (B) Overview of the pleiotropic functions of cGMP-mediated ANP signaling including its effect on anxiety. *Panel A was drawn with PyMOL (DeLano Scientific) using data from the Protein Database file 1T34 (Ogawa, H., Qiu, Y., Ogata, C. M., & Misono, K. S. (2004). Crystal structure of hormone-bound atrial natriuretic peptide receptor extracellular domain: Rotation mechanism for transmembrane signal transduction.* Journal of Biological Chemistry, 279, *28625–28631).*

4. PHYSIOLOGICAL ACTIONS OF NATRIURETIC PEPTIDES IN THE BRAIN

The expression of natriuretic peptides and their receptors in specific regions of the central nervous system, particularly in circumventricular areas, may explain why these peptides are engaged in integrating signals and executing responses from different neural pathways (Hodes & Lichtstein, 2014). Levin and colleagues reported that administration of ANP into the fourth ventricle caused a significant decrease in mean arterial blood pressure through an interaction with the central α_2-adrenergic system (Levin, Weber, & Mills, 1988). However, other studies could not confirm this observation. Since intracerebroventricular infusion of ANP in euhydrated conscious sheep inhibits vasopressin secretion, no change in mean arterial blood pressure was monitored (Charles, Richards, & Espiner, 1992; Lee, Malvin, Claybaugh, & Huang, 1987). In contrast, intracerebroventricularly administered CNP induced a prompt fall in arterial pressure in normal sheep without increasing the heart rate, while plasma cortisol and aldosterone levels increased abruptly during the first hour. There is evidence from laboratory studies in rats which suggests that CNP, whose expression and direct actions are mainly confined to the vascular endothelium, not only modulates the adrenocortical response, but also results in improved consolidation of learning in a passive avoidance paradigm (Telegdy, Kokavszky, & Nyerges, 1999). Schultz and colleagues reported that in anesthetized rats with intact arterial baroreceptors, intracerebroventricular injection of ANP into the third ventricle did not alter the heart rate or the mean arterial pressure, unless sinoaortic denervation was performed (Schultz, Steele, & Gardner, 1990).

Centrally infused human ANP in anesthetized dogs resulted in a significant decrease in the glomerular filtration rate, while blood pressure, heart rate, and plasma aldosterone were unchanged (Shoji et al., 1987). While injection of ANP into the brains of conscious, unrestrained rats had no detectable effect on the basal plasma corticosterone level, it attenuated dose dependently the increase in plasma corticosterone induced by centrally injected angiotensin II. This observation suggests that local expression of ANP in the brain is involved in the regulation of the hypothalamo–pituitary–adrenal (HPA) axis (Itoh et al., 1986). Intracerebroventricular administration of human ANP in restrained euhydrated rabbits neither affected blood pressure nor heart rate, but prevented the increase in plasma arginine vasopressin, aldosterone, and corticosterone levels in dehydrated animals

(Kallaras et al., 2004). Likewise, basal secretion of arginine vasopressin was inhibited by centrally administered BNP in conscious rats in a manner comparable to that of ANP (Yamada et al., 1988).

There is evidence which suggests that arginine vasopressin functionally interacts with ANP at the level of the central nervous system to significantly attenuate the central pressor effects of vasopressin (Stepniakowski, Budzikowski, Loń, & Szczepańska-Sadowska, 1991, 1994). One study reports on hyperthermic effects of centrally administered natriuretic peptides in the rat brain which were abolished upon intramuscular injection of the antipyretic drug noraminophenazone. This finding suggests that natriuretic peptides play a role in thermoregulation mediated by the cyclooxygenase pathway (Pataki, Jászberényi, & Telegdy, 1999). In summary, all three natriuretic peptides are linked to different vegetative functions and modulate the secretion of other hormones.

5. NEUROPROTECTIVE EFFECTS OF NATRIURETIC PEPTIDES

Growing evidence from laboratory studies suggests that natriuretic peptides exert distinct neuromodulatory roles in various disease models (Hodes & Lichtstein, 2014). Wiggins et al. reported that *ANP* mRNA expression in the rat neocortex, but not expression of *CNP* transcripts, was increased following an acute, unilateral episode of cortical spreading depression (CSD), which is characterized by slow, self-propagating waves of cellular depolarization in the gray matter (Wiggins, Shen, & Gundlach, 2003). Since preconditioning with CSD protects against effects of ischemia, the authors suggested that ANP may contribute to CSD-mediated neuroprotection, possibly via effects on cGMP production. Kuribayashi and colleagues reported that ANP ameliorated *N*-methyl-D-aspartate (NMDA)-induced neurotoxicity in the retina following intravitreal injection of rat eyes with NMDA, possibly through binding to NPR-A (Kuribayashi et al., 2006).

Proton nuclear magnetic resonance imaging demonstrated inhibition of edema formation in ischemic brain regions in a rat model of intravenous ANP injection following occlusion of the left middle cerebral artery as compared to saline treatment (Naruse, Aoki, Takei, Horikawa, & Ueda, 1991). Even 4 h after onset of a hemorrhagic injury induced by intracerebral injection of bacterial collagenase into the rat brain, the infusion of ANP in the peritoneum resulted in reduced edema formation (Rosenberg & Estrada, 1995). Similarly, intravenous injection of human BNP was associated with improved long-term neurological recovery after acute brain injury (James et al., 2010). However, in human patients with spontaneous subarachnoid or intracerebral hemorrhage, increased levels of MR-proANP were associated with increased mortality and poor functional outcome after 180 days (Fischer et al., 2014). Likewise, BNP plasma levels were higher among those patients who deteriorated in their neurological status during the acute phase of stroke (Montaner et al., 2012). However, these findings may be due to neurocardiogenic dysfunction leading to adverse prognosis and do not necessarily contradict a potential neuroprotective role of natriuretic peptides.

6. EVIDENCE OF ANXIOLYTIC-LIKE EFFECTS OF ANP

In the early 1990s, Kellner, Wiedemann, and Holsboer published an important paper demonstrating that, in humans, intravenously administered ANP inhibited the corticotropin-releasing hormone (CRH)-induced secretion of adrenocorticotropic hormone (ACTH), also known as corticotrophin (Kellner, Wiedemann, & Holsboer, 1992). The discovery that ANP is a peptidergic antagonist of pituitary–adrenocortical activity shed light on the central function of natriuretic peptides (Fig. 3B). Whereas arginine vasopressin significantly amplifies the CRH-stimulated ACTH and cortisol release (Watabe et al., 1988), ANP infusion significantly decreases vasopressin levels in healthy male subjects and antagonized the stimulatory action of vasopressin. This observation suggests a complex interplay between the two peptide hormones on the regulation of the HPA axis and may explain the lack of HPA activation during lactate-induced panic attacks (Kellner, Herzog, Holsboer, & Wiedemann, 1995; Kellner, Herzog, Yassouridis, Holsboer, & Wiedemann, 1995). The same authors found increased circulating ANP levels in lactate-induced panic attacks which confirmed the hypothesis that ANP release is an intrinsic mechanism contributing to the unresponsiveness of the HPA axis in emotionally anxious conditions (Kellner, Knaudt, Jahn, Holsboer, & Wiedemann, 1998). However, intravenous administration of CNP in healthy male volunteers augmented the increase of ACTH during challenge by the panicogenic cholecystokinin tetrapeptide (CCK-4), suggesting that ANP and CNP exert opposing effects on anxiety behavior (Jahn et al., 2001; Kellner et al., 2002). In patients with panic disorder, administration of 150 μg of ANP reduced the occurrence of CCK-4-induced panic attacks as compared to placebo treatment (Ströhle, Kellner, Holsboer, & Wiedemann,

2001). Among the 10 participants studied on two different days, 7 in the placebo condition and only 2 in the ANP treatment condition experienced symptoms of a panic attack.

Biró, Tóth, and Telegdy (1995) examined the behavioral consequences of centrally administered ANP in rats and found that ANP dose dependently abolished the normal preference for the closed arms in an elevated plus-maze model while increasing the percentage of time spent on the open arms, which is consistent with an anxiolytic-like effect. The authors reported, in another publication, that multiple neurotransmitter systems, most likely dopaminergic, α-adrenergic and β-adrenergic synapses, might be involved in the natriuretic peptide-induced anxiolysis (Biró, Tóth, & Telegdy, 1996).

Given the fact that peripherally infused ANP-reduced panic symptoms in humans and intracerebral ANP displayed anxiolytic effects in animal experiments, Herrmann-Lingen et al. (2003) analyzed associations between NT-proANP and anxiety in patients with and without congestive heart failure and found an independent association of higher NT-proANP levels with lower anxiety. Recently, we assessed the relationship between circulating MR-proANP concentrations and symptoms of anxiety in patients with cardiovascular risk factors (Meyer et al., 2015). Data from the DIAST-CHF study showed a significant inverse association between elevated scores on the anxiety subscale of the Hospital Anxiety and Depression Scale (HADS) and both log-transformed plasma mid-regional proANP (MR-proANP) and amino-terminal proBNP (NT-proBNP) concentrations. The significant association between HADS anxiety scores and plasma MR-proANP remained stable after adjustment for sex, age, body-mass index, and Framingham score as a measure for the severity of heart failure symptoms, while NT-proBNP lost its significance in independent prediction of anxiety.

The higher ANP levels typically measured in the plasma of heart failure patients may protect their hearts against the surge of catecholamines which usually accompanies panic attacks and anxiety symptoms. Given that anxiety is a known trigger for ventricular tachyarrhythmia and sudden cardiac death, the anxiolytic-like effects of ANP may be particularly beneficial in the context of comorbid heart diseases (Meyer, Buss, & Herrmann-Lingen, 2010). Therefore, it will be interesting to see whether both circulating and brain-expressed ANP may contribute to the buffering of the aversive emotions of anxiety in selected patient groups, such as those after myocardial infarction with reduced left ventricular function who are particularly prone to the adverse effects of heightened sympathoadrenal activity.

REFERENCES

Abdelalim, E. M., Osman, A. H., Takada, T., Torii, R., & Tooyama, I. (2007). Immunohistochemical mapping of natriuretic peptide receptor-A in the brainstem of *Macaca fascicularis*. *Neuroscience*, *145*, 1087–1096.

Anand-Srivastava, M. B. (2005). Natriuretic peptide receptor-C signaling and regulation. *Peptides*, *26*, 1044–1059.

Biró, E., Tóth, G., & Telegdy, G. (1995). Involvement of neurotransmitters in the 'anxiolytic-like' action of atrial natriuretic peptide in rats. *Neuropeptides*, *29*, 215–220.

Biró, E., Tóth, G., & Telegdy, G. (1996). Effect of receptor blockers on brain natriuretic peptide and C-type natriuretic peptide caused anxiolytic state in rats. *Neuropeptides*, *30*, 59–65.

de Bold, A. J., Borenstein, H. B., Veress, A. T., & Sonnenberg, H. (1981). A rapid and potent natriuretic response to intravenous injection of atrial myocardial extract in rats. *Life Sciences*, *28*, 89–94.

Brown, J., & Czarnecki, A. (1990). Autoradiographic localization of atrial and brain natriuretic peptide receptors in rat brain. *American Journal of Physiology*, *258*, R57–R63.

Charles, C. J., Richards, A. M., & Espiner, E. A. (1992). Central C-type natriuretic peptide but not atrial natriuretic factor lowers blood pressure and adrenocortical secretion in normal conscious sheep. *Endocrinology*, *131*, 1721–1726.

Dunsmoor, J. E., & Paz, R. (2015). Fear generalization and anxiety: Behavioral and neural mechanisms. *Biological Psychiatry*, *78*, 336–343.

Fischer, M., Katan, M., Morgenthaler, N. G., Seiler, M., Müller, B., Lackner, P., et al. (2014). The prognostic value of midregional proatrial natriuretic peptide in patients with hemorrhagic stroke. *Cerebrovascular Diseases*, *37*, 128–133.

Flynn, T. G., de Bold, M. L., & de Bold, A. J. (1983). The amino acid sequence of an atrial peptide with potent diuretic and natriuretic properties. *Biochemical and Biophysical Research Communications*, *117*, 859–865.

Flynn, T. G., Davies, P. L., Kennedy, B. P., de Bold, M. L., & de Bold, A. J. (1985). Alignment of rat cardionatrin sequences with the preprocardionatrin sequence from complementary DNA. *Science*, *228*, 323–325.

Gardner, D. G., Vlasuk, G. P., Baxter, J. D., Fiddes, J. C., & Lewicki, J. A. (1987). Identification of atrial natriuretic factor gene transcripts in the central nervous system of the rat. *Proceedings of the National Academy of Sciences of the United States of America*, *84*, 2175–2179.

Hansson, M. (2002). Natriuretic peptides in relation to the cardiac innervation and conduction system. *Microscopy Research and Technique*, *58*, 378–386.

Herman, J. P., Dolgas, C. M., Rucker, D., & Langub, M. C., Jr. (1996). Localization of natriuretic peptide-activated guanylate cyclase mRNAs in the rat brain. *Journal of Comparative Neurology*, *369*, 165–187.

Herrmann-Lingen, C., Binder, L., Klinge, M., Sander, J., Schenker, W., Beyermann, B., et al. (2003). High plasma levels of N-terminal pro-atrial natriuretic peptide associated with low anxiety in severe heart failure. *Psychosomatic Medicine*, *65*, 517–522.

Hodes, A., & Lichtstein, D. (2014). Natriuretic hormones in brain function. *Frontiers in Endocrinology (Lausanne)*, *5*, 201.

Itoh, H., Nakao, K., Katsuura, G., Morii, N., Yamada, T., Shiono, S., et al. (1986). Possible involvement of central atrial natriuretic polypeptide in regulation of hypothalamo–pituitary–adrenal axis in conscious rats. *Neuroscience Letters*, *69*, 254–258.

Jahn, H., Montkowski, A., Knaudt, K., Ströhle, A., Kiefer, F., Schick, M., et al. (2001). Alpha-helical-corticotropin-releasing hormone reverses anxiogenic effects of C-type natriuretic peptide in rats. *Brain Research*, *893*, 21–28.

James, M. L., Wang, H., Venkatraman, T., Song, P., Lascola, C. D., & Laskowitz, D. T. (2010). Brain natriuretic peptide improves long-term functional recovery after acute CNS injury in mice. *Journal of Neurotrauma*, *27*, 217–228.

Kallaras, C., Angelopoulos, N., Bountzioukas, S., Mavroudis, K., Karamouzis, M., & Guiba-Tziampiri, O. (2004). Intracerebroventricular administration of atrial natriuretic peptide prevents increase of plasma ADH, aldosterone and corticosterone levels in restrained conscious dehydrated rabbits. *Journal of Endocrinological Investigation, 27*, 844–853.

Kaneko, T., Shirakami, G., Nakao, K., Nagata, I., Nakagawa, O., Hama, N., et al. (1993). C-type natriuretic peptide (CNP) is the major natriuretic peptide in human cerebrospinal fluid. *Brain Research, 612*, 104–109.

Kawata, M., Nakao, K., Morii, N., Kiso, Y., Yamashita, H., Imura, H., et al. (1985). Atrial natriuretic polypeptide: Topographical distribution in the rat brain by radioimmunoassay and immunohistochemistry. *Neuroscience, 16*, 521–546.

Kellner, M., Herzog, L., Holsboer, F., & Wiedemann, K. (1995a). Circadian changes in the sensitivity of the corticotropin-releasing hormone-stimulated HPA system after arginine vasopressin and atrial natriuretic hormone in human male controls. *Psychoneuroendocrinology, 20*, 515–524.

Kellner, M., Herzog, L., Yassouridis, A., Holsboer, F., & Wiedemann, K. (1995b). Possible role of atrial natriuretic hormone in pituitary–adrenocortical unresponsiveness in lactate-induced panic. *The American Journal of Psychiatry, 152*, 1365–1367.

Kellner, M., Knaudt, K., Jahn, H., Holsboer, F., & Wiedemann, K. (1998). Atrial natriuretic hormone in lactate-induced panic attacks: Mode of release and endocrine and pathophysiological consequences. *Journal of Psychiatric Research, 32*, 37–48.

Kellner, M., Wiedemann, K., & Holsboer, F. (1992). Atrial natriuretic factor inhibits the CRH-stimulated secretion of ACTH and cortisol in man. *Life Sciences, 50*, 1835–1842.

Kellner, M., Yassouridis, A., Hua, Y., Wendrich, M., Jahn, H., & Wiedemann, K. (2002). Intravenous C-type natriuretic peptide augments behavioral and endocrine effects of cholecystokinin tetrapeptide in healthy men. *Journal of Psychiatric Research, 36*, 1–6.

Koller, K. J., Lowe, D. G., Bennett, G. L., Minamino, N., Kangawa, K., Matsuo, H., et al. (1991). Selective activation of the B natriuretic peptide receptor by C-type natriuretic peptide (CNP). *Science, 252*, 120–123.

Kuribayashi, K., Kitaoka, Y., Kumai, T., Munemasa, Y., Kitaoka, Y., Isenoumi, K., et al. (2006). Neuroprotective effect of atrial natriuretic peptide against NMDA-induced neurotoxicity in the rat retina. *Brain Research, 1071*, 34–41.

Langub, M. C., Jr., Watson, R. E., Jr., & Herman, J. P. (1995). Distribution of natriuretic peptide precursor mRNAs in the rat brain. *Journal of Comparative Neurology, 356*, 183–199.

Lee, J., Malvin, R. L., Claybaugh, J. R., & Huang, B. S. (1987). Atrial natriuretic factor inhibits vasopressin secretion in conscious sheep. *Proceedings of the Society for Experimental Biology and Medicine, 185*, 272–276.

Levin, E. R., Weber, M. A., & Mills, S. (1988). Atrial natriuretic factor-induced vasodepression occurs through central nervous system. *American Journal of Physiology, 255*, H616–H622.

Matsukawa, N., Grzesik, W. J., Takahashi, N., Pandey, K. N., Pang, S., Yamauchi, M., et al. (1999). The natriuretic peptide clearance receptor locally modulates the physiological effects of the natriuretic peptide system. *Proceedings of the National Academy of Sciences of the United States of America, 96*, 7403–7408.

Matsuo, H., & Kangawa, K. (1984). Human and rat atrial natriuretic polypeptides (hANP & rANP) purification, structure and biological activity. *Clinical and Experimental Hypertension. Part A, 6*, 1717–1722.

McKenzie, J. C., Berman, N. E., Thomas, C. R., Young, J. K., Compton, L. Y., Cothran, L. N., et al. (1994). Atrial natriuretic peptide-like (ANP-LIR) and ANP prohormone immunoreactive astrocytes and neurons of human cerebral cortex. *Glia, 12*, 228–243.

McKenzie, J. C., Juan, Y. W., Thomas, C. R., Berman, N. E., & Klein, R. M. (2001). Atrial natriuretic peptide-like immunoreactivity in neurons and astrocytes of human cerebellum and inferior olivary complex. *Journal of Histochemistry and Cytochemistry*, *49*, 1453–1467.

Meyer, T., Buss, U., & Herrmann-Lingen, C. (2010). Role of cardiac disease severity in the predictive value of anxiety for all-cause mortality. *Psychosomatic Medicine*, *72*, 9–15.

Meyer, T., Herrmann-Lingen, C., Chavanon, M. L., Nolte, K., Pasedach, C. A., Binder, L., et al. (2015). Higher plasma levels of MR-pro-atrial natriuretic peptide are linked to less anxiety: Results from the observational DIAST-CHF study. *Clinical Research in Cardiology*, *104*, 574–581.

Montaner, J., García-Berrocoso, T., Mendioroz, M., Palacios, M., Perea-Gainza, M., Delgado, P., et al. (2012). Brain natriuretic peptide is associated with worsening and mortality in acute stroke patients but adds no prognostic value to clinical predictors of outcome. *Cerebrovascular Diseases*, *34*, 240–245.

Naruse, S., Aoki, Y., Takei, R., Horikawa, Y., & Ueda, S. (1991). Effects of atrial natriuretic peptide on ischemic brain edema in rats evaluated by proton magnetic resonance method. *Stroke*, *22*, 61–65.

Nemer, M., Chamberland, M., Sirois, D., Argentin, S., Drouin, J., Dixon, R. A., et al. (1984). Gene structure of human cardiac hormone precursor, pronatriodilatin. *Nature*, *312*, 654–656.

Ogawa, Y., Itoh, H., Yoshitake, Y., Inoue, M., Yoshimasa, T., Serikawa, T., et al. (1994). Molecular cloning and chromosomal assignment of the mouse C-type natriuretic peptide (CNP) gene (*Nppc*): Comparison with the human CNP gene (*NPPC*). *Genomics*, *24*, 383–387.

Oikawa, S., Imai, M., Ueno, A., Tanaka, S., Noguchi, T., Nakazato, H., et al. (1984). Cloning and sequence analysis of cDNA encoding a precursor for human atrial natriuretic polypeptide. *Nature*, *309*, 724–726.

Pataki, I., Jászberényi, M., & Telegdy, G. (1999). Hyperthermic effect of centrally administered natriuretic peptides in the rat. *Peptides*, *20*, 193–197.

Rosenberg, G. A., & Estrada, E. Y. (1995). Atrial natriuretic peptide blocks hemorrhagic brain edema after 4-hour delay in rats. *Stroke*, *26*, 874–877.

Ryan, M. C., & Gundlach, A. L. (1995). Anatomical localisation of preproatrial natriuretic peptide mRNA in the rat brain by in situ hybridisation histochemistry: Novel identification in olfactory regions. *Journal of Comparative Neurology*, *356*, 168–182.

Saavedra, J. M., & Kurihara, M. (1991). Autoradiography of atrial natriuretic peptide (ANP) receptors in the rat brain. *Canadian Journal of Physiology and Pharmacology*, *69*, 1567–1575.

Schultz, H. D., Steele, M. K., & Gardner, D. G. (1990). Central administration of atrial peptide decreases sympathetic outflow in rats. *American Journal of Physiology*, *258*, R1250–R1256.

Shoji, M., Kimura, T., Matsui, K., Ota, K., Iitake, K., Inoue, M., et al. (1987). Effects of centrally administered atrial natriuretic peptide on renal functions. *Acta Endocrinologica (Copenhagen)*, *115*, 433–440.

Sonnenberg, H., Cupples, W. A., de Bold, A. J., & Veress, A. T. (1982). Intrarenal localization of the natriuretic effect of cardiac atrial extract. *Canadian Journal of Physiology and Pharmacology*, *60*, 1149–1152.

Standaert, D. G., Needleman, P., & Saper, C. B. (1986). Organization of atriopeptin-like immunoreactive neurons in the central nervous system of the rat. *Journal of Comparative Neurology*, *253*, 315–341.

Steinhelper, M. E. (1993). Structure, expression, and genomic mapping of the mouse natriuretic peptide type-B gene. *Circulation Research*, *72*, 984–992.

Stepniakowski, K., Budzikowski, A., Loń, S., & Szczepańska-Sadowska, E. (1991). Central ANP attenuates pressor responses to central AVP in WKY and SHR. *Brain Research Bulletin*, *27*, 247–249.
Stepniakowski, K., Budzikowski, A., Loń, S., & Szczepańska-Sadowska, E. (1994). Central cardiovascular effects of AVP and ANP in normotensive and spontaneously hypertensive rats. *Journal of the Autonomic Nervous System*, *47*, 33–43.
Ströhle, A., Kellner, M., Holsboer, F., & Wiedemann, K. (2001). Anxiolytic activity of atrial natriuretic peptide in patients with panic disorder. *The American Journal of Psychiatry*, *158*, 1514–1516.
Tanaka, I., Misono, K. S., & Inagami, T. (1984). Atrial natriuretic factor in rat hypothalamus, atria and plasma: determination by specific radioimmunoassay. *Biochemical and Biophysical Research Communications*, *124*, 663–668.
Telegdy, G., Kokavszky, K., & Nyerges, A. (1999). Action of C-type natriuretic peptide (CNP) on passive avoidance learning in rats: Involvement of transmitters. *European Journal of Neuroscience*, *11*, 3302–3306.
Thuerauf, D. J., Hanford, D. S., & Glembotski, C. C. (1994). Regulation of rat brain natriuretic peptide transcription. A potential role for GATA-related transcription factors in myocardial cell gene expression. *Journal of Biological Chemistry*, *269*, 17772–17775.
Watabe, T., Tanaka, K., Kumagae, M., Itoh, S., Kogure, M., Hasegawa, M., et al. (1988). Role of endogenous arginine vasopressin in potentiating corticotropin-releasing hormone-stimulated corticotropin secretion in man. *The Journal of Clinical Endocrinology and Metabolism*, *66*, 1132–1137.
Wiggins, A. K., Shen, P. J., & Gundlach, A. L. (2003). Atrial natriuretic peptide expression is increased in rat cerebral cortex following spreading depression: Possible contribution to sd-induced neuroprotection. *Neuroscience*, *118*, 715–726.
Winquist, R. J., Faison, E. P., Waldman, S. A., Schwartz, K., Murad, F., & Rapoport, R. M. (1984). Atrial natriuretic factor elicits an endothelium-independent relaxation and activates particulate guanylate cyclase in vascular smooth muscle. *Proceedings of the National Academy of Sciences of the United States of America*, *81*, 7661–7664.
Yamada, T., Nakao, K., Itoh, H., Shirakami, G., Kangawa, K., Minamino, N., et al. (1988). Intracerebroventricular injection of brain natriuretic peptide inhibits vasopressin secretion in conscious rats. *Neuroscience Letters*, *95*, 223–228.
Yan, W., Wu, F., Morser, J., & Wu, Q. (2000). Corin, a transmembrane cardiac serine protease, acts as a pro-atrial natriuretic peptide-converting enzyme. *Proceedings of the National Academy of Sciences of the United States of America*, *97*, 8525–8529.
Yang-Feng, T. L., Floyd-Smith, G., Nemer, M., Drouin, J., & Francke, U. (1985). The pronatriodilatin gene is located on the distal short arm of human chromosome 1 and on mouse chromosome 4. *The American Journal of Human Genetics*, *37*, 1117–1128.

CHAPTER SIX

Role of Nitric Oxide in Stress-Induced Anxiety: From Pathophysiology to Therapeutic Target

A. Kumar[1], P. Chanana
University Institute of Pharmaceutical Sciences, UGC Centre of Advanced Study, Panjab University, Chandigarh, India
[1]Corresponding author: e-mail address: kumaruips@yahoo.com

Contents

Abstract

Stress is often marked by a state of hyperarousal to aid the initiation of necessary stress response for the successful management of stressful stimuli. It can be manifested as a challenge (stimulus) that requires behavioral, psychological, and physiological adaptations for the maintenance of a state of homeostasis in response to stressful stimuli. In an organism, miscellaneous stressors trigger a wide spectrum of alterations in hormonal and neuronal physiologies, resulting in behavioral (anxiety and depression disorders, diminished food intake and gastrointestinal dysfunctions, decline in sexual behavior, diabetes, and loss of cognitive function) and other physiological responses. Stress serves as a potent etiological link to development of several neuropsychiatric diseases such as depression, anxiety, and cognitive impairments. Exposure to stressful stimuli has been found to be associated with activation of nitric oxide synthase and generation of NO which reacts with spontaneous oxygen species to aid formation of active nitrogen radicals. High concentrations of reactive nitrogen radicals may cause damage to intracellular proteins, in addition to causing impairment to components of the mitochondrial transport chain, leading to cellular energy deficiency. This may further serve as an

Vitamins and Hormones, Volume 103
ISSN 0083-6729
http://dx.doi.org/10.1016/bs.vh.2016.09.004

etiological link to the development of secondary neurological diseases associated with chronic stress. Also, during stress exposure, pharmacological inhibition of nitric oxide production displays reduction in indicators of anxiety- and depressive-like behavior in animal models. Therefore, the purpose of this chapter is to present an overview on the role of NO in stress-evoked emergence of secondary neurological disorders like anxiety as well as citing examples where NO has been used as a therapeutic target for the management of stress-induced anxiety-like behavior.

1. INTRODUCTION

1.1 Stress

Stress is a common experience of daily life characterized by exposure to externally and internally generated stress causing stimuli against which all organisms maintain a homeostasis. Stress can be manifested as a challenge (stimulus) that requires behavioral, psychological, and physiological adaptations for the maintenance of a state of homeostasis in response to stressful stimuli. It is often marked by a state of hyperarousal for the initiation of necessary stress response for the successful management of stressful stimuli. The initiation and continuation of hyperaroused stress response is often controlled by two pivotal physiological systems, i.e., the hypothalamic–pituitary–adrenal (HPA) axis and the sympathoadrenal medullary (SAM) system; these components of stress response manifest their control via cardinal molecules like cortisol and norepinephrine (NE)/epinephrine (Negrao, Deuster, Gold, Singh, & Chrousos, 2000). A stressor can be defined as any stimulus, either internal or external, that innervates the HPA axis and the sympathetic nervous system, resulting in physiological changes that aid in coping up with the stressful conditions (Maier & Watkins, 1998). When stressors challenge an organism's integrity, a set of physiologic reactions involving the HPA axis, the SAM system, and the inflammatory cascades counteract the threat and adjust the organism to the new situation often termed as "the stress response." Stress can be classified as *systemic* or *physiologic*, where conscious interpretation of the stimulus is not required and the physical characteristics of the stimulus have enough innate capacity to initiate the responsive elements to stressful stimuli in the organism. The other type is the *processive* or *psychogenic* where quality of the stressful stimulus depends on the interpretation of the stimuli by subject, and therefore the physiologic impact as well as the response to the stressor varies among individuals.

In an organism, diverse stressors trigger a myriad of interacting hormonal and neuronal physiologies, resulting in behavioral (anxiety and depression disorders, depleted food intake and gastrointestinal dysfunctions, decrease in sexual behavior, diabetes, and loss of cognitive function) as well as physiological responses (innervation of pituitary–adrenal axis and release of glucocorticoids into the blood stream) (Henry & Stephens, 1977). Fig. 1 describes the diseases precipitated by chronic stress. The alterations in various physiologies in response to stressful stimuli as well as the generation of various molecular intermediates designate stress as an important precipitant variant of various chronic metabolic, endocrinological, immunomodulatory, as well as inflammatory diseases. Due to its high sensitivity to stress-induced degenerative conditions as well as the presence of large amounts of polyunsaturated fatty acids, brain serves as a vulnerable target for different stressors as well as free radical damage (Gutteridge, 1995; Şahin & Gümüşlü, 2007). Stress serves as a potent etiological link to the development of several neuropsychiatric diseases such as depression, anxiety, and cognitive impairments. The pathological molecular pathways converged by stressful stimuli-like HPA axis activation and release of corticosteroids, alteration of neurotransmitter release and synthesis, free radical generation, microglial

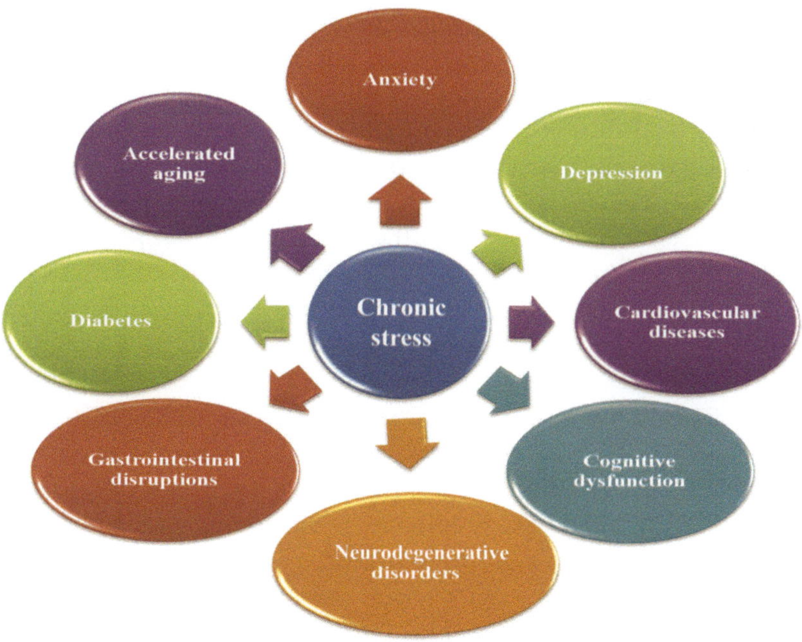

Fig. 1 Flowchart depicting secondary disorders associated with chronic stress.

activation as well as resultant neuronal damage have been found to be responsible for the interplay between stress and development of other neurobehavioral disorders (Connor & Leonard, 1998; Fukui, Onodera, Shinkai, Suzuki, & Urano, 2001; Mitra, Vyas, Chatterjee, & Chattarji, 2005; Rinwa & Kumar, 2014).

1.2 Anxiety as Response to Stress

According to previous studies stress induces anxiety-like behavior in both humans and animals (Liezmann, Klapp, & Peters, 2011). Different kinds of stress may it be chronic unpredictable, postnatal, prenatal, restraint, or even repetitive social defeat stress, all have shown to be precipitants of anxiety-like behavior. An important highlight in the field of stress research concerns delineating underlying mechanisms and pathophysiological pathways by which stress possibly triggers the onset of diseases like anxiety (Esch, Stefano, Fricchione, & Benson, 2002). The pathophysiological mechanisms involved in the induction or worsening of anxiety-like behavior with stress conditions can be attributed to HPA axis and sympathetic nervous system activation, thereby aiding the release of glucocorticosteroids and catecholamines and the induction of oxidative stress. The modulatory effects of nitric oxide synthase (NOS), neuronal activation characterized by elevated levels of c-Fos, direct and indirect microglial activation via induction of distinct patterns of monocyte/macrophage trafficking in specific brain areas controlling anxiety, and thus bringing out enhanced release of proinflammatory cytokines detrimental to neuronal cascades have all shown to be cardinal mechanisms in pathophysiology of stress-induced anxiety. Each of the precipitating mechanisms has been illustrated in detail as follows:

1. *The HPA axis activation*: Stress has been characterized by various physiologic changes in the body. Stress activates sympathoadrenomedullary system (SAS) and HPA axis, causing the release of catecholamines and stress hormones, namely, glucocorticoids and corticosterone from the adrenal gland (Dieleman et al., 2015; McCarty & Gold, 1996; Negrao et al., 2000; Tsigos & Chrousos, 2002). The paraventricular nucleus (PVN) of the hypothalamus is considered the culminating point of the HPA stress response. In response to stress, release of corticotropin-releasing hormone (CRH) from the parvocellular neurosecretory neurons occurs which stimulates the anterior pituitary to release adrenocorticotropic hormone (ACTH) into the circulation. This eventually sensitizes the adrenal gland to ACTH, resulting in corticosterone release

(Buijs et al., 1999; Lowry, 2002; Weiser, Osterlund, & Spencer, 2011). The stress-responsive systems are interconnected and in certain regions of the brain such as hippocampus and amygdala, the activation of the HPA axis is further facilitated by stress-induced NE release (Plotsky, 1987). It has been evident that PVN also facilitates direct release of corticosterone through the hypothalamic–spinal–adrenal stress axis as well as the SAS via involvement of catecholamines. This net corticosterone release and surge as a response to stress causes accumulation of pathological cellular and metabolic paradigms of stress, giving way to secondary neurological disorders.

Corticosteroids further act through specific mineralocorticoid and glucocorticoid receptors (GRs) localized in hippocampus and amygdala that are involved in the regulation of fear and anxiety-like behavior. The HPA axis activation and elevated release of corticosterone have also been found to induce oxidative stress conditions in specific brain regions controlling anxiety and depression. The accumulation of oxidative free radicals generated as a consequence of oxidative stress provides impetus to neuronal damage and thus may serve another mechanism of stress-induced anxiety (Zafir & Banu, 2009).

2. *Oxidative stress*: Encephalic regions have found to be vulnerable to oxidative free radicals, thereby oxidative stress contributes considerably toward neuronal degeneration in the central nervous system (Hovatta, Juhila, & Donner, 2010). The relative generation of reactive oxygen species (ROS) characterized by increased lipid peroxidation and nitrite levels is greatly augmented under conditions of chronic stress. Stress has been shown to be responsible for the depletion of several free radical scavenging antioxidants such as glutathione peroxidase, catalase, and superoxide dismutase (Matsumoto et al., 1999). Consequently a situation of oxidative burden is triggered, which has been implicated in stress as well as in the pathogenesis of several disease states. Since polyunsaturated fatty acids are found in abundance in encephalic regions and one of the important consequences of oxidative free radicals is peroxidation of membranal lipids, this reaction produces marked damage to the structure and function of neuronal membranes (Fontella et al., 2005). Therefore, lipid peroxidation was supposed as the major biochemical alteration and consequence of oxidant-induced neuronal injury, thereby giving way to development of neurobehavioral disorders like anxiety.
3. *Neuronal activation and c-Fos levels*: Exposure to stressful stimuli has been shown to cause neuronal activation often characterized by raised c-Fos

expression in specific brain regions controlling fear, emotion, and anxiety. c-Fos is an endogenous protein that acts as an indirect marker of neuronal activity, as its elevated levels are often linked with prolonged neuronal firing. c-Fos expression, as determined by immunohistochemical evaluations, was used as an indicator of regional neuronal activation (Wohleb et al., 2011) wherein discrete patterns of increased neuronal c-Fos expression within specific regions of the brain were observed, for example, in the prefrontal cortex (PFC), lateral septum (LS), bed nucleus of the stria terminalis (BNST), PVN, and medial amygdala (Wohleb et al., 2011). These are distinct encephalic regions associated with fear, anxiety, and threat appraisal. The stress-triggered neuronal activation and elevated c-Fos levels are under the direct control of β-adrenergic receptor activation. The same could be justified as proponalol when administered to stressed subjects was found to inhibit the corresponding neuronal activation in specific brain regions regulating anxiety-like behavior (Wohleb et al., 2011).

4. *Indirect and direct microglial activation and release of proinflammatory cytokines*: Stressful conditions trigger direct as well as indirect microglial activation and thus innervate release of detrimental proinflammatory cytokines. Neuronal activation as a consequence of stress is regionally associated with variations in microglial morphology. It causes changes like increased cell size, as well as thickening and shortening of cell processes, a process defined as deramification of microglia which thereby aids direct microglial activation (Wohleb et al., 2011). These activated microglia appear to be more inflammatory and phagocytic (Walker et al., 2014). In line with this observation, enriched microglia isolated from the brains of stressed mice have been found to be associated with increased proinflammatory cytokine gene expression as well as amplified expression of inflammatory markers (Tynan et al., 2010; Wohleb et al., 2011).

   Stressful stimuli are also associated with distinct patterns of monocyte/macrophage trafficking to the brain, and these macrophages further start behaving as activated microglia. Brain macrophages can be defined as peripherally derived monocytes that either accumulate in the perivascular space ($CD11b^{+}/CD45^{hi}$) or cross the blood–brain barrier into the brain parenchyma. Consequently under stressful conditions, the cluster of differentiated (CD45) expression is altered and these cells become indistinguishable from resident ramified and activated microglia both morphologically and phenotypically (Mildner et al., 2007).

In addition, these cells were also found to be associated with increased expression of inflammatory markers like Iba1 (Wohleb, Powell, Godbout, & Sheridan, 2013). The direct microglial activation and consequent triggering of neuroinflammatory responses can thereby precipitate conditions of stress-induced anxiety.

Like neuronal activation these macrophages getting differentiated into activated ramified microglial cells were also specifically observed in discrete brain regions associated with fear, anxiety, and threat appraisal including the PFC, PVN, LS, BNST, and amygdala. This plausibly outlines the etiological link between stress and anxiety-like behavior.

1.3 Nitric Oxide Mechanism in Stress-Induced Anxiety-Like Behavior

Nitric oxide (NO), a gaseous free radical belonging to the category of reactive nitrogen species (RNS), is synthesized through the transformation of L-arginine to L-citrulline by NOS in the presence of oxygen, NADPH (nicotinamide adenine dinucleotide phosphate), and pivotal cofactors such as tetrahydrobiopterin (Andrew & Mayer, 1999). There are three main isoforms of NOS, neuronal NOS (nNOS, type I), inducible NOS (iNOS, type II), and endothelial NOS (eNOS, type III) whereby each one has a discrete expression profile (Stuehr, 1999). Both nNOS and eNOS can be included under cNOS (constitutive NOS), i.e., calcium-dependent NOS, whereas iNOS is calcium independent. Though nNOS is principally active in the cytosol of central and peripheral neurons for the purpose of signaling and regulation, it has also been found in the sarcolemma and cytoplasm of all muscle fibers (Frandsen, Lopez-Figueroa, & Hellsten, 1996). Interestingly, nNOS is also found to be present in the hippocampus, hypothalamus, and pituitary and adrenal gland, signifying colocalization with the HPA axis and a role in pathogenesis of stress and related comorbidities (Gądek-Michalska, Tadeusz, Rachwalska, Spyrka, & Bugajski, 2012; Lai et al., 2005). Furthermore, transcriptional regulation of nNOS by glucocorticoids has also been evident in the hippocampus, implicating its importance in the stress response (Lopez-Figueroa, Itoi, & Watson, 1998; Reagan, McKittrick, & McEwen, 1999; Zhou et al., 2011).

It has been usually recognized that psychophysiological stress is associated with upregulation of NOS mRNA expression as well as increased enzymatic activity. The high amounts of NO thus generated interfere with various components and fundamental mechanisms of the stress response on different levels, thereby exerting detrimental effects to cellular components such as proteins, lipids, and DNA. This leads to precipitation of several stress-induced secondary neurological disorders like anxiety, depression, and cognitive deficits. The major points of highlight here seem to be delineating the mechanisms by which stress leads to NO surge as well as the pathophysiological pathways via which NO serves as a mediator for the development of secondary neurological disorders associated with stress.

1.4 Mechanisms Leading to Stress-Induced NO Surge

(i) *Direct activation of microglial iNOS*: One of the three isoforms of NOS is iNOS which unlike the cNOSs (eNOS and nNOS) is found to be calcium insensitive, likely due to its tight noncovalent interaction with calmodulin (CaM) and Ca^{2+}. iNOS is not expressed in resting cells and it produces NO only in response to a wide range of stimuli, mostly endotoxins and endogenous proinflammatory mediators (Nathan, 1992). The encephalic localization of the iNOS is attributed to glial cells. Psychophysiological stress is associated with extensive microglial activation brought about via the enhanced release of corticosterone, norepinephrine, and adrenaline (Sugama, 2009). The consecutive release of proinflammatory cytokines following microglial activation leads to direct activation of glial or astrocytic iNOS and thus aids enhanced NO synthesis from L-arginine. Once induced, iNOS facilitates extensive production of NO for prolonged periods (until substrate depletion) (Hickey, Granger, & Kubes, 2001). It has been evident that a single 6 h acute immobilization stress induces upregulation of iNOS expression and activity in the cerebral cortex. This iNOS expression surge is intermediated by the NMDA receptor and subsequent innervation of the transcriptional factor, nuclear factor kappa light chain enhancer of activated B cells (NF-κB) (Madrigal et al., 2001). The acute stress-induced activation of the NMDA receptor also increases levels of tumor necrosis factor-alpha (TNFα) via upregulation of TNFα-convertase. It has been demonstrated that by antagonizing TNFα-convertase, the stress-induced translocation of NF-κB and subsequent iNOS expression can be meticulously controlled. This further confirms the role of TNFα in iNOS expression surge under stressful conditions (Madrigal et al., 2002). Recently, researchers found a degree of encephalic regional specificity associated with this chronic stress-induced iNOS expression. For example, chronic unpredictable mild stress increases plasma nitrite levels and iNOS mRNA expression specifically in the cortex as well as hippocampus (Peng et al., 2012; Wang, An, & Zhang, 2008).

(ii) *Indirect as well as direct activation of nNOS via calcium modulation*: nNOS belongs to the constitutive NOS category, since it is a calcium-CaM-controlled isoenzyme. The relationship between nNOS release and components of HPA axis has been put forward by several researchers. nNOS is present in the hippocampus, hypothalamus, and pituitary and adrenal gland, suggesting its colocalization with the HPA axis (Gądek-Michalska et al., 2012; Lai et al., 2005). Furthermore, several studies have depicted the transcriptional regulation of nNOS by glucocorticoids in the hippocampus, implicating the role of nNOS in pathogenesis of stress and related comorbidities. Another hypothesis states that the activation of HPA axis with stressful stimuli leads to elevated release of corticosterone, and this corticosterone via direct action over postsynaptic mineralocorticoid receptors directly potentiates nNOS activity (Zhou et al., 2011). NO produced by the inducible isoform of nitric oxide synthase (iNOS) can act as a neurotransmitter on presynaptic neurons. This presynaptic NO causes glutamate release, thereby innervating postsynaptic NMDA receptors (NMDAR) to increase calcium (Ca^{2+}) concentration and in the presence of CaM, culminating nNOS release. Studies have also reflected that increased release of proinflammatory cytokines like TNF-α following microglial activation brings about an increase in quinolinic acid to kynurenic acid ratio as well as inhibition of excitatory amino acid (glutamate) removal via focal loss of astroglial EAAT1/2 (excitatory amino acid transporter) leading to a net NMDA agonism (Bal-Price & Brown, 2001; Rai, Kamat, Nath, & Shukla, 2013). It can be hypothesized from the above-mentioned reports that this net NMDA agonism via involvement of various pathways may activate not only release of proinflammatory cytokines but may also regulate nNOS activation. The above steps lead to enhanced nNOS release as well as activity and thus contribute to NO load in the brain as a consequence of stress.

Studies reflect that a single 2 h acute restraint stress significantly increases the density of neurons expressing nNOS in the amygdaloid nucleus, but for the neuronal density to increase to the same level in hippocampus and entorhinal cortex it took nearly 5 days (Echeverry, Guimaraes, & Del Bel, 2004). Significant increases in nNOS mRNA and activity have also been described in the anterior pituitary and adrenal gland (Kishimoto, Tsuchiya, Emson, & Nakayama, 1996) of rats subjected to neurogenic stress. This signifies that stress causes a localized NO surge in the brain areas targeted toward discrete areas that control fear, emotion, behavior, etc.

The NO surge produced as a consequence of stressful stimuli as per above-mentioned mechanisms is associated with elemental detrimental changes to neuronal physiologies via DNA damage, protein and lipids degradation, altered neurotransmitter release, as well as activation of direct apoptotic and neurodegenerative cascades. These mechanisms may be responsible for the precipitation of stress-induced anxiety behaviors. This can be supported by studies where it has been evident that infusion of iNOS as well as nNOS inhibitors in specific brain areas has been associated with attenuation of anxiety-like behaviors associated with stress. The ethanol withdrawal-induced anxiogenic-like behavior observed in rats is inhibited by administration of the selective iNOS inhibitor, 1400 W, into the dorsolateral periaqueductal gray region of the brain (Bonassoli, Contardi, Milani, & de Oliveira, 2013). The data with control of anxiety via NOS inhibition are further supported by studies using systemic treatment. Intraperitoneal injection of 1400 W has also shown to increase survival of cortical neurons and decrease the depressive-like behavior in mice (Peng et al., 2012). The administration of another nNOS inhibitor 1-(-2-trifluoromethylphenyl)-imidazole (TRIM) systemically 30 min prior to testing induces anxiolytic-like behavior as depicted by increased time spent in the light compartment of a light–dark compartment test (Volke, Wegener, Bourin, & Vasar, 2003).

The etiological link between NO surge and precipitation of anxiety-like behavior as a consequence of stress can be attributed to certain mechanisms like effect of NO on HPA axis, mitochondrial insufficiencies and consequent energy deficits, oxidative stress enhancement, as well as neurodegenerative responses associated with nitrosylation and nitrotyrosination. Each mechanism has been explained in detailed as follows:

1. *HPA axis activation*: The NO load observed as a response to stress has been shown to modify HPA axis signaling. The stress research was diverted to exploration of this pathway when it became evident through immunohistochemical studies that increased levels of nNOS mRNA and protein and enhanced nNOS activity are present in the PVNs of rats exposed to stressful conditions, such as forced swimming, immobilization, and endotoxin administration (Bonassoli et al., 2013; Kostoglou-Athanassiou et al., 1998; Lee, Barbanel, & Rivier, 1995; Sanchez, Moreno, Vacas, Carretero, & Vazquez, 1999; Tsuchiya, Kishimoto,

Koyama, & Ozawa, 1997). Interestingly, simultaneous exposure to neurogenic stress (e.g., restraint) and immunoinflammatory stress (e.g., administration of LPS) stimulated the NO system in the rat PVN to a greater extent than either stress alone (Yang, Oskin, & Krukoff, 1999).

NO has been found to directly aid HPA axis activation which has been depicted in several stress models. NO plays a stimulatory role on release of CRH from rat mediobasal hypothalamus, as well as ACTH release in response to IL-2- and LPS-induced stress–anxiety comorbid (Karanth, Lyson, & McCann, 1993). When the stress is neurogenic in origin, NO enhances stress axis activation by increasing the release of CRH and AVP from the hypothalamus and ACTH from the pituitary gland. This HPA axis activation further leads to enhanced glucocorticoid release from adrenal glands; these glucocorticosteroids thereby aid neuronal damage via elevating oxidative free radical levels, neuronal DNA damage and bringing alterations to dendritic structures, etc.

The exact mechanisms that lead to HPA axis activation and elevated release of CRH as a consequence of NO surge are still under investigation. Many hypotheses have been put forward till date. The first hypothesis underlines the importance of NO/sGC/cGMP system in the modulation of the HPA axis by NO. It was evident in a double-blind, placebo-controlled, crossover clinical trial, in which healthy male athletes were treated with tadalafil (a long-acting type-V phosphodiesterase inhibitor that increases cGMP concentrations), that exercise-induced activation of the HPA axis, assessed by measurement of salivary cortisol concentrations, was significantly potentiated by tadalafil (Di Luigi et al., 2008). This clearly depicts that one of the pathways by which NO causes potentiation of CRH release is via activation of NO/sGC/cGMP pathway. Another hypothesis states that elevated NO levels may be responsible for downregulation of hippocampal GRs. The downregulation of local GR expression via NO surge was accorded through either the soluble guanylyl cyclase/cGMP or peroxynitrite/extracellular signal-regulated kinase signaling pathways (Zhou et al., 2011). This subsequently leads to an increase in levels of hypothalamic CRH which ultimately through the enhanced corticosteroid release induces secondary disorders associated with stress.

2. *Oxidative stress, mitochondrial insufficiencies, and consequent energy deficits*: High levels of NO and its derivatives are destructive to cellular components such as proteins, lipids, and DNA. Nitric oxide can react directly with ROS and correspondingly arise to a situation of reactive nitrosative

stress. A primary reaction in the production of RNS is the combination of NO and superoxide anions to form the highly reactive metabolite, peroxynitrite ($ONOO^-$), a potent neurotoxin that may be detrimental to several cellular components as well as macromolecules (Lipton et al., 1993). It has been suggested that NO and peroxynitrite can impair almost all components of the mitochondrial respiratory chain and thus alter the production of adenosine triphosphate (Almeida & Bolaños, 2001; Sarti, Forte, Mastronicola, Giuffrè, & Arese, 2012). These RNS reversibly or irreversibly inhibit mitochondrial oxygen consumption, particularly at complex IV (also known as cytochrome *c* oxidase), which may lead to cellular energy insufficiency and ultimately cell death in pathological conditions (Sarti et al., 2012). Cytochrome *c* is the most vulnerable mitochondrial target for RNS, and it has also been evident that inhibition of cytochrome *c* oxidase by NO and peroxynitrite causes neuronal dysfunction as well.

In addition NO has also shown to modulate mitochondrial membrane potential in neuronal cells which leads to mitochondrial permeability transition pore (MPTP) opening. MPTP opening can be held responsible for the exchange of solutes and several small proteins between the mitochondrial matrix and the cytosol (Tatton & Olanow, 1999). These phenomena may lead to mitochondrial swelling, rupture of the outer mitochondrial membrane and activation of caspases, and thus initiation of neuronal apoptotic pathways.

3. *Activation of nitrosylation and nitrotyrosination responses*: S-nitrosylation is the process of covalent binding of NO to the thiol side chain of the amino acid cysteine, leading to the formation of another NO derivative called *S*-nitroso-protein. Under physiological conditions, NO S-nitrosylates the NMDA receptor, thereby averting glutamate excitotoxicity by blocking calcium influx, and thus assisting cell survival (Lipton & Stamler, 1994). Excessive production of NO and thus resultant abnormal S-nitrosylation of proteins such as apolipoprotein E, cyclin-dependent kinase 5, dynamin-related protein 1, parkin, peroxiredoxin 2, protein disulfide isomerase, heat shock protein 90, and X-linked inhibitor of apoptosis have all being associated with the development of neurodegenerative conditions, giving way to secondary neurological pathologies associated with stress (Anand & Stamler, 2012). Additionally, peroxynitrite also incites another process called protein nitrotyrosination. This process is an irreversible chemical transformation wherein addition of a nitro group to the tyrosine residue in target

proteins occurs, thereby generating 3-nitrotyrosine. This posttranslational alteration usually degrades standard physiological function of the proteins, thereby, serving as a potential indicator in several neurodegenerative conditions (Peluffo et al., 2004).

Fig. 2 clearly depicts the pathways leading to NO load during chronic stress as well as the pathophysiological role played by NO in accounting neuronal damage. Thus it can be concluded that a complex cascade of events involving the above-mentioned mechanisms may be held responsible for the etiological interplay between stress and NO load in specific encephalic regions and precipitation of secondary neurological disorders like anxiety.

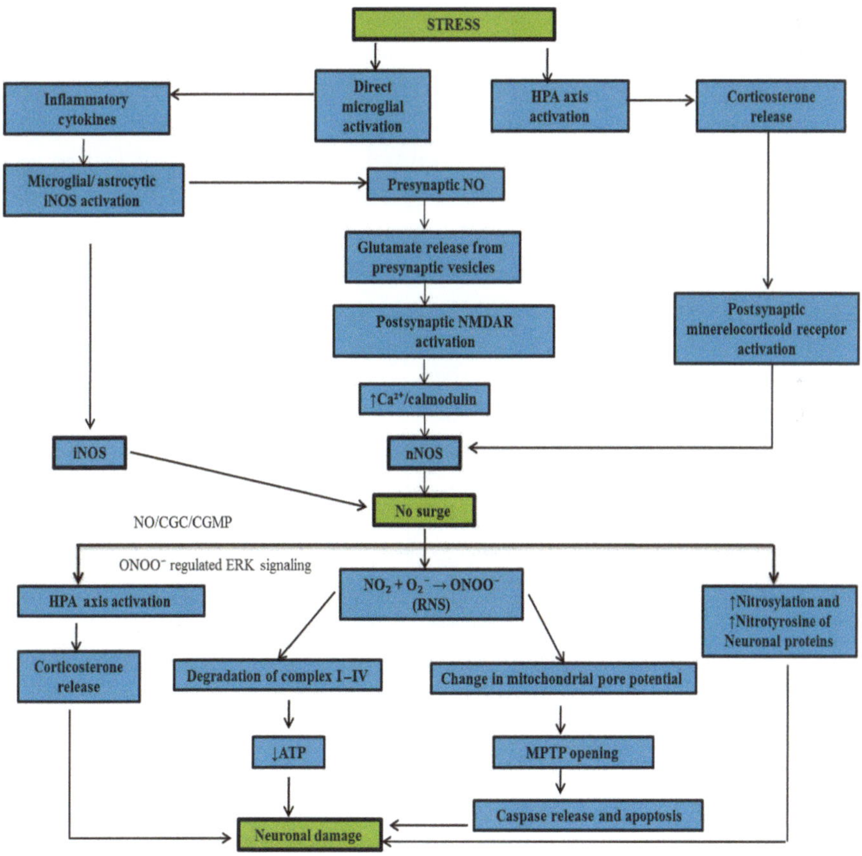

Fig. 2 Pathways leading to NO load during chronic stress as well as the pathophysiological role played by NO in accounting neuronal damage.

1.5 Therapeutics Targeting NO Mechanisms in Stress-Induced Anxiety-Like Behavior

As discussed previously NO plays a cardinal role in stress-induced secondary neurological disorders of which anxiety being a major consequence. A number of pathophysiological manifestations associated with stress-induced anxiety are either found to be initiated or worsened by NO signaling, for example, HPA axis activation, oxidative stress, mitochondrial impairment, and activation of neuroapoptotic or neurodegenerative pathways. Therefore research converged to the development of agents that potentially target the NO signaling pathways so as to manage comorbidities associated with stress at a basal level. Two major categories of agents developed in this regard are plant extracts or biogenic flavonoids and triglyceride saponins as well as the synthetically derived or chemical antidepressants and NOS modulators.

(a) *Biogenic therapeutics targeting NO mechanisms in management of stress-induced anxiety*: Plant extracts or herbal interventions are lately being utilized in the management of stress and thus neurological disorders related to it. *Withania somnifera* known as ashwagandha or Indian ginseng has been commonly used in Indian traditional medicines. When investigated in an animal model of restraint stress-induced anxiety-like behavior, the drug demonstrated dose-dependent antianxiety-like behavior as well as depletion in total encephalic nitrite levels. In addition authors depicted that the antianxiety effect of *W. somnifera* was comparable to that produced by nonspecific NOS inhibitor *N*-nitro-arginine methyl ester (L-NAME) as well as the anxiolytic effect was abolished with L-arginine pretreatment, confirming that this reported antianxiety profile of *W. somnifera* could be attributed to the inhibition of NOS (Khan & Ghosh, 2010, 2011). St. John's Wort (*Hypericum perforatum*) is a perennial herb, the major phytoconstituents of which are flavonoids. Kumar et al. explored the anxiolytic potential St. John's Wort in 6 h restraint stress-induced anxiety-like behavior (Kumar, Garg, & Prakash, 2010). It was evident from studies that in addition to anxiolytic properties the herbal intervention also exhibited antinociceptive, antioxidant, and memory consolidating properties against stress-induced hyperalgesia, oxidative damage, and spatial memory losses (Kumar, Garg, & Prakash, 2010; Prakash, Arulkumar, & Sabesan, 2010). Furthermore, one of the mechanisms of action of St. John's Wort has been reported to be inhibition of iNOS activity

(Tedeschi et al., 2003); therefore, it can be proposed that the plausible neuroprotective potential of St. John's Wort against stress-induced anxiety can also be attributed to modulation of NOS activity. The anti-anxiety and mitoprotective activity of some other flavonoids, like hesperidin and naringin against immobilization stress-induced anxiety-like behavior, was also found to be significantly regulated by NO modulators. It was evident that pretreatment with L-arginine and L-NAME potentially reversed and potentiated, respectively, the protective effect of both hesperidin and naringin against immobilization stress-induced anxiety-like behavior and mitochondrial impairments signifying NO pathways as plausible targets of the above-mentioned flavonoids (Viswanatha et al., 2011; Viswanatha, Shylaja, Sandeep Rao, Santhosh Kumar, & Jagadeesh, 2012). Similarly, *Panax quinquefolium* (glycosidic saponin) was also found to show its activity via NO modulatory pathways against chronic unpredictable stress-induced cognitive deficits (Rinwa & Kumar, 2014). The above-stated examples clearly depict that NO mechanisms portray a cardinal role in pathogenesis of stress-induced secondary neurological disorders and thereby it can be treated as a therapeutic target. Sleep deprivation has also been postulated as a type of stress, and an etiological link between sleep deprivation-induced stress and anxiety-like behavior has also been established. Another phytopolyphenol whose anxiolytic potential against 72-h sleep deprivation-induced anxiety-like behavior and oxidative damage was found to be mediated through NOS modulation is curcumin. Kumar and Singh demonstrated that the neuroprotective effect of curcumin against sleep deprivation-induced anxiety was potentiated and reversed by pretreatment with L-NAME and L-arginine, respectively (Kumar & Singh, 2008).

(b) *Antidepressants and synthetic drugs targeting NO mechanisms in management of stress-induced anxiety*: Many antidepressants and other synthetic moieties have also shown to target NO pathogenic pathways for their potential antianxiety and antioxidant effects against stressful stimuli. Antidepressants, for example, serotonin–norepinephrine reuptake inhibitors (venlafaxine), selective serotonin reuptake inhibitors (citalopram), serotonin antagonist and reuptake inhibitor (trazodone), and tricyclic antidepressants (imipramine), have demonstrated potential antianxiety properties against 6 h immobilized stress as well as sleep deprivation-induced anxiety-like behavior. In addition, these agents exhibited potent antioxidant profile by attenuating raised MDA levels (malondialdehyde) and nitrite levels, improving depleted GSH as well as antioxidant enzymes like catalase contents. Furthermore, nitric oxide mechanism in protective effect of imipramine and venlafaxine against acute immobilization stress-induced anxiety-like behavior was highlighted wherein the protective effect of above-mentioned antidepressants was found to be regulated by pretreatment with NOS inhibitors (L-NAME and methylene blue) as well as NO synthesis precursors (L-arginine) (Kumar & Garg, 2008; Kumar, Garg, Gaur, & Kumar, 2009; Kumar, Garg, Gaur, & Kumar, 2010). A plausible signaling pathway responsible for the antianxiety-like potential of antidepressants against stress-induced anxiety could be attributed to the potential activation of serotonin type IA receptors. Agonism of these receptors via an unknown mechanism mediates the downregulation of hippocampal nNOS, leading to a decrease in NO and subsequent inhibition of cAMP response element-binding protein phosphorylation (Zhang et al., 2010).

Stress-induced pathologies have also shown to be benefitted by direct administration of synthetic iNOS and nNOS inhibitors, for example, aminoguanidine as well as TRIM in stress-induced secondary neuropathologies. Aminoguanidine demonstrated a direct NO-cGMP pathway modification as antianxiety mechanism in stress-induced anxiety-like behavior. When combined with sildenafil (a cGMP inhibitor), the positive effect of aminoguanidine was further potentiated, signifying a NO regulatory potential (Gilhotra & Dhingra, 2009). Similarly, TRIM which is a selective iNOS and nNOS inhibitor, when evaluated in mice subjected to chronic unpredictable stress brought about beneficial effects which were found to be comparable to fluoxetine in terms of improvement in grooming behavior (Mutlu, Ulak, Laugeray, & Belzung, 2009). Therefore NO mechanistics have proved to be a potent therapeutic target for the management of stress-induced anxiety-like behavior.

2. SUMMARY

Stress has long been observed to play a role in the precipitation of neuronal pathologies and thus development of secondary disorders like anxiety, depression, cognitive impairments, and other mental disorders. A growing body of evidence suggests that nitric oxide serves a pivotal etiological link to the interplay between stress and anxiety. The nitrergic system has been implicated in regulating a number of physiological pathways, for example, HPA axis activation, oxidative stress, mitochondrial insufficiency, activation of neuroinflammatory responses, as a consequence of exposure to both short- and long-term stressful stimuli, all of which contribute to extensive damage to neurological physiologies and thus may induce precipitation of secondary neurological disorders as a direct consequence of stress. Therefore, stress, both physical and psychological, is attracting increasing attention among neuroresearchers. Targeting the same, NOS inhibitors demonstrating potent anxiolytic and antidepressant activity have been tried in stress-induced anxiety and depression. But still, the intrinsic cross talk between neuroendocrine stress and nitrergic system activation has become a topic of immediate research consideration and should be looked upon intricately in near future.

ACKNOWLEDGMENT

The authors would like to thank the University Grants Commission, New Delhi, for providing the financial support under the UGC-RFSMS [award file number: F.5-94/2007(BSR)] scheme for carrying out this research work.

REFERENCES

Almeida, A., & Bolaños, J. P. (2001). A transient inhibition of mitochondrial ATP synthesis by nitric oxide synthase activation triggered apoptosis in primary cortical neurons. *Journal of Neurochemistry*, 77(2), 676–690.

Anand, P., & Stamler, J. S. (2012). Enzymatic mechanisms regulating protein S-nitrosylation: Implications in health and disease. *Journal of Molecular Medicine*, *90*(3), 233–244.

Andrew, P. J., & Mayer, B. (1999). Enzymatic function of nitric oxide synthases. *Cardiovascular Research*, *43*(3), 521–531.

Bal-Price, A., & Brown, G. C. (2001). Inflammatory neurodegeneration mediated by nitric oxide from activated glia-inhibiting neuronal respiration, causing glutamate release and excitotoxicity. *The Journal of Neuroscience*, *21*(17), 6480–6491.

Bonassoli, V. T., Contardi, E. B., Milani, H., & de Oliveira, R. M. W. (2013). Effects of nitric oxide synthase inhibition in the dorsolateral periaqueductal gray matter on ethanol withdrawal-induced anxiety-like behavior in rats. *Psychopharmacology*, *228*(3), 487–498.

Buijs, R. M., Wortel, J., Van Heerikhuize, J. J., Feenstra, M. G., Ter Horst, G. J., Romijn, H. J., et al. (1999). Anatomical and functional demonstration of a multisynaptic suprachiasmatic nucleus adrenal (cortex) pathway. *European Journal of Neuroscience*, *11*(5), 1535–1544.

Connor, T. J., & Leonard, B. E. (1998). Depression, stress and immunological activation: The role of cytokines in depressive disorders. *Life Sciences*, *62*(7), 583–606.

Dieleman, G. C., Huizink, A. C., Tulen, J. H., Utens, E. M., Creemers, H. E., van der Ende, J., et al. (2015). Alterations in HPA-axis and autonomic nervous system functioning in childhood anxiety disorders point to a chronic stress hypothesis. *Psychoneuroendocrinology*, *51*, 135–150.

Di Luigi, L., Baldari, C., Sgro, P., Emerenziani, G. P., Gallotta, M. C., Bianchini, S., et al. (2008). The type 5 phosphodiesterase inhibitor tadalafil influences salivary cortisol, testosterone, and dehydroepiandrosterone sulphate responses to maximal exercise in healthy men. *The Journal of Clinical Endocrinology & Metabolism, 93*(9), 3510–3514.

Echeverry, M. B., Guimaraes, F. S., & Del Bel, E. A. (2004). Acute and delayed restraint stress-induced changes in nitric oxide producing neurons in limbic regions. *Neuroscience, 125*(4), 981–993.

Esch, T., Stefano, G. B., Fricchione, G. L., & Benson, H. (2002). The role of stress in neurodegenerative diseases and mental disorders. *Neuroendocrinology Letters, 23*(3), 199–208.

Fontella, F. U., Siqueira, I. R., Vasconcellos, A. P. S., Tabajara, A. S., Netto, C. A., & Dalmaz, C. (2005). Repeated restraint stress induces oxidative damage in rat hippocampus. *Neurochemical Research, 30*(1), 105–111.

Frandsen, U., Lopez-Figueroa, M., & Hellsten, Y. (1996). Localization of nitric oxide synthase in human skeletal muscle. *Biochemical and Biophysical Research Communications, 227*(1), 88–93.

Fukui, K., Onodera, K., Shinkai, T., Suzuki, S., & Urano, S. (2001). Impairment of learning and memory in rats caused by oxidative stress and aging, and changes in antioxidative defense systems. *Annals of the New York Academy of Sciences, 928*(1), 168–175.

Gądek-Michalska, A., Tadeusz, J., Rachwalska, P., Spyrka, J., & Bugajski, J. (2012). Effect of repeated restraint on homotypic stress-induced nitric oxide synthases expression in brain structures regulating HPA axis. *Pharmacological Reports, 64*(6), 1381–1390.

Gilhotra, N., & Dhingra, D. (2009). Involvement of NO–cGMP pathway in anti-anxiety effect of aminoguanidine in stressed mice. *Progress in Neuro-Psychopharmacology and Biological Psychiatry, 33*(8), 1502–1507.

Gutteridge, J. M. (1995). Lipid peroxidation and antioxidants as biomarkers of tissue damage. *Clinical Chemistry, 41*(12), 1819–1828.

Henry, J. P., & Stephens, P. M. (1977). *Stress, health, and the social environment: A sociobiologic approach to medicine* (pp. 245–263). New York: Springer-Verlag.

Hickey, M. J., Granger, D. N., & Kubes, P. (2001). Inducible nitric oxide synthase (iNOS) and regulation of leucocyte/endothelial cell interactions: Studies in iNOS-deficient mice. *Acta Physiologica Scandinavica, 173*(1), 119–126.

Hovatta, I., Juhila, J., & Donner, J. (2010). Oxidative stress in anxiety and comorbid disorders. *Neuroscience Research, 68*(4), 261–275.

Karanth, S., Lyson, K., & McCann, S. M. (1993). Role of nitric oxide in interleukin 2-induced corticotropin-releasing factor release from incubated hypothalami. *Proceedings of the National Academy of Sciences of the United States of America, 90*(8), 3383–3387.

Khan, Z. A., & Ghosh, A. R. (2010). Possible nitric oxide modulation in protective effects of withaferin A against stress induced neurobehavioural changes. *Journal of Medicinal Plants Research, 4*(6), 490–495.

Khan, Z. A., & Ghosh, A. R. (2011). L-arginine abolishes the anxiolytic-like effect of withaferin A in the elevated plus-maze test in rats. *African Journal of Pharmacy and Pharmacology, 5*(2), 234–237.

Kishimoto, J., Tsuchiya, T., Emson, P. C., & Nakayama, Y. (1996). Immobilization-induced stress activates neuronal nitric oxide synthase (nNOS) mRNA and protein in hypothalamic-pituitary-adrenal axis in rats. *Brain Research, 720*(1), 159–171.

Kostoglou-Athanassiou, I., Costa, A., Navarra, P., Nappi, G., Forsling, M. L., & Grossman, A. B. (1998). Endotoxin stimulates an endogenous pathway regulating corticotropin-releasing hormone and vasopressin release involving the generation of nitric oxide and carbon monoxide. *Journal of Neuroimmunology, 86*(1), 104–109.

Kumar, A., & Garg, R. (2008). A role of nitric oxide mechanism involved in the protective effects of venlafaxine in sleep deprivation. *Behavioural Brain Research, 194*(2), 169–173.

Kumar, A., Garg, R., Gaur, V., & Kumar, P. (2009). Nitric oxide mechanism in protective effect of imipramine and venlafaxine against acute immobilization stress-induced behavioral and biochemical alteration in mice. *Neuroscience Letters*, *467*(2), 72–75.

Kumar, A., Garg, R., Gaur, V., & Kumar, P. (2010). Possible role of NO modulators in protective effect of trazodone and citalopram (antidepressants) in acute immobilization stress in mice. *Indian Journal of Experimental Biology*, *48*, 1131–1135.

Kumar, A., Garg, R., & Prakash, A. K. (2010). Effect of St. John's Wort (Hypericum perforatum) treatment on restraint stress-induced behavioral and biochemical alteration in mice. *BMC Complementary and Alternative Medicine*, *10*(1), 18.

Kumar, A., & Singh, A. (2008). Possible nitric oxide modulation in protective effect of (Curcuma longa, Zingiberaceae) against sleep deprivation-induced behavioral alterations and oxidative damage in mice. *Phytomedicine*, *15*(8), 577–586.

Lai, F. J., Huang, S. S., Hsieh, M. C., Hsin, S. C., Wu, C. H., Hsin, Y. C., et al. (2005). Upregulation of neuronal nitric oxide synthase mRNA and protein in adrenal medulla of water-deprived rats. *Journal of Histochemistry & Cytochemistry*, *53*(1), 45–53.

Lee, S., Barbanel, G., & Rivier, C. (1995). Systemic endotoxin increases steady-state gene expression of hypothalamic nitric oxide synthase: Comparison with corticotropin-releasing factor and vasopressin gene transcripts. *Brain Research*, *705*(1), 136–148.

Liezmann, C., Klapp, B., & Peters, E. (2011). Stress, atopy and allergy: A re-evaluation from a psychoneuroimmunologic perspective. *Dermatoendocrinology*, *3*(1), 37–40.

Lipton, S. A., Choi, Y. B., Pan, Z. H., Lei, S. Z., Chen, H. S. V., Sucher, N. J., et al. (1993). A redox-based mechanism for the neuroprotective and neurodestructive effects of nitric oxide and related nitroso-compounds. *Nature*, *364*, 626–632.

Lipton, S. A., & Stamler, J. S. (1994). Actions of redox-related congeners of nitric oxide at the NMDA receptor. *Neuropharmacology*, *33*(11), 1229–1233.

Lopez-Figueroa, M. O., Itoi, K., & Watson, S. J. (1998). Regulation of nitric oxide synthase messenger RNA expression in the rat hippocampus by glucocorticoids. *Neuroscience*, *87*(2), 439–446.

Lowry, C. A. (2002). Functional subsets of serotonergic neurones: Implications for control of the hypothalamic-pituitary-adrenal axis. *Journal of Neuroendocrinology*, *14*(11), 911–923.

Madrigal, J. L., Hurtado, O., Moro, M. A., Lizasoain, I., Lorenzo, P., Castrillo, A., et al. (2002). The increase in TNF-α levels is implicated in NF-kB activation and inducible nitric oxide synthase expression in brain cortex after immobilization stress. *Neuropsychopharmacology*, *26*, 155–163.

Madrigal, J. L., Moro, M. A., Lizasoain, I., Lorenzo, P., Castrillo, A., Boscá, L., et al. (2001). Inducible nitric oxide synthase expression in brain cortex after acute restraint stress is regulated by nuclear factor kB-mediated mechanisms. *Journal of Neurochemistry*, *76*(2), 532–538.

Maier, S. F., & Watkins, L. R. (1998). Cytokines for psychologists: Implications of bidirectional immune-to-brain communication for understanding behavior, mood, and cognition. *Psychological Reviews*, *105*, 83–107.

Matsumoto, K., Yobimoto, K., Huong, N. T. T., Abdel-Fattah, M., Van Hien, T., & Watanabe, H. (1999). Psychological stress-induced enhancement of brain lipid peroxidation via nitric oxide systems and its modulation by anxiolytic and anxiogenic drugs in mice. *Brain Research*, *839*(1), 74–84.

McCarty, R., & Gold, P. (1996). Catecholamines, stress, and disease: A psychobiological perspective. *Psychosomatic Medicine*, *58*, 590–597.

Mildner, A., Schmidt, H., Nitsche, M., Merkler, D., Hanisch, U. K., Mack, M., et al. (2007). Microglia in the adult brain arise from Ly-6ChiCCR2+ monocytes only under defined host conditions. *Nature Neuroscience*, *10*(12), 1544–1553.

Mitra, R., Vyas, A., Chatterjee, G., & Chattarji, S. (2005). Chronic-stress induced modulation of different states of anxiety-like behavior in female rats. *Neuroscience Letters*, *383*(3), 278–283.

Mutlu, O., Ulak, G., Laugeray, A., & Belzung, C. (2009). Effects of neuronal and inducible NOS inhibitor 1-[2-(trifluoromethyl) phenyl] imidazole (TRIM) in unpredictable chronic mild stress procedure in mice. *Pharmacology, Biochemistry, and Behavior*, *92*(1), 82–87.

Nathan, C. (1992). Nitric oxide as a secretory product of mammalian cells. *The FASEB Journal*, *6*(12), 3051–3064.

Negrao, A. B., Deuster, P. A., Gold, P. W., Singh, A., & Chrousos, G. P. (2000). Individual reactivity and physiology of the stress response. *Biomedicine and Pharmacotherapy*, *54*, 122–128.

Peluffo, H., Shacka, J. J., Ricart, K., Bisig, C. G., Martìnez-Palma, L., Pritsch, O., et al. (2004). Induction of motor neuron apoptosis by free 3-nitro-l-tyrosine. *Journal of Neurochemistry*, *89*(3), 602–612.

Peng, Y. L., Liu, Y. N., Liu, L., Wang, X., Jiang, C. L., & Wang, Y. X. (2012). Inducible nitric oxide synthase is involved in the modulation of depressive behaviors induced by unpredictable chronic mild stress. *Journal of Neuroinflammation*, *9*, 75.

Plotsky, P. M. (1987). Facilitation of immunoreactive corticotropin-releasing factor secretion into the hypophysial-portal circulation after activation of catecholaminergic pathways or central norepinephrine injection. *Endocrinology*, *121*(3), 924–930.

Prakash, D. J., Arulkumar, S., & Sabesan, M. (2010). Effect of nanohypericum (Hypericum perforatum gold nanoparticles) treatment on restraint stress induced behavioral and biochemical alteration in male albino mice. *Pharmacognosy Research*, *2*(6), 330.

Rai, S., Kamat, P. K., Nath, C., & Shukla, R. (2013). A study on neuroinflammation and NMDA receptor function in STZ (ICV) induced memory impaired rats. *Journal of Neuroimmunology*, *254*(1), 1–9.

Reagan, L. P., McKittrick, C. R., & McEwen, B. S. (1999). Corticosterone and phenytoin reduce neuronal nitric oxide synthase messenger RNA expression in rat hippocampus. *Neuroscience*, *91*(1), 211–219.

Rinwa, P., & Kumar, A. (2014). Modulation of nitrergic signalling pathway by American ginseng attenuates chronic unpredictable stress-induced cognitive impairment, neuroinflammation, and biochemical alterations. *Naunyn-Schmiedeberg's Archives of Pharmacology*, *387*(2), 129–141.

Şahin, E., & Gümüşlü, S. (2007). Immobilization stress in rat tissues: Alterations in protein oxidation, lipid peroxidation and antioxidant defense system. *Comparative Biochemistry and Physiology, Part C: Toxicology & Pharmacology*, *144*(4), 342–347.

Sanchez, F., Moreno, M. N., Vacas, P., Carretero, J., & Vazquez, R. (1999). Swim stress enhances the NADPH–diaphorase histochemical staining in the paraventricular nucleus of the hypothalamus. *Brain Research*, *828*(1), 159–162.

Sarti, P., Forte, E., Mastronicola, D., Giuffrè, A., & Arese, M. (2012). Cytochrome c oxidase and nitric oxide in action: Molecular mechanisms and pathophysiological implications. *Biochimica et Biophysica Acta (BBA)—Bioenergetics*, *1817*(4), 610–619.

Stuehr, D. J. (1999). Mammalian nitric oxide synthases. *Biochimica et Biophysica Acta (BBA)—Bioenergetics*, *1411*(2), 217–230.

Sugama, S. (2009). Stress-induced microglial activation may facilitate the progression of neurodegenerative disorders. *Medical Hypotheses*, *73*(6), 1031–1034.

Tatton, W. G., & Olanow, C. W. (1999). Apoptosis in neurodegenerative diseases: The role of mitochondria. *Biochimica et Biophysica Acta (BBA)—Bioenergetics*, *1410*(2), 195–213.

Tedeschi, E., Menegazzi, M., Margotto, D., Suzuki, H., Förstermann, U., & Kleinert, H. (2003). Anti-inflammatory actions of St. John's wort: Inhibition of human inducible nitric-oxide synthase expression by down-regulating signal transducer and activator of transcription-1α (STAT-1α) activation. *Journal of Pharmacology and Experimental Therapeutics*, *307*(1), 254–261.

Tsigos, C., & Chrousos, G. P. (2002). Hypothalamic–pituitary–adrenal axis, neuroendocrine factors and stress. *Journal of Psychosomatic Research*, *53*(4), 865–871.

Tsuchiya, T., Kishimoto, J., Koyama, J., & Ozawa, T. (1997). Modulatory effect of L-NAME, a specific nitric oxide synthase (NOS) inhibitor, on stress-induced changes in plasma adrenocorticotropic hormone (ACTH) and corticosterone levels in rats: Physiological significance of stress-induced NOS activation in hypothalamic–pituitary–adrenal axis. *Brain Research*, *776*(1), 68–74.

Tynan, R. J., Naicker, S., Hinwood, M., Nalivaiko, E., Buller, K. M., Pow, D. V., et al. (2010). Chronic stress alters the density and morphology of microglia in a subset of stress-responsive brain regions. *Brain, Behavior, and Immunity*, *24*(7), 1058–1068.

Viswanatha, G. L., Shylaja, H., Rao, K. S., Ashwini, Y., Kumar, V. R., Mohan, C. G., et al. (2011). Amelioration of immobilization stress-induced biochemical and behavioral alterations and mitochondrial dysfunction by naringin in mice: Possible mechanism of nitric oxide modulation. *Journal of Chinese Integrative Medicine*, *9*(11), 1254–1263.

Viswanatha, G. L., Shylaja, H., Sandeep Rao, K. S., Santhosh Kumar, V. R., & Jagadeesh, M. (2012). Hesperidin ameliorates immobilization-stress-induced behavioral and biochemical alterations and mitochondrial dysfunction in mice by modulating nitrergic pathway. *ISRN Pharmacology*, *2012*, 479570.

Volke, V., Wegener, G., Bourin, M., & Vasar, E. (2003). Antidepressant-and anxiolytic-like effects of selective neuronal NOS inhibitor 1-(2-trifluoromethylphenyl)-imidazole in mice. *Behavioural Brain Research*, *140*(1), 141–147.

Walker, F. R., Beynon, S. B., Jones, K. A., Zhao, Z., Kongsui, R., Cairns, M., et al. (2014). Dynamic structural remodelling of microglia in health and disease: A review of the models, the signals and the mechanisms. *Brain, Behavior, and Immunity*, *37*, 1–14.

Wang, D., An, S. C., & Zhang, X. (2008). Prevention of chronic stress-induced depression-like behavior by inducible nitric oxide inhibitor. *Neuroscience Letters*, *433*(1), 59–64.

Weiser, M. J., Osterlund, C., & Spencer, R. L. (2011). Inhibitory effects of corticosterone in the hypothalamic paraventricular nucleus (PVN) on stress-induced adrenocorticotrophic hormone secretion and gene expression in the PVN and anterior pituitary. *Journal of Neuroendocrinology*, *23*(12), 1231–1240.

Wohleb, E. S., Hanke, M. L., Corona, A. W., Powell, N. D., La'Tonia, M. S., Bailey, M. T., et al. (2011). β-Adrenergic receptor antagonism prevents anxiety-like behavior and microglial reactivity induced by repeated social defeat. *The Journal of Neuroscience*, *31*(17), 6277–6288.

Wohleb, E. S., Powell, N. D., Godbout, J. P., & Sheridan, J. F. (2013). Stress-induced recruitment of bone marrow-derived monocytes to the brain promotes anxiety-like behavior. *The Journal of Neuroscience*, *33*(34), 13820–13833.

Yang, W., Oskin, O., & Krukoff, T. L. (1999). Immune stress activates putative nitric oxide-producing neurons in rat brain: Cumulative effects with restraint. *Journal of Comparative Neurology*, *405*(3), 380–387.

Zafir, A., & Banu, N. (2009). Induction of oxidative stress by restraint stress and corticosterone treatments in rats. *Indian Journal of Biochemistry & Biophysics*, *46*(1), 53.

Zhang, J., Huang, X. Y., Ye, M. L., Luo, C. X., Wu, H. Y., Hu, Y., et al. (2010). Neuronal nitric oxide synthase alteration accounts for the role of 5-HT1A receptor in modulating anxiety-related behaviors. *The Journal of Neuroscience*, *30*(7), 2433–2441.

Zhou, Q. G., Zhu, L. J., Chen, C., Wu, H. Y., Luo, C. X., Chang, L., et al. (2011). Hippocampal neuronal nitric oxide synthase mediates the stress-related depressive behaviors of glucocorticoids by downregulating glucocorticoid receptor. *The Journal of Neuroscience*, *31*, 7579–7590.

CHAPTER SEVEN

Nitric Oxide and Anxiety

K. Gulati[1], N. Rai, A. Ray
Vallabhbhai Patel Chest Institute, University of Delhi, Delhi, India
[1]Corresponding author: e-mail address: kavgul2002@yahoo.com

Contents

Abstract

Anxiety is a common neuropsychiatric disorder which affects both physical and mental health. Complex neurobiological mechanisms are involved in the genesis of anxiety, and the drugs used to date, though effective, are not free from shortcomings. Conventional agents like the classical benzodiazepines and the atypical nonbenzodiazepine agents like buspirone have their own limitations. There is thus need to explore newer neurochemical pathways to develop efficacious and safer drugs for the disorder. Nitric oxide (NO) is a unique neuromodulator substance, with the ability to influence and modulate several other conventional messengers which play an important role in anxiety. The currently available experimental and clinical data indicate that NO may be involved in the regulation of anxiety-like behavior induced by a variety of stimuli. These studies have explored the pharmacological and biochemical basis of nitrergic mechanisms in anxiety, and the data available are equivocal. This chapter reviews the research data available in this specific area and suggests that in view of the nature of the existing data, there is considerable scope for future research in this field.

1. INTRODUCTION

Anxiety disorders are among the most frequent psychiatric conditions in humans and cause loss of quality of life. Anxiety disorders represent a huge burden in terms of both their social impact and their economic cost

Vitamins and Hormones, Volume 103
ISSN 0083-6729
http://dx.doi.org/10.1016/bs.vh.2016.09.001

(Cryan & Sweeney, 2011). Anxiety is accompanied by emotional, cognitive, behavioral, and physical components. These disorders are commonly associated with persistent and uncontrollable nervousness, stress, and worry that are triggered by anticipation of future events and memories of past events (Chansky, 2004). The anxiety response is an important mechanism by which we adapt and respond to real dangers. Dysregulation of this healthy response results in "marked, persistent, and excessive or unreasonable fear," culminating in a significant interference in normal life. Anxiety is often accompanied by autonomic symptoms, such as headache, perspiration, palpitations, tightness in the chest, and mild stomach discomfort (Sadock & Sadock, 2007).

At a physiological level, anxiety is an adaptive emotional reaction to both innate and conditioned stimuli and helps the survival of species by warning the individual of potential dangers through a series of neurochemical, neuroendocrine, and behavioral responses. However, when these reactions become constant and intense on a chronic basis, with inadequate defensive responses, it may lead to pathological anxiety (Marques et al., 2016).

Anxiety disorders could be commonly categorized as: generalized anxiety disorder, panic disorder, social phobia, obsessive–compulsive disorder, and posttraumatic stress disorder (PTSD). Anxiety may also exist with other comorbid psychiatric conditions like depression as well as non-neuropsychiatric disorders, viz., gastrointestinal disease, bronchial asthma, cardiovascular disease, migraine, cancer, and chronic pain (Harter, Conway, & Merikangas, 2003; Kessler, Chiu, Demler, Merikangas, & Walters, 2005; Merikangas, 2003; Roy-Byrne et al., 2008). Anxiety could also be classified as "trait" and "state" anxiety—the former being related to interindividual differences in anxiety proneness wherein the temperament and personality decides the susceptibility or resistance to anxiety, and the latter being a transitory emotional state characterized by subjective, psychological, and autonomic dysfunction. Irrespective of the type and nature, it is well documented that anxiety results from stress—both emotional or environmental or a combination of the two, and both central and peripheral mechanisms/manifestations are possible. For example, exposure to stress during surgery upregulates proinflammatory cytokines which are also involved in mood and memory disturbances (Shapira-Lichter et al., 2008). On the other hand, preoperative waiting before cardiac surgery increases plasma IL-6 levels (Ai, Kronfol, Seymour, & Bolling, 2005).

At the neurobiological level, the basis of anxiety can be conceived of as a disruption in the fundamental mechanism of fight and flight responses associated with stressful situations and regulated by the hypothalamic–pituitary–adrenal (HPA) axis. Stress, fear, and anxiety, therefore, involve brain structures participating in the regulation of the HPA axis, viz., limbic structures such as the amygdala, the hypothalamus, the periaqueductal gray, and the hippocampus.

Anxiety disorders are characterized by a variety of neuroendocrine, neurotransmitter, and neuroanatomical disruptions. The limbic system via its interconnections with lower brain stem and higher cortical areas regulates the stress- and fear-related anxiety responses. In addition to emotional factors, environmental factors may also influence the functioning of the limbic system and its connections and increase the risk for such psychopathology. Complex neurotransmitter systems in the brain regulate the genesis of the anxiety response. Biogenic amines (catecholamines, serotonin, histamine), amino acids (GABA, glutamate), and neuropeptides may interact at different levels in the brain to influence anxiety-related responses. Increased activity in emotion-processing brain regions in patients who have an anxiety disorder could result from decreased inhibitory signaling by γ-amino butyric acid (GABA) or increased excitatory neurotransmission by glutamate. The regulatory role of other neurotransmitters, viz., norepinephrine (NE), dopamine (DA), and serotonin (5-HT), and their interactions has also been reported during pathological anxiety. In the central nervous system, classic neurotransmitters are colocalized and coreleased with neuropeptides, many of which are expressed in limbic regions where they can influence stress and emotion circuitry, and their functional significance has been suggested. Neuropeptides with particularly strong links to psychopathology include cholecystokinin (CCK), galanin, neuropeptide Y, vasopressin (AVP), oxytocin, and corticotropin-releasing factor (CRF), among others. Well-documented antianxiety agents are known to target the NE, DA, and 5-HT systems in the brain, and newer agents exploring the interactions with these biogenic agents with other neurotransmitters are being explored. CRF and interactions with AVP and NE have been suggested in the limbic brain with influence on stress- and anxiety-related responses (Marques et al., 2016; Martin, Ressler, Binder, & Nemeroff, 2009).

Anxiety has become an important area of research interest in neuropsychopharmacology. However, in spite of the several preclinical and clinical studies in the area, the causative factors, etiology, and underlying mechanisms of anxiety disorders, as for most psychiatric disorders, remain relatively poorly understood. Further, though several drugs are available for treating anxiety acting through different mechanisms, the pharmacological treatment of this neuropsychiatric condition is still far from satisfactory and there is constant look out for newer mechanisms and strategies for treatment.

2. NITRIC OXIDE

Nitric oxide (NO) discovered by John Dalton is a 32-Da small, free radical molecule. Its reactivity is due to the unpaired electron in the outer valence orbital of its oxygen constituent. In the atmosphere, it is formed from oxygen and nitrogen during lighting and thunderstorms. Initially it was recognized as a constituent of cigarette smoke and a common air pollutant which causes acid rain and destroys the ozone layer. Now it is stated as the most important signaling molecule in mammalian physiology with complex actions. It is a short-lived, ubiquitous, chemical transmitter which decays within 6–10 s after its synthesis and quickly gets turned into nitrates and nitrites, or gets iron bound in hemoglobin and enzymes.

In 1992, NO was described as the "Molecule of the Year" by *Science*. In 1998, the importance of the nitric oxide as a signaling molecule in the cardiovascular system was realized by the valuable discoveries of the three US scientists, Robert F. Furchgott, Louis J. Ignarro, and Ferid Murad, and they were awarded a Nobel Prize in Physiology and Medicine. Murad and his coworkers found that nitrates release nitric oxide which stimulate soluble guanylate cyclase (sGC) causing an increase in cGMP, which produces smooth muscle cells relaxation resulting in vasodilation (Arnold, Mittal, Katsuki, & Murad, 1977). He was the first to propose that the colorless, odorless gas could function as a signaling molecule. Then in 1980, Furchgott identified nitric oxide as a biological agent and published that endothelial cells produce an unknown signaling molecule that induce relaxation of vascular smooth muscle cells. He named the molecule as endothelium-derived relaxing factor (EDRF) (Furchgott & Zawadzki, 1980). In 1986, Ignarro and coworkers through a series of analyses went on to conclude that EDRF from the artery and vein was either NO or a related compound identical to nitric oxide (SoRelle, 1998). As a result of these path-breaking discoveries, we now know that NO regulates a wide variety of physiological and pathological processes especially in the cardiovascular, immune, and nervous systems (Fig. 1).

2.1 Biosynthesis of Nitric Oxide

NO is produced from L-arginine by five electron oxidation to nitric oxide and L-citrulline, and this reaction is catalyzed by nitric oxide synthase (NOS) enzyme. This enzyme simultaneously requires several cofactors, viz.,

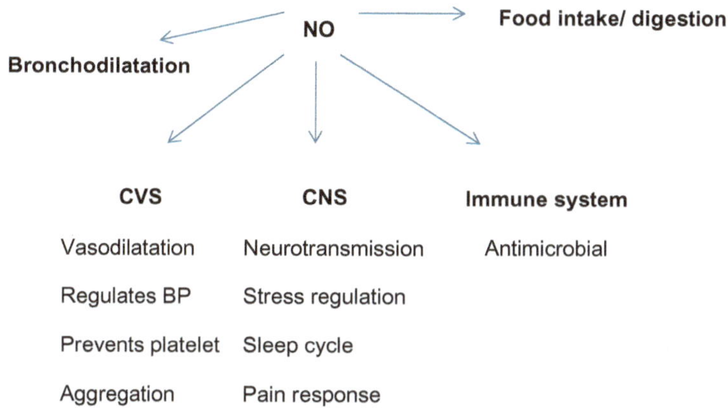

Fig. 1 A schematic representation of the physiological role of NO.

tetrahydrobiopterin (BH_4), nicotinamide adenine dinucleotide phosphate (NADPH), flavin adenine dinucleotide (FAD), flavin mononucleotide (FMN), heme, and Ca^{2+}-calmodulin, for its proper function, and hence NOS is one of the most regulated enzymes in biology (Xia & Zweier, 1997). The NOS enzyme is a homodimer, and each monomer is comprised of a heme and tetrahydrobiopterin-containing "oxygenase domain" at its amino terminal end and "reductase domain" at its carboxy terminal end (Liu & Gross, 1996; Xia & Bredt, 1996). The oxygenase domain binds and oxidizes L-arginine, whereas the reductase domain containing FAD and FMN facilitates the transfer of electrons from NADPH to heme (Abu-Soud, Yoho, & Stuehr, 1994). The three different isoforms of NOS commonly are referred to as: NOS-I (nNOS), NOS-II (iNOS), and NOS-III (eNOS) based on the order in which they were first discovered or identified in human beings and other organisms. All the NOSs share between them 50–60% sequence homology (Stuehr, 1999). The two constitutive isoforms, initially identified in neurons (nNOS) and vascular endothelial cells (eNOS), are regulated by calcium and calmodulin. The third isoform is an inducible, transcriptionally regulated isoform (iNOS), which was initially identified in macrophages but can be expressed in response to inflammatory cytokines and endotoxins in number of other cells (Hobbs, Fukuto, & Ignarro, 1994). Both endothelial NOS (eNOS) and neuronal NOS (nNOS) are constitutively expressed in mammalian cells, and they are well characterized in the cardiovascular system and nervous system, respectively. Nitric oxide derived from iNOS plays both regulatory and cytotoxic roles in the immune response. However, it has been shown that all the three NOSs are existent in CVS and CNS, and there is overlapping of their functions. NO once synthesized has a half-life of a few seconds only, and due to both hydrophilic and lipophilic nature diffuses readily through cytoplasm and plasma membranes. It acts within cells where produced and also adjacent cells, thus having auto- and paracrine functions. Thus, an innocuous gas, consisting of one molecule of oxygen and one molecule of nitrogen, with the unique ability of being formed in one cell and diffusing into another cell to exert an action, with immense translational significance, was indeed a new concept. Later, with the discovery of the pathophysiological significance of carbon monoxide (CO) and hydrogen sulfide (H_2S), the three (i.e., NO, CO, and H_2S) collectively were labeled as gasotransmitters.

2.2 NO and Signal Transduction Mechanism

The role of NO in cellular signaling has become one of the most rapidly growing areas in biology in the past three decades. Nitric oxide plays a pivotal role as a bioregulator through both cGMP-dependent and cGMP-independent mechanisms (Lucas et al., 2000). In health, NO diffuses back to the presynaptic terminal, increases cGMP levels through the activation of sGC, and regulates vascular tone (Ignarro, 2002; Moncada, Palmer, & Higgs, 1991). NO also signals through ion channels including sodium, voltage-gated calcium, calcium-activated and ATP-sensitive potassium, and cyclic nucleotide-gated channels, as well as AMPA receptors to modulate synaptic strength and intrinsic postsynaptic neuronal excitability. NO can also directly modify nuclear transcription factors and can profoundly affect gene expression (Nott & Riccio, 2009). Exposure to excess reactive nitrogen species (RNS) produced in response to iNOS induction results in the proinflammatory response and is a common feature of several neurodegenerative disorders.

However, other important effects of NO such as cytoprotection and antiadhesion appear to occur independent of cGMP signaling. Likewise, $NO^{\bullet}$ regulation of inflammation has frequently been associated with signal transduction events that do not involve the second messenger. Cyclic GMP production and phosphorylation of protein via activation of soluble guanyl cyclase is the main signaling mechanism of NO (Pfeilschifter, Eberhardt, & Huwiler, 2003). In the neuronal, cardiovascular, gastrointestinal, and other systems, NO mediates its biological effects through guanylate cyclase (GC) and stimulates cGMP synthesis. GC is a heterodimeric NO receptor and composed of two subunits, α and β. To date, four isoforms of sGC have been distinguished, i.e., α1, α2, β1, and β2. Only two heterodimers α1/β1 and α2/β1 are activated by NO (Russwurm, Behrends, Harteneck, & Koesling, 1998). Isoform α1/β1 is profusely distributed in various tissues with the eminent mRNA levels in brain, lung, heart, kidney, spleen, and muscle (Budworth, Meillerais, Charles, & Powell, 1999). The cGMP synthesized then acts as a secondary messenger and regulates various physiological effects such as vascular smooth muscle tone and motility, phototransduction, and maintaining fluid and electrolyte homeostasis. The family of cGMP-dependent protein kinases, cGMP-regulated phosphodiesterases, and cyclic nucleotide-gated channels also act as downstream effectors for producing various physiological actions (Biel et al., 1999; Boolell et al., 1996; Lucas et al., 2000).

In biological systems, the mode and rate of NO metabolism is dependent on its own concentration, its diffusibility, and the surrounding concentration of other bioreactants. Diffusion is essential for understanding the ability of NO to act as local modulator and its major determinant of biological life time (Rassaf et al., 2002). NO can travel significant distances to reach target cells surrounding the NO generating cell. During this migration, NO can interact with molecular oxygen, thiols, and reduced hemoproteins (Dejam et al., 2003). The extent of either of these reactions largely depends on the microenvironmental conditions under which NO is released, most important, the concentration of NO and other bioreactants, and this is responsible for the large variation in half-lives of EDRF and NO (seconds to minutes) (Butler & Rhodes, 1997). The heme group acts as a NO scavenger. NO diffuses throughout the membranes to the lumen of the vessels, where it reacts with hemoglobin thus forming nitrates and methemoglobin (metHb) (Kelm, 1999). Furthermore, NO can form dinitrosyl complexes with ferrum and binds to proteins containing a heme group, such as hemoglobin. Thus, erythrocytes can transport NO by nitrosation and transnitrosation to release it in other tissues. Another mechanism proposed to store NO is the production of *S*-nitroso-L-glutathione (GSNO) by reduced glutathione (GSH) nitrosylation. NO is released from GSNO by several enzymes, such as GSH peroxidase, thioredoxin reductase, and γ-glutamyl transpeptidase. It has been demonstrated that the reaction of NO with O_2 within the hydrophobic regions of membranes is approximately 300 times more rapid than in the surrounding aqueous medium. This implies that biological membrane and other hydrophobic tissue compartments are important sites for NO metabolism. Several reactive oxygen-derived species, such as superoxide anion, hydrogen peroxide, and hydroxyl radical, also interact with NO and are suggested to be involved in the breakdown of NO. Thus the half-life of NO and the ratio of the NO metabolites (nitrite and nitrate) in aqueous solutions depend on such surrounding conditions of oxygen-derived radicals, the pH, and the concentration of transition metals and ions (Tannenbaum, 1979).

2.3 Nitric Oxide and Central Nervous System

The discovery of NO as a signaling molecule in the nervous system has radically changed the concept of neural communication (Bruhwyler, Chleide, Liegeois, & Carreer, 1993). The physical properties of NO prevent its storage in lipid-lined vesicles and metabolism by hydrolytic enzymes. Therefore, unlike conventional neurotransmitters, NO is neither stored in synaptic vesicles nor released by exocytosis, but synthesized on demand in an optimum amount. It simply diffuses to the structure in the vicinity of the producing cell, both neuronal and nonneuronal, and regulates local brain metabolism and cerebral blood flow (Esplugues, 2002), gene expression, and release of various neurotransmitters. It also plays an important role in morphogenesis and synaptic plasticity, smooth muscle tone, platelet aggregation, and host response to infection, and regulates various other physiological and pathophysiological processes. On the other hand, its excessive generation, in a variety of neuronal disorders, produces neurotoxicity and initiates a neurotoxic cascade similar to glutamate (Esplugues, 2002).

NO was first characterized in the CNS as the intercellular messenger mediating the increase in cyclic GMP levels that follows activation of glutamate receptor. Subsequently, it was shown that NO acts as a neurotransmitter in both the CNS and PNS by mechanisms that are dependent on cyclic GMP (Bredt, 1999). The majority of information deals with nNOS, which, although present in some cerebral vessels and in glial cells, is predominantly found in neurons. nNOS-containing neurons are present in many areas of the CNS with the highest densities occurring in the accessory olfactory bulb and granule cells of the cerebellum. As mentioned above, eNOS is mainly involved in the regulation of vascular function and, although present in some of neurons and glia, is predominantly located in the endothelial cells of cerebral vessels. Induction of iNOS in glial cells has been implicated in the unspecific immune response of the brain and is usually associated with the pathological conditions (Calabrese et al., 2007; Zhang & Snyder, 1995).

This gasotransmitter regulates the release of "classical neurotransmitters" in many brain areas. In fact NO has been shown to indirectly stimulate the release of acetylcholine in the nucleus accumbens by stimulating adjacent glutamatergic neurons (Krumenacker, Hanafy, & Murad, 2004). Basal NO concentrations have been shown to reduce the release of GABA in Ca^{2+}- and Na^{+}-dependent manner, whereas high levels of NO increase GABA release. NO donors stimulated the release of noradrenaline and glutamate in the hippocampus. In rat medial preoptic area, NO increased the release of both dopamine and serotonin in a sGC–cGMP-dependent way (Guix, Uribesalgo, Coma, & Munoz, 2005).

NO has complex influences on brain development, memory formation, and behavior through regulation of synaptic plasticity (Joca, Guimaraes, & Del-Bel, 2007). Inhibition of NO synthesis produces amnesia, and disrupts spatial learning and olfactory memory. Immunohistochemical studies reveal transient expression of nNOS in discrete areas of the developing nervous system. In the embryonic sensory ganglia virtually all neuronal cells are nNOS positive, whereas in adult life only 1% express NOS (Nakane, 2003).

NO has been implicated in various levels of the nociceptive neural pathways, both peripherally (primary afferent neurons and dorsal root ganglia) and centrally (brain stem and several sensory structures of the thalamus). Functionally most nociceptive reflexes involve interaction of NO and NMDA receptors, and it is established that synthesis of NO enhances spinal facilitation of the afferent input conveyed to the cortex and subsequently manifested in behavioral responses. However, NO has differential regulatory role in nociception depending on the type of pain stimuli. Inhibition of NO has antinociceptive effects when pain stems from chemically stimulated peripheral nerve terminals and in models of thermal hyperalgesia or visceral pain while intrathecal administration of L-arginine induces allodynia by converting a nonnoxious to a noxious mechanical stimulus. In contrast, blockade of NO synthesis exacerbates pain in models of mechanical hyperalgesia. The implication of NO in the antinociceptive effects of drugs is also controversial. For instance, inhibitors of NO/GC system potentiate the antinociceptive actions of morphine while attenuating the antinociceptive effects of beta-endorphin. Further there is evidence which suggests that the splice variant of nNOS, nNOS-2, modulates morphine analgesia but not morphine tolerance (He et al., 2007).

Studies indicate that NO may be involved in excitotoxicity. Nitric oxide has been implicated as an endogenous anticonvulsant on one hand, and is shown to sensitize/promote the convulsive effects on the other stimuli.

For example, L-arginine increased seizure severity in response to subconvulsive dose of NMDA, suggesting NO to be a proconvulsant, whereas NOS inhibitor doubled the duration of seizures in bicuculline-induced model. Similarly, NOS inhibition has been found to increase the severity of seizures and mortality in response to kainic acid, whereas attenuation of convulsion in response to tacrine by L-NAME has also been reported (Tutka, Klonowski, Dzieciuch, Kleinrok, & Czuczwar, 1996; Urbanska et al., 1996). Studies from our laboratory have shown that NO synthase inhibitors, L-NAME and 7-nitroindazole, significantly blocked the incidence of seizures and postictal mortality in response to aminophylline administration, suggesting that NO plays a role in drug-induced seizures. Interestingly, pretreatment with the NO mimetics, L-arginine and glyceryltrinitrate, potentiated the proconvulsant effects of subthreshold doses of aminophylline and precipitated the seizures in most animals. Nevertheless, other studies have also reported protective effects for NO synthase inhibitors, i.e., attenuation of seizures induced by a combination of subthreshold electroshock and subconvulsive dose aminophylline, and these were closely paralleled by changes in brain NO activity. In support of the proconvulsant role for NO, another recent study has shown that potentiation of subthreshold PTZ seizures was also blocked by NO synthase inhibitors. Further, in the PTZ kindling model, the onset of kindling was delayed by L-NAME coadministration. In another study, seizures induced by the antitubercular agent, INH, and INH + electroshock were also antagonized by pretreatment with L-NAME or 7-nitroindazole. It thus appears that the NO synthase inhibitors have a clear-cut profile in relation to various seizure models, with NO-lowering agents being protective in most situations (Gulati, Ray, Pal, & Vijayan, 2005; Ray & Gulati, 2007; Ray, Gulati, Anand, & Vijayan, 2005).

The involvement of nitrosative stress in the development of neurodegenerative disorders is also being increasingly documented. In these diseases, NO is produced in excess by iNOS induction owing to the proinflammatory response, which is a common feature of neurodegenerative disorders. Moreover NO is much more harmful under pathological conditions that involve the production of reactive oxygen species, such as superoxide anions and the formation of peroxynitrite. The formation of nitrotyrosine, a marker of nitrosative stress, has been documented in patients with Alzheimer's disease and Parkinson's disease. Furthermore, NO has been shown to activate both the constitutive and inducible isoforms of cyclooxygenase, which are upregulated in brain cells under proinflammatory conditions. During the catalytic cycle of cyclooxygenase, the release of free radicals and formation of prostaglandins occur, two events that are closely related to the development of neuroinflammation. Interestingly, inducible cyclooxygenase is upregulated in the brain of patients affected by Alzheimer's disease and is considered a marker of the progression of dementia in this disease. Keeping this in mind, the activation of inducible cyclooxygenase can be considered as an indirect way for NO to exert neurotoxicity. Another target for NO-induced neurotoxicity is protein disulfide isomerase (PDI), which catalyzes thiol–disulfide exchange, thereby promoting the formation of disulfide bonds and protein rearrangement. During the course of neurodegenerative diseases and cerebral ischemia, immature and denatured proteins accumulate, the latter being toxic to neurons. Under normal circumstances, PDI upregulation would reduce the abnormal accumulation of misfolded proteins to protect neurons. However, in brains of patients with Alzheimer's disease or Parkinson's disease, this protective role is lost owing to the S-nitrosylation of crucial cysteines and inhibition of PDI enzymatic activity (Calabrese et al., 2007; Chanrion et al., 2007).

2.4 Nitric Oxide—Cytotoxicity vs Cytoprotection: A Question of Balance

NO can function as a beneficial physiological agent utilized for essential functions or can mediate several cytotoxic and pathological effects (Mason & Cockcroft, 2006). Whether NO is harmful or helpful depends on a variety of factors such as (a) the cellular environment in which NO is released, (b) the rate of NO flux as determined by which NOS enzyme is activated, and (c) the array of second messenger cascades available for beneficial/toxic cell signaling (Hummel, Fischer, Martin, Schafer, & Buettner, 2006). Interactions between RNS and ROS generate potentially cytotoxic agents, which mediate some of the pathology associated with Parkinson's diseases, chronic inflammation, and atherosclerosis (Kroncke, Fehsel, & Kolb-Bachofen, 1997; Pfeilschifter, Eberhardt, & Beck, 2001). On the other hand, NO can serve as a chain-terminating antioxidant by reacting with chain carrying peroxyl radicals (Joshi, Ponthier, & Lancaster, 1999). In the cell the antioxidant properties of NO can be greatly amplified by the activation of signal transduction pathways that lead to increased synthesis of endogenous antioxidants or downregulate responses to proinflammatory stimuli (Niziolek, Korytowski, & Girotti, 2003). NO has been shown to be cytoprotective through its reaction with lipid radicals as well as with iron. Wink and Mitchell (1998) have shown that either H_2O_2 or *t*-butyl hydroperoxide (*t*BH)-mediated cytotoxicity can be attenuated by the presence of NO releasing diazen-1-ium-1,2-diolats (NONOates). Nitroglycerin, a NO generator, attenuates ferric nitrilotriacetate-induced renal oxidative stress, hyperproliferative response, and necrosis in ddY mice. Exogenous NO can also act as a hydroxyl radical scavenger and protect neurons from oxidative stress. Nitroglycerine can attenuate MPTP-induced dopamine depletion in the striatum by virtue of its OH scavenging action (Darley-Usmar, Wiseman, & Halliwell, 1995). It has been suggested that NO terminates oxidant stress in brain by (i) suppressing iron-induced generation of hydroxyl radicals via Fenton reaction, (ii) interrupting the chain reaction of lipid peroxidation, (iii) augmenting the antioxidative potency of reduced glutathione, and (iv) inhibiting cysteine proteases. Nitric oxide has also been found to act as a powerful antioxidant for astrocytes preserving mitochondrial and cellular integrity during oxidative stress (Pagliaro, 2003; Patel, Levonen, Crawford, & Darley-Usmar, 2000).

The ratio of NO to ROS may be all-important, and it has been shown that a 1:1 ratio of superoxide to NO generates peroxynitrite and induces lipid peroxidation, whereas an excess of NO can inhibit lipid peroxidation by scavenging peroxyl radicals (Jourd'heuil, Kang, & Grisham, 1997). The major differences between cNOS and iNOS activities do not reside in the concentration of NO generated per enzyme but rather in the duration of NO produced. Cytotoxicity usually correlates with the product of iNOS and not with the product of two cNOSs (possible exception of brain injury). Thus, regulated vs constant unregulated NO synthesis differentiates between the messenger and killer properties of NO.

2.5 Nitric Oxide, Stress, and Anxiety

Stress is any external or internal stimulus which disrupts the physiological milieu and results from emotional and environmental demands on the organism. The ability or the lack of it to cope with such aversive and challenging situations is a crucial determinant of health and disease. Such stressors manifest as physiological and behavioral responses that are determined by genetic, behavioral, and environmental factors, and complex neurobiological mechanisms have been implicated. The brain and its neural circuitry is a target for such stress effects and limbic brain areas like hypothalamus, hippocampus, and amygdala along with their connections to higher and lower brain areas play an important role (McEwen & Gianaros, 2010).

These brain areas are also crucial for anxiety, having high levels of NOS enzyme.

Experimental evidence has implicated the possible role of NO in anxiety-like behavior. However, in view of the equivocal nature of data available, the role of this gasotransmitter is still not clearly defined. While some studies have proposed an anxiolytic role for NO, others have indicated to the contrary, and hence studies are still ongoing to delineate its role in this psychopathology (Kalouda & Pitsikas, 2015).

NO production may be considerably modified in stress and adaptation to diverse conditions that tend to suggest that NO plays an important role in stress and adaptive response of the organism (Esch, Stefano, Fricchione, & Benson, 2002). Early studies showed that NO did not affect basal CRH and AVP release, but did inhibit the release of these neuropeptides from rat hypothalamic explants that had been stimulated with potassium chloride or cytokine interleukin beta. By contrast, other studies showed that NO stimulated the release of CRH from rat mediobasal hypothalamus in vitro. On the basis of in vivo studies, the current view proposes that NO has differential effects on the different components of the stress axis (Malyshev & Manukhina, 1998). The NOS inhibitor L-NAME has been shown to decrease the ACTH response in response to shock and to decrease upregulation of CRH and AVP expression in the hypothalamic paraventricular nucleus following exposure to neurogenic stressors; accordingly, the intracerebroventricular administration of NO increased the number of CRH and AVP transcripts in the rat paraventricular nucleus (Nelson, Kriegsfeld, Dawson, & Dawson, 1997), thus suggesting that NO has a stimulatory role in the hypothalamus. By contrast, L-NAME augmented the release of ACTH from the pituitary in response to AVP and circulating proinflammatory cytokines, which implies an inhibitory role for NO at the level of the median eminence or the pituitary. The effect of NO during stress regulation may depend upon the fine balance between the opposing effects of the molecule on the paraventricular nucleus and the pituitary gland, which may be governed by the type and duration of stress stimulus (Kim & Rivier, 2000; Salamon, Esch, & Stefano, 2005).

In chronic mild stress (CMS) model the participation of hippocampal NO production and their regulation by protein kinase C (PKC) was investigated on memory impairment in mice. The mice showed a poor learning performance in both open field and passive avoidance inhibitory task as compared to controls (Palumbo et al., 2007). This was corroborated by histological studies, showing a morphological alteration in the hippocampus of these mice. On the other hand, chronic stress induced a diminished NO production by nNOS. Partial restoration of nNOS activity was obtained after PKC activity blockade. Treatment of mice with nNOS inhibitors resulted in memory deficit as after chronic stressed animals suggesting a showing protective role of nNOS against memory impairments following stress (Joca et al., 2007). Harvey, Oosthuizen, Brand, Wegener, and Stein (2004) reported the role of NO in neuronal toxicity which is regulated by glutamate and GABA and has important implications in stress-related hippocampal degeneration. NO has been shown to be an important modulator of the febrile response to pyrogens and to psychological stress. Colonic temperature was measured in unanesthetized rats before and after treatment with the more selective nNOS inhibitor 7-nitroindazole or with the selective iNOS inhibitor aminoguanidine (AG) under unrestrained or restrained conditions (Pokk & Vali, 2002). Intraperitoneal injection of AG did not affect restraint fever, indicating that iNOS is unlikely to be involved in restraint fever. On the other hand, intraperitoneal injection of 7-nitroindazole significantly attenuated the rise in the colonic temperature caused by restraint stress (RS), indicating that NO produced by nNOS plays an important role in the genesis of RS-induced fever (Sanches, Steiner, & Branco, 2002). Using genetic and pharmacological methods, Zhou et al. (2007) investigated the roles of nNOS in depression induced by CMS in mice. The mice exposed to CMS exhibited behavioral changes typical of depression, and impaired neurogenesis in the hippocampus. The CMS-induced behavioral despair and hippocampal neurogenesis impairment were prevented and reversed in the null mutant mice lacking nNOS gene (nNOS−/−) and in the mice receiving nNOS inhibitor. Moreover, nNOS−/− mice exhibited antidepressant-like properties, suggesting that nNOS overexpression in the hippocampus is essential for chronic stress-induced depression and inhibiting nNOS signaling in brain may represent a novel approach for the treatment of depressive disorders (Echeverry, Guimaraes, & Del Bel, 2004). It has been shown that NOS inhibition in rostral ventrolateral medulla attenuates pressor response to psychological stress in rabbits (Pshennikova, Bondarenko, & Shimkovich, 2001). A sustained overproduction of NO via iNOS expression may be responsible, at least in part, for some of the neurodegenerative changes caused by stress and support a possible neuroprotective role for specific iNOS inhibitors in this situation (Matsumoto et al., 1999). The amygdaloid complex plays a central role in neuroanatomical circuits that coordinate defensive responses. A high density of NO-producing neurons is observed in the medial amygdala, and these neurons are activated after exposure to threatening stimuli.

Systemic treatment with L-NAME, L-NOARG, and 7-NI produced anxiolytic-like behavior in rats in the elevated plus maze (EPM) (Volke, Wegener, & Vasar, 2003). On the other hand, it has been reported that administration of the NO donor 3-morpholinosydnonimine (SIN-1) evoked an anxiolytic-like response in the light/dark exploration test (Perez-Nievas, Garcia-Bueno, Caso, Menchen, & Leza, 2007; Tsuchiya, Kishimoto, Koyama, & Ozawa, 1997). In contrast to this, nitric oxide inhibition has been found to intensify cold restraint-induced gastric ulceration in rats. L-Arginine has been shown to protect against stress-induced gastric mucosal lesions by preserving gastric mucus (Leza, Salas, Sawicki, Russell, & Radomski, 1998). Madrigal et al. (2001) have suggested that excitatory amino acids and subsequent activation of NF-kB account for stress-induced iNOS expression in cerebral cortex. An acute and delayed RS-induced change in nitric oxide-producing neurons in limbic regions has also been reported (Smriga & Torii, 2003).

Pharmacological and biochemical data from our laboratory also indicated an important role of NO in stress susceptibility and adaptation (Gulati, Ray, Masood, & Vijayan, 2006; Masood, Banerji, Vijayan, & Ray, 2004). Further, a recent study was conducted to evaluate the neurobehavioral and biochemical effects of acute and repeated RS for 1 or 6 h, and its modulation by nitrergic agents in rats. Acute RS (1 or 6 h) reduced open arm entries (OAE) and open arm time (OAT) in the EPM test—which were attenuated by the NO precursor, L-arginine, but not influenced appreciably by the NO synthase inhibitor, L-NAME. This neurobehavioral suppression was associated with differential changes in brain NO metabolites (NO_x) but consistently reduced GSH and raised MDA levels in comparison to the control group. Following RS of 1 h daily for 10 days the neurobehavioral suppression and changes in brain oxidative stress markers were less pronounced as compared to the group exposed to single RS for 1 h, indicating adaptation on repeated exposure. L-Arginine pretreatment facilitated this adaptation to repeated RS (1 h). Interestingly, RS of 6 h daily for 10 days induced severe behavioral suppression and aggravation of MDA and NO_x levels, and L-NAME pretreatment tended to protect against these chronic RS-induced aggravations. These results suggest that acute and chronic RS induces duration/intensity-dependent neurobehavioral and oxidative injury which are under the differential regulatory control of NO (Gulati, Chakraborti, & Ray, 2009; Gulati & Ray, 2014).

Studies were also conducted to investigate the influence of age and sex on stress-induced anxiogenesis and the involvement of NO. The effect of RS on neurobehavioral and brain oxidative/nitrosative stress markers and their modulation by antioxidants and nitrergic agents was evaluated in both young and old rats. RS-exposed rats induced anxiogenesis when tested on EPM and open field (OF) tests, and such changes were more marked in older rats. These behavioral alterations were associated with increased levels of MDA and reduced levels of GSH, catalase, and stable NO metabolites (NO_x)—the intensity of these effects being greater in older rats. Administration of α-tocopherol and *N*-acetylcysteine consistently reversed the RS-induced behavioral deficits and biochemical alterations in both age groups of rats. Similar attenuations of RS-induced changes were seen after pretreatment with NO precursor, L-arginine, attenuated, while the NOS inhibitor, L-NAME, tended to aggravate the RS-induced effects in both age groups of rats. The results suggested that susceptibility to stress-induced anxiogenesis may increase with age and that NO and its interactions of reactive oxygen species in brain exerted a regulatory influence in such responses to stress (Chakraborti, Gulati, & Ray, 2008).

In another study, the effects of stress (RS) were investigated on anxiety-like behavior in male and female rats. NO metabolites (NO_x) and asymmetric dimethyl arginine, an endogenous NOS inhibitor, were measured in brain homogenates of these rats for corroborative purposes. Interestingly, RS induced anxiogenesis in the EPM test with the effects being greater in magnitude in males as compared to females. The behavioral alterations were associated with enhanced levels of ADMA and reductions in levels of NO_x in brain homogenates—the intensity of these effects was greater in males as compared to females. Administration of NO precursor, L-arginine, reversed the RS-induced behavioral and biochemical changes, while NOS inhibitor, L-NAME, had opposite effects. Further, pretreatment with formestane (50 mg/kg), an aromatase inhibitor and estrogen synthesis blocker, aggravated stress-induced anxiogenesis which was accompanied with increase in ADMA and decrease in NO_x levels in the females. It was hypothesized that brain NO and its interactions with estrogens play a crucial modulatory role in the differential anxiogenic response to stress among males and females (Chakraborti, Gulati, & Ray, 2014). There is also some evidence of sex-related differences in nitrergic activity in certain areas of CNS (Pires, Costa, Saraiva, Bonikovski, & Futuro Neto, 2003).

Recently studies evaluated the possible involvement of nitric oxide during modulatory effects of some centrally acting and anxiolytic agents.

In a pioneering study, the involvement of nitric oxide in the effects of the neurosteroid, dehydroepiandrosterone sulfate (DHEAS), on stress-induced neurobehavioral and brain oxidative/nitrosative stress markers was investigated in rats. Exposure of rats to RS suppressed behavioral activity in the EPM, and this was associated with increase in malondialdehyde (MDA) and decrease in reduced glutathione (GSH) and brain NO metabolite (NO_x) levels in brain homogenates. Pretreatment with DHEAS reversed the stress-induced changes in behavioral and oxidative stress markers and also brain NO_x levels. The anxiolytic effect of DHEAS was blocked by pretreatment with NOS inhibitor, L-NAME, while pretreatment of rats with L-arginine produced potentiation of action of subeffective dose of DHEAS. The DHEAS effects were stress specific as these behavioral and biochemical parameters were not much influenced in nonstressed rats. These observations suggested that pretreatment with DHEAS attenuated stress-induced and NO-dependent mechanisms may be involved in this effect (Chakraborti, Gulati, & Ray, 2011).

A similar study evaluated the effects of the opioid, morphine, on stress (RS)-induced anxiogenesis in rats. In agreement with our earlier observations, RS consistently induced an anxiety-like response in the EPM test, i.e., reduced number of OAE and OAT spent in the open arms as compared to controls. Prior administration of morphine attenuated this RS-induced anxiogenic response in a dose-related manner. The RS-induced neurobehavioral suppression was associated with reductions in end-metabolites of NO oxidation (NO_x) in brain homogenates, which were also reversed with morphine. Coadministration of subeffective doses of morphine and L-arginine had a synergistic anxiolytic effects in the EPM and on brain NO_x levels, whereas L-NAME neutralized these effects of morphine. Repeated RS for 5 days induced adaptative changes as evidenced by normalization of behavioral suppression (simulating anxiolysis) and elevations in brain NO_x, as compared to acute RS. Morphine potentiated behavioral adaptation in the EPM and elevated brain NO_x levels, and the NOS inhibitor, L-NAME, blocked the morphine effect of stress adaptation. These results suggested opioid–NO interactions during stress-induced anxiogenesis (Anand, Gulati, & Ray, 2012). This hypothesis was further extended when Joshi, Ray, and Gulati (2015) showed that the anxiolytic effects of morphine were dependent on NO and Hsp70. Acute and chronic RS showed differential effects on anxiogenesis in the EPM and changes in brain NO_x and Hsp70 levels. Pretreatment with morphine attenuated acute RS effects on behavioral and biochemical parameters, without much influencing chronic RS effects. Morphine effects were potentiated after L-arginine and attenuated after L-NAME. It was inferred that acute and chronic RS-induced anxiety modulations were differentially influenced by morphine and L-arginine and that complex interactions involving brain NO and Hsp70 could regulate such effects (Joshi et al., 2015). A recent review has also given an elaborate description of the suggested role of NO in stress and anxiety (Gulati, Joshi, & Ray, 2015).

PTSD has received considerable attention in recent years. It is characterized by anxiogenesis, cognitive deficits, hyperarousal, increased startle, and cognitive deficits, and is caused by exposure to a severe traumatic event. Multiple exposures to trauma are now increasingly being recognized as crucial in predicting the onset and severity of this disorder. In a pioneering study, Harvey, Bothma, Nel, Wegener, and Stein (2005) investigated the role of NO and NO-mediated signaling pathways in PTSD. They used time-dependent sensitization stress, an animal model of repeated trauma and a putative animal model of PTSD, to study the accumulation of nitrogen oxides (NO_x), the stable oxidative metabolites of NO, in the hippocampus of rats 7 days after exposure to stress. Earlier studies have found the model to closely reproduce key behavioral, neurochemical, and endocrine manifestations of PTSD. The effect of stress was investigated on the hippocampal NO-cGMP signaling pathway, and specifically the release of nitrogen oxides (NO_x) and their modulation by NMDA receptor, NO, cGMP, and NF-kB-selective agents. Immediately after stress, rats received the glutamate NMDA receptor antagonist, memantine, the NO synthase inhibitor, 7-nitroindazole sodium salt (7-NINA), the cGMP-specific PDE inhibitor, sildenafil (SIL), or the NF-kB antagonist, pyrrolidinedithiocarbamate (PDTC), for 7 days. The results revealed that stress significantly increased hippocampal NO_x on day 7 poststress, which was blocked by either 7-NINA or PDTC, while MEM was without effect. SIL, however, significantly augmented stress-induced NO_x accumulation. They inferred that elevated cGMP acts as a protagonist in driving stress-related events, while both nNOS (neuronal NOS) and iNOS (inducible/immunological NOS) could represent a therapeutic target in preventing the effects of severe stress. Recent clinical evidence has also indicated that PTSD may involve neurodegenerative pathology, and brain imaging studies have revealed a marked reduction in hippocampal volume and these findings also positively correlated with similar findings in rodents. Such hippocampal atrophy was accompanied by stress-induced increases in plasma cortisol. Interestingly, marked suppression of plasma cortisol levels was also noted in patients of

PTSD. It was hypothesized that such stress and glucocorticoids increased glutamate release, which could be an important mediator of glucocorticoid-induced neurotoxicity. Preclinical studies have also shown that the glutamate and nitric oxide could play a putative role in anxiety-related behaviors. It was suggested that by virtue of its role in neuronal toxicity, cellular memory processes, and as a neuromodulator, NO and nitrergic pathways could be a decisive factor in stress-related hippocampal degenerative pathology and cognitive deficits seen in patients with PTSD (Oosthuizen, Wegener, & Harvey, 2005).

3. CONCLUSION

In conclusion, since the discovery of the role of NO as a neuromodulator in the brain, several behavioral and cognitive effects have been proposed. The role of NO in anxiety and related disorders has been extensively researched, and both preclinical and clinical data are available. However, most of the data generated to date are equivocal to say the least, which leaves the researcher with the scope to investigate various other possibilities involved. Additional research in the area is required in order to arrive at a consensus.

REFERENCES

Abu-Soud, H. M., Yoho, L. L., & Stuehr, D. J. (1994). Calmodulin controls neuronal nitric-oxide synthase by a dual mechanism. Activation of intra- and interdomain electron transfer. *The Journal of Biological Chemistry*, *269*, 32047–32050.

Ai, A. L., Kronfol, Z., Seymour, E., & Bolling, S. F. (2005). Effects of mood state and psychosocial functioning on plasma interleukin-6 in adult patients before cardiac surgery. *International Journal of Psychiatry in Medicine*, *35*, 363–376.

Anand, R., Gulati, K., & Ray, A. (2012). Pharmacological evidence for the role of nitric oxide in the modulation of stress-induced anxiety by morphine in rats. *European Journal of Pharmacology*, *676*, 71–74.

Arnold, W. P., Mittal, C. K., Katsuki, S., & Murad, F. (1977). Nitric oxide activates guanylate cyclase and increases guanosine 3′:5′-cyclic monophosphate levels in various tissue preparations. *Proceedings of the National Academy of Sciences of the United States of America*, *74*, 3203–3207.

Biel, M., Seeliger, M., Pfeifer, A., Kohler, K., Gerstner, A., Ludwig, A., et al. (1999). Selective loss of cone function in mice lacking the cyclic nucleotide-gated channel CNG3. *Proceedings of the National Academy of Sciences of the United States of America*, *96*, 7553–7557.

Boolell, M., Allen, M. J., Ballard, S. A., Gepi-Attee, S., Muirhead, G. J., Naylor, A. M., et al. (1996). Sildenafil: An orally active type 5 cyclic GMP-specific phosphodiesterase inhibitor for the treatment of penile erectile dysfunction. *International Journal of Impotence Research*, *8*, 47–52.

Bredt, D. S. (1999). Endogenous nitric oxide synthesis: Biological functions and pathophysiology. *Free Radical Research*, *31*, 577–596.

Bruhwyler, J., Chleide, E., Liegeois, J. F., & Carreer, F. (1993). Nitric oxide: A new messenger in the brain. *Neuroscience and Biobehavioral Reviews*, *17*, 373–384.
Budworth, J., Meillerais, S., Charles, I., & Powell, K. (1999). Tissue distribution of the human soluble guanylate cyclases. *Biochemical and Biophysical Research Communications*, *263*, 696–701.
Butler, A. R., & Rhodes, P. (1997). Chemistry, analysis, and biological roles of S-nitrosothiols. *Analytical Biochemistry*, *249*, 1–9.
Calabrese, V., Mancuso, C., Calvani, M., Rizzarelli, E., Butterfield, D. A., & Stella, A. M. (2007). Nitric oxide in the central nervous system: Neuroprotection versus neurotoxicity. *Nature Reviews. Neuroscience*, *8*, 766–775.
Chakraborti, A., Gulati, K., & Ray, A. (2008). Age related differences in stress-induced neurobehavioral responses in rats: Modulation by antioxidants and nitrergic agents. *Behavioural Brain Research*, *194*, 86–91.
Chakraborti, A., Gulati, K., & Ray, A. (2011). Involvement of nitric oxide in the protective effects of dehydroepiandrosterone sulphate on stress induced neurobehavioral suppression and brain oxidative injury in rats. *European Journal of Pharmacology*, *652*, 55–59.
Chakraborti, A., Gulati, K., & Ray, A. (2014). Possible role of nitric oxide (NO) in the regulation of gender related differences in stress induced anxiogenesis in rats. *Nitric Oxide*, *43*, 74–80.
Chanrion, B., Mannoury la Cour, C., Bertaso, F., Lerner-Natoli, M., Freissmuth, M., Millan, M. J., et al. (2007). Physical interaction between the serotonin transporter and neuronal nitric oxide synthase underlies reciprocal modulation of their activity. *Proceedings of the National Academy of Sciences of the United States of America*, *104*, 8119–8124.
Chansky, T. E. (2004). *Freeing your child from anxiety: Powerful, practical solutions to overcome your child's fears, worries, and phobias*. New York: Broadway Books.
Cryan, J. F., & Sweeney, F. F. (2011). The age of anxiety: Role of animal models of anxiolytic action in drug discovery. *British Journal of Pharmacology*, *164*, 1129–1161.
Darley-Usmar, V., Wiseman, H., & Halliwell, B. (1995). Nitric oxide and oxygen radicals: A question of balance. *FEBS Letters*, *369*, 131–135.
Dejam, A., Kleinbongard, P., Rassaf, T., Hamada, S., Gharini, P., Rodriguez, J., et al. (2003). Thiols enhance NO formation from nitrate photolysis. *Free Radical Biology and Medicine*, *35*, 1551–1559.
Echeverry, M. B., Guimaraes, F. S., & Del Bel, E. A. (2004). Acute and delayed restraint stress-induced changes in nitric oxide producing neurons in limbic regions. *Neuroscience*, *125*, 981–993.
Esch, T., Stefano, G. B., Fricchione, G. L., & Benson, H. (2002). Stress-related diseases—A potential role for nitric oxide. *Medical Science Monitor*, *8*, RA103–RA118.
Esplugues, J. V. (2002). NO as a signalling molecule in the nervous system. *British Journal of Pharmacology*, *135*, 1079–1095.
Furchgott, R. F., & Zawadzki, J. V. (1980). The obligatory role of endothelial cells in the relaxation of arterial smooth muscle by acetylcholine. *Nature*, *288*, 373–376.
Guix, F. X., Uribesalgo, I., Coma, M., & Munoz, F. J. (2005). The physiology and pathophysiology of nitric oxide in the brain. *Progress in Neurobiology*, *76*, 126–152.
Gulati, K., Chakraborti, A., & Ray, A. (2009). Differential role of nitric oxide (NO) in acute and chronic stress induced neurobehavioral modulation and oxidative injury in rats. *Pharmacology, Biochemistry, and Behavior*, *92*, 272–276.
Gulati, K., Joshi, J. C., & Ray, A. (2015). Recent advances in stress research: Focus on nitric oxide. *European Journal of Pharmacology*, *765*, 406–414.
Gulati, K., & Ray, A. (2014). Differential neuromodulatory role of NO in anxiety and seizures: An experimental study. *Nitric Oxide*, *43*, 55–61.
Gulati, K., Ray, A., Masood, A., & Vijayan, V. K. (2006). Involvement of nitric oxide (NO) in the regulation of stress susceptibility and adaptation in rats. *Indian Journal of Experimental Biology*, *44*, 809–815.

Gulati, K., Ray, A., Pal, G., & Vijayan, V. K. (2005). Possible role of free radicals in theophylline-induced seizures in mice. *Pharmacology, Biochemistry, and Behavior, 82*, 241–245.

Harter, M. C., Conway, K. P., & Merikangas, K. R. (2003). Associations between anxiety disorders and physical illness. *European Archives of Psychiatry and Clinical Neuroscience, 253*, 313–320.

Harvey, B. H., Bothma, T., Nel, A., Wegener, G., & Stein, D. J. (2005). Involvement of the NMDA receptor, NO-cyclic GMP and nuclear factor K-beta in an animal model of repeated trauma. *Human Psychopharmacology, 20*, 367–373.

Harvey, B. H., Oosthuizen, F., Brand, L., Wegener, G., & Stein, D. J. (2004). Stress-restress evokes sustained iNOS activity and altered GABA levels and NMDA receptors in rat hippocampus. *Psychopharmacology, 175*, 494–502.

He, J., Wang, T., Wang, P., Han, P., Yin, Q., & Chen, C. (2007). A novel mechanism underlying the susceptibility of neuronal cells to nitric oxide: The occurrence and regulation of protein S-nitrosylation is the checkpoint. *Journal of Neurochemistry, 102*, 1863–1874.

Hobbs, A. J., Fukuto, J. M., & Ignarro, L. J. (1994). Formation of free nitric oxide from l-arginine by nitric oxide synthase: Direct enhancement of generation by superoxide dismutase. *Proceedings of the National Academy of Sciences of the United States of America, 91*, 10992–10996.

Hummel, S. G., Fischer, A. J., Martin, S. M., Schafer, F. Q., & Buettner, G. R. (2006). Nitric oxide as a cellular antioxidant: A little goes a long way. *Free Radical Biology and Medicine, 40*, 501–506.

Ignarro, L. J. (2002). Nitric oxide as a unique signaling molecule in the vascular system: A historical overview. *Journal of Physiology and Pharmacology, 53*, 503–514.

Joca, S. R., Guimaraes, F. S., & Del-Bel, E. (2007). Inhibition of nitric oxide synthase increases synaptophysin mRNA expression in the hippocampal formation of rats. *Neuroscience Letters, 421*, 72–76.

Joshi, M. S., Ponthier, J. L., & Lancaster, J. R., Jr. (1999). Cellular antioxidant and pro-oxidant actions of nitric oxide. *Free Radical Biology and Medicine, 27*, 1357–1366.

Joshi, J. C., Ray, A., & Gulati, K. (2015). Effects of morphine on stress induced anxiety in rats: Role of nitric oxide and Hsp70. *Physiology and Behavior, 139*, 393–396.

Jourd'heuil, D., Kang, D., & Grisham, M. B. (1997). Interactions between superoxide and nitric oxide: Implications in DNA damage and mutagenesis. *Frontiers in Bioscience, 2*, d189–d196.

Kalouda, T., & Pitsikas, N. (2015). The nitric oxide donor molsidomine induces anxiolytic-like behavior in two different rat models of anxiety. *Pharmacology, Biochemistry, and Behavior, 138*, 111–116.

Kelm, M. (1999). Nitric oxide metabolism and breakdown. *Biochimica et Biophysica Acta, 1411*, 273–289.

Kessler, R. C., Chiu, W. T., Demler, O., Merikangas, K. R., & Walters, E. E. (2005). Prevalence, severity, and comorbidity of 12-month DSM-IV disorders in the National Comorbidity Survey Replication. *Archives of General Psychiatry, 62*, 617–627.

Kim, C. K., & Rivier, C. L. (2000). Nitric oxide and carbon monoxide have a stimulatory role in the hypothalamic-pituitary-adrenal response to physico-emotional stressors in rats. *Endocrinology, 141*, 2244–2253.

Kroncke, K. D., Fehsel, K., & Kolb-Bachofen, V. (1997). Nitric oxide: Cytotoxicity versus cytoprotection—How, why, when, and where? *Nitric Oxide, 1*, 107–120.

Krumenacker, J. S., Hanafy, K. A., & Murad, F. (2004). Regulation of nitric oxide and soluble guanylyl cyclase. *Brain Research Bulletin, 62*, 505–515.

Leza, J. C., Salas, E., Sawicki, G., Russell, J. C., & Radomski, M. W. (1998). The effects of stress on homeostasis in JCR-LA-cp rats: The role of nitric oxide. *The Journal of Pharmacology and Experimental Therapeutics, 286*, 1397–1403.

Liu, Q., & Gross, S. S. (1996). Binding sites of nitric oxide synthases. *Methods in Enzymology*, *268*, 311–324.
Lucas, K. A., Pitari, G. M., Kazerounian, S., Ruiz-Stewart, I., Park, J., Schulz, S., et al. (2000). Guanylyl cyclases and signaling by cyclic GMP. *Pharmacological Reviews*, *52*, 375–414.
Madrigal, J. L., Moro, M. A., Lizasoain, I., Lorenzo, P., Castrillo, A., Bosca, L., et al. (2001). Inducible nitric oxide synthase expression in brain cortex after acute restraint stress is regulated by nuclear factor kappaB-mediated mechanisms. *Journal of Neurochemistry*, *76*, 532–538.
Malyshev, I. Y., & Manukhina, E. B. (1998). Stress, adaptation, and nitric oxide. *Biochemistry*, *63*, 840–853.
Marques, A. A., Bevilaqua, M. C., da Fonseca, A. M., Nardi, A. E., Thuret, S., & Dias, G. P. (2016). Gender differences in the neurobiology of anxiety: Focus on adult hippocampal neurogenesis. *Neural Plasticity*, *2016*, 5026713.
Martin, E. I., Ressler, K. J., Binder, E., & Nemeroff, C. B. (2009). The neurobiology of anxiety disorders: Brain imaging, genetics, and psychoneuroendocrinology. *The Psychiatric Clinics of North America*, *32*, 549–575.
Mason, R. P., & Cockcroft, J. R. (2006). Targeting nitric oxide with drug therapy. *Journal of Clinical Hypertension (Greenwich)*, *8*, 40–52.
Masood, A., Banerji, B., Vijayan, V. K., & Ray, A. (2004). Pharmacological and biochemical studies on the possible role of nitric oxide in stress adaptation in rats. *European Journal of Pharmacology*, *493*, 111–115.
Matsumoto, K., Yobimoto, K., Huong, N. T., Abdel-Fattah, M., Van Hien, T., & Watanabe, H. (1999). Psychological stress-induced enhancement of brain lipid peroxidation via nitric oxide systems and its modulation by anxiolytic and anxiogenic drugs in mice. *Brain Research*, *839*, 74–84.
McEwen, B. S., & Gianaros, P. J. (2010). Central role of the brain in stress and adaptation: Links to socioeconomic status, health, and disease. *Annals of the New York Academy of Sciences*, *1186*, 190–222.
Merikangas, K. R. (2003). Implications of genomics for public health: The role of genetic epidemiology. *Cold Spring Harbor Symposia on Quantitative Biology*, *68*, 359–364.
Moncada, S., Palmer, R. M., & Higgs, E. A. (1991). Nitric oxide: Physiology, pathophysiology, and pharmacology. *Pharmacological Reviews*, *43*, 109–142.
Nakane, M. (2003). Soluble guanylyl cyclase: Physiological role as an NO receptor and the potential molecular target for therapeutic application. *Clinical Chemistry and Laboratory Medicine*, *41*, 865–870.
Nelson, R. J., Kriegsfeld, L. J., Dawson, V. L., & Dawson, T. M. (1997). Effects of nitric oxide on neuroendocrine function and behavior. *Frontiers in Neuroendocrinology*, *18*, 463–491.
Niziolek, M., Korytowski, W., & Girotti, A. W. (2003). Chain-breaking antioxidant and cytoprotective action of nitric oxide on photodynamically stressed tumor cells. *Photochemistry and Photobiology*, *78*, 262–270.
Nott, A., & Riccio, A. (2009). Nitric oxide-mediated epigenetic mechanisms in developing neurons. *Cell Cycle*, *8*, 725–730.
Oosthuizen, F., Wegener, G., & Harvey, B. H. (2005). Nitric oxide as inflammatory mediator in post-traumatic stress disorder (PTSD): Evidence from an animal model. *Neuropsychiatric Disease and Treatment*, *1*, 109–123.
Pagliaro, P. (2003). Differential biological effects of products of nitric oxide (NO) synthase: It is not enough to say NO. *Life Sciences*, *73*, 2137–2149.
Palumbo, M. L., Fosser, N. S., Rios, H., Zorrilla Zubilete, M. A., Guelman, L. R., Cremaschi, G. A., et al. (2007). Loss of hippocampal neuronal nitric oxide synthase contributes to the stress-related deficit in learning and memory. *Journal of Neurochemistry*, *102*, 261–274.

Patel, R. P., Levonen, A., Crawford, J. H., & Darley-Usmar, V. M. (2000). Mechanisms of the pro- and anti-oxidant actions of nitric oxide in atherosclerosis. *Cardiovascular Research, 47*, 465–474.

Perez-Nievas, B. G., Garcia-Bueno, B., Caso, J. R., Menchen, L., & Leza, J. C. (2007). Corticosterone as a marker of susceptibility to oxidative/nitrosative cerebral damage after stress exposure in rats. *Psychoneuroendocrinology, 32*, 703–711.

Pfeilschifter, J., Eberhardt, W., & Beck, K. F. (2001). Regulation of gene expression by nitric oxide. *Pflügers Archiv, 442*, 479–486.

Pfeilschifter, J., Eberhardt, W., & Huwiler, A. (2003). Nitric oxide and mechanisms of redox signaling. *Journal of the American Society of Nephrology, 14*, S237–S240.

Pires, J. G., Costa, P. G., Saraiva, F. P., Bonikovski, V., & Futuro Neto, H. A. (2003). Gender-related differences in the effects of nitric oxide donors on neuroleptic-induced catalepsy in mice. *Brazilian Journal of Medical and Biological Research, 36*, 239–245.

Pokk, P., & Vali, M. (2002). The effects of the nitric oxide synthase inhibitors on the behaviour of small-platform-stressed mice in the plus-maze test. *Progress in Neuro-Psychopharmacology and Biological Psychiatry, 26*, 241–247.

Pshennikova, M. G., Bondarenko, N. A., & Shimkovich, M. V. (2001). Nitric oxide as a factor of genetically determined resistance to stress damages and adaptive protection. *Bulletin of Experimental Biology and Medicine, 132*, 1048–1050.

Rassaf, T., Preik, M., Kleinbongard, P., Lauer, T., Heiss, C., Strauer, B. E., et al. (2002). Evidence for in vivo transport of bioactive nitric oxide in human plasma. *The Journal of Clinical Investigation, 109*, 1241–1248.

Ray, A., & Gulati, K. (2007). Nitric oxide and central nervous system: Perspective and trends. In A. Ray & K. Gulati (Eds.), *Current trends in pharmacology* (pp. 367–382). New Delhi: I.K. International Publishing House Pvt. Ltd.

Ray, A., Gulati, K., Anand, S., & Vijayan, V. K. (2005). Pharmacological studies on mechanisms of aminophylline-induced seizures in rats. *Indian Journal of Experimental Biology, 43*, 849–853.

Roy-Byrne, P. P., Davidson, K. W., Kessler, R. C., Asmundson, G. J., Goodwin, R. D., Kubzansky, L., et al. (2008). Anxiety disorders and comorbid medical illness. *General Hospital Psychiatry, 30*, 208–225.

Russwurm, M., Behrends, S., Harteneck, C., & Koesling, D. (1998). Functional properties of a naturally occurring isoform of soluble guanylyl cyclase. *The Biochemical Journal, 335*, 125–130.

Sadock, B. J., & Sadock, V. A. (2007). *Kaplan and Sadock's synopsis of psychiatry: Behavioral sciences/clinical psychiatry*. Philadelphia: Lippincott Williams and Wilkins.

Salamon, E., Esch, T., & Stefano, G. B. (2005). Role of amygdala in mediating sexual and emotional behavior via coupled nitric oxide release. *Acta Pharmacologica Sinica, 26*, 389–395.

Sanches, D. B., Steiner, A. A., & Branco, L. G. (2002). Involvement of neuronal nitric oxide synthase in restraint stress-induced fever in rats. *Physiology and Behavior, 75*, 261–266.

Shapira-Lichter, I., Beilin, B., Ofek, K., Bessler, H., Gruberger, M., Shavit, Y., et al. (2008). Cytokines and cholinergic signals co-modulate surgical stress-induced changes in mood and memory. *Brain, Behavior, and Immunity, 22*, 388–398.

Smriga, M., & Torii, K. (2003). Prolonged treatment with L-lysine and L-arginine reduces stress-induced anxiety in an elevated plus maze. *Nutritional Neuroscience, 6*, 125–128.

SoRelle, R. (1998). Nobel prize awarded to scientists for nitric oxide discoveries. *Circulation, 98*, 2365–2366.

Stuehr, D. J. (1999). Mammalian nitric oxide synthases. *Biochimica et Biophysica Acta, 1411*, 217–230.

Tannenbaum, S. R. (1979). Nitrate and nitrite: Origin in humans. *Science, 205*, 1332–1335.

Tsuchiya, T., Kishimoto, J., Koyama, J., & Ozawa, T. (1997). Modulatory effect of L-NAME, a specific nitric oxide synthase (NOS) inhibitor, on stress-induced changes in plasma adrenocorticotropic hormone (ACTH) and corticosterone levels in rats: Physiological significance of stress-induced NOS activation in hypothalamic-pituitary-adrenal axis. *Brain Research, 776*, 68–74.

Tutka, P., Klonowski, P., Dzieciuch, J., Kleinrok, Z., & Czuczwar, S. J. (1996). NG-nitro-L-arginine differentially affects glutamate- or kainate-induced seizures. *Neuroreport, 7*, 1605–1608.

Urbanska, E. M., Drelewska, E., Borowicz, K. K., Blaszczak, P., Kleinrok, Z., & Czuczwar, S. J. (1996). NG-nitro-L-arginine, a nitric oxide synthase inhibitor, and seizure susceptibility in four seizure models in mice. *Journal of Neural Transmission, 103*, 1145–1152.

Volke, V., Wegener, G., & Vasar, E. (2003). Augmentation of the NO-cGMP cascade induces anxiogenic-like effect in mice. *Journal of Physiology and Pharmacology, 54*, 653–660.

Wink, D. A., & Mitchell, J. B. (1998). Chemical biology of nitric oxide: Insights into regulatory, cytotoxic, and cytoprotective mechanisms of nitric oxide. *Free Radical Biology and Medicine, 25*, 434–456.

Xia, H., & Bredt, D. S. (1996). Cloned and expressed nitric oxide synthase proteins. *Methods in Enzymology, 268*, 427–436.

Xia, Y., & Zweier, J. L. (1997). Direct measurement of nitric oxide generation from nitric oxide synthase. *Proceedings of the National Academy of Sciences of the United States of America, 94*, 12705–12710.

Zhang, J., & Snyder, S. H. (1995). Nitric oxide in the nervous system. *Annual Review of Pharmacology and Toxicology, 35*, 213–233.

Zhou, Q. G., Hu, Y., Hua, Y., Hu, M., Luo, C. X., Han, X., et al. (2007). Neuronal nitric oxide synthase contributes to chronic stress-induced depression by suppressing hippocampal neurogenesis. *Journal of Neurochemistry, 103*, 1843–1854.

CHAPTER EIGHT

The Endocannabinoid System and Anxiety

S.F. Lisboa*,†,[1], F.V. Gomes‡,[2], A.L.B. Terzian*,†,[2], D.C. Aguiar§, F.A. Moreira§, L.B.M. Resstel*,†, F.S. Guimarães*,†,[1]

*Medical School of Ribeirão Preto, University of São Paulo, Ribeirão Preto, Brazil
†Center for Interdisciplinary Research on Applied Neurosciences (NAPNA), University of São Paulo (USP), Ribeirão Preto, Brazil
‡University of Pittsburgh, Pittsburgh, PA, United States
§Institute of Biological Sciences, Universidade Federal de Minas Gerais, Belo Horizonte, Brazil
[1]Corresponding authors: e-mail address: sa_lisboa@hotmail.com; fsguimar@fmrp.usp.br

Contents

[2] These authors contributed equally to this work.

Vitamins and Hormones, Volume 103
ISSN 0083-6729
http://dx.doi.org/10.1016/bs.vh.2016.09.006

Abstract

The medical properties of *Cannabis sativa* is known for centuries. Since the discovery and characterization of the endogenous cannabinoid system, several studies have evaluated how cannabinoid compounds and, particularly, how the modulation of the endocannabinoid (eCB) system influences a wide range of functions, from metabolic to mental disorders. Cannabinoids and eCB system often exert opposite effects on several functions, such as anxiety. Although the mechanisms are not completely understood, evidence points to different factors influencing those effects. In this chapter, the recent advances in research about the relationship between eCB system and anxiety disorders in humans, as well as in animal models, will be discussed. The recent data addressing modulation of the eCBs in specific brain areas, such as the medial prefrontal cortex, amygdaloid complex, bed nucleus of *stria terminalis*, hippocampus, and dorsal periaqueductal gray, will be summarized. Finally, data from animal models addressing the mechanisms through which the eCB system modulates anxiety-related behavior dependent on stressful situations, such as the involvement of different receptors, distinct eCBs, modulation of neurotransmitters release, HPA axis and immune system activation, and plastic mechanisms, will also be discussed.

1. INTRODUCTION

1.1 Neurobiology of Anxiety, Defense Levels, and Animal Models

Fear and anxiety are adaptive defensive responses that have evolved to protect animals from dangerous stimuli (Blanchard, Blanchard, Griebel, & Nutt, 2008; Gross & Canteras, 2012; McNaughton & Corr, 2004; Tovote, Fadok, & Luthi, 2015). The distinction between these two emotions is not always clear. They are often associated with the actual (fear) vs potential (anxiety) presence of threat (Blanchard et al., 2008; Tovote et al., 2015). Gray and McNaughton (2000), however, have argued that the critical difference between anxiety and fear rests on what they called "defensive direction." Whereas fear would lead the animal to leave a dangerous situation, anxiety occurs when the animal has to enter (cautions risk assessment) or prevent entrance (passive avoidance) to such situation (McNaughton & Corr, 2004; McNaughton & Zangrossi, 2008). From these different perspectives, it is clear that the behavioral defensive responses to danger will vary considerably depending on the characteristics of the threatening stimuli, such as potential vs real, innate vs learned, proximal vs distal, and escape availability (McNaughton & Corr, 2004).

Although fear and anxiety are normal emotions, they can be considered pathological when triggered by inappropriate stimuli or are too extreme in terms of intensity and/or duration (McNaughton & Zangrossi, 2008). These defensive responses are organized by partially distinct brain systems (Canteras, Resstel, Bertoglio, Carobrez, & Guimaraes, 2010; McNaughton & Corr, 2004). A defensive circuitry, responsive to innate threat such as predators, includes the amygdala, the medial hypothalamus, and the dorsal periaqueductal gray (dPAG). Another important circuitry is related to aversive conditioning. Aversive Pavlovian conditioning allows the learning and extinction of associations between neutral and unpleasant stimuli. The neural substrate of this process is relatively well known in rodents. Acquisition and retention of fear conditioning are likely to occur in the lateral nucleus of the amygdala (Canteras et al., 2010), which projects to the central nucleus of the amygdala (CeA) by direct or indirect pathways (via the basolateral nucleus of the amygdala, BLA). The CeA, the primary output of the system, sends direct projection to the ventrolateral PAG and other structures that led to the behavioral (freezing) and autonomic changes associated to fear conditioning (LeDoux, 2000). These systems are regulated by the prefrontal cortex (PFC) and the hippocampal formation. The orbitofrontal PFC, particularly its medial part, coordinates defensive behavioral responses in a complex way, as will be discussed later (Bishop, 2007; Canteras et al., 2010). The hippocampal formation provides context information during threatening situations (Canteras et al., 2010). It also plays a significant role on fear extinction, providing the contextual information needed to acquisition and retrieval of this process (Tovote et al., 2015). In addition, a septum-hippocampus system, according to Gray's proposal, would generate anxiety in response to conflict situations. As a consequence, it increases arousal and selective attention and interrupts ongoing behaviors (Gray & McNaughton, 2000; Oehrn et al., 2015). Another brain structure that has been associated with defensive responses is the bed nucleus of the *stria terminalis* (BNST). It receives major inputs from the central (CeA) and basolateral (BLA) nucleus of the amygdala and plays an important role in anxiety responses (Tovote et al., 2015). Within these structures, moreover, different subsets of neurons can play even distinct roles. For example, independent circuitries in the medial hypothalamus are involved in social vs predator fear (Silva et al., 2013) and stimulation of distinct BNST subnuclei can result in either anxiolytic or anxiogenic responses (Tovote et al., 2015).

As will be discussed later, cannabinoid type 1 (CB_1) receptors and the enzymes responsible for the synthesis and metabolism of endocannabinoids (eCBs) are highly expressed in all these brain regions (Hu & Mackie, 2015; Tsou, Brown, Sanudo-Pena, Mackie, & Walker, 1998). The eCB system, therefore, is in a critical position to influence defensive responses.

1.2 Cannabinoids and the eCB System

The herb *Cannabis sativa* (marijuana) has been used for centuries as a drug of abuse or as a medicine. Therefore, it is surprising that the scientific knowledge on its chemical constituents only emerged by the second half of the 20th century. This herb produces more than a hundred compounds termed cannabinoids, from which the main responsible for the typical effects of marijuana, including its abuse potential, "high," amnesia, sedation, and analgesia, is Δ^9-tetrahydrocannabinol (Δ^9-THC). Apart from Δ^9-THC, other phytocannabinois are of major medical interest, particularly cannabidiol (CBD) (Mechoulam, 1970). After the identification of Δ^9-THC as the main responsible for marijuana effects, several synthetic analogs have been obtained, which have contributed to the understanding of the pharmacological effects of marijuana. Δ^9-THC and its synthetic analogs induce typical pharmacological effects in rodents, termed the cannabinoid tetrad, consisting of hypolocomotion, catalepsy, analgesia, and hypothermia in mice (Martin et al., 1991).

The characterization of a pharmacological assay for cannabinoids enabled a better understanding of the cannabinoid pharmacology, culminating with the discovery of a receptor specific for cannabinoid compounds (Devane, Dysarz, Johnson, Melvin, & Howlett, 1988; Matsuda, Lolait, Brownstein, Young, & Bonner, 1990). This, in turn, raised the hypothesis that the body might produce its own cannabinoid compounds. Accordingly, a brain cannabinoid was isolated and characterized as an arachidonic acid derivative, arachidonoyl ethanolamide, also termed anandamide (AEA) (Devane et al., 1992). Another endogenous cannabinoid agonist, 2-arachidonoyl glycerol (2-AG) was later described (Mechoulam et al., 1995), as well as a second cannabinoid receptor (Munro, Thomas, & Abu-Shaar, 1993). These endogenous cannabinoid agonists have been termed eCBs, and the receptors have been named by their order of discovery as CB_1 and CB_2 (Howlett et al., 2002).

AEA and 2-AG are thought to be synthesized on-demand from the membrane of postsynaptic neurons after an increase in neuronal activity and to act mainly through CB_1 receptors located in presynaptic terminals. Their actions are terminated by the enzymes fatty acid amide hydrolase (FAAH) and monoacylglycerol lypase (MAGL), respectively. Altogether, the eCBs, their receptors, and the enzymes responsible for their metabolism constitute a chemical messenger system termed the eCB system (Howlett et al., 2002; Pertwee et al., 2010). In addition, other endogenous compounds have been proposed as eCBs. Moreover, other receptors have also been included in this system, such as the transient receptor potential vanilloid type-1 (TRPV1) channel, a cation-permeable ion channel that is also activated by AEA (Di Marzo & De Petrocellis, 2012).

2. eCBs AND ANXIETY IN HUMANS

2.1 Evidence From Behavioral and Brain Image Studies

2.1.1 Pharmacological Manipulation of the Cannabinoid System

Cannabis use is generally associated with stress relief and mood elevation. However, in some cases, it produces dysphoric effects that include panic or heightened anxiety (Hall & Solowij, 1998). These bidirectional effects of cannabinoids observed in humans can be mimicked in laboratory animals, with low doses being anxiolytic whereas high doses are anxiogenic (Viveros, Marco, & File, 2005).

Additionally, some aversive effects that may result from cannabis smoking, such as anxiety and panic (Hall & Solowij, 1998), are rarely observed after oral administration of Δ^9-THC. For instance, functional magnetic resonance imaging revealed that oral Δ^9-THC reduces amygdala reactivity in healthy volunteers exposed to social signals of threat (Gorka, Fitzgerald, de Wit, & Phan, 2015; Phan et al., 2008), an effect similar to anxiolytic drugs such as benzodiazepines, although contradictory results can be also found (Bhattacharyya et al., 2010). Furthermore, an early placebo-controlled study showed that nabilone, a synthetic derivative of Δ^9-THC, reduces anxiety in patients (Fabre & McLendon, 1981). Dronabinol, another synthetic derivative of Δ^9-THC, facilitates extinction of learned fear in humans and increases activation in the ventromedial PFC and hippocampus upon presentation of conditioned stimulus (CS) (Rabinak et al., 2014, 2013). Recent evidence indicates that nabilone and Δ^9-THC relieve insomnia, nightmares, and other symptoms in individuals with post-traumatic stress disorder (PTSD) (Cameron, Watson, & Robinson, 2014; Fraser, 2009; Jetly, Heber, Fraser, & Boisvert, 2015; Roitman, Mechoulam, Cooper-Kazaz, & Shalev, 2014; Table 1). Large and well-designed controlled trials, however, are needed to better delineate the potential role of cannabinoids as an adjunct or alternative treatment to conventional approaches to PTSD management.

Table 1 Effect of Drugs Interfering With Endocannabinoid System in Humans

Drug (Dose)	Subject Condition	Model	Behavioral Effect	Physiological Alteration (If Evaluated)	References
Dronabinol (7.5 mg; oral; single administration)	Healthy volunteers	Fear extinction	↓ Recovery of fear	↓ AMY and ↑ HIP and PFC reactivity to a CS	Rabinak et al. (2014, 2013)
Nabilone (from 2 to 8 mg a day; oral; 4 weeks)	Patients suffering from anxiety		Anxiolytic		Fabre and McLendonn (1981)
Nabilone (from 0.5 to 3 mg; oral; 7 weeks)	Military personal with PTSD		↓ Nightmares		Jetly et al. (2015)
Nabilone (from 0.5 to 6 mg; oral; range, 1 day to 36 weeks)	PTSD patients		↓ Insomnia, nightmares, and PTSD symptoms		Cameron et al. (2014)
Nabilone (from 0.5 to 6 mg; oral; range, 4–12 months)	PTSD patients		↓ Nightmares, flashbacks		Fraser (2009)
CBD (300 mg; oral; single administration)	Healthy volunteers	Simulated public-speaking	Anxiolytic		Zuardi et al. (1993)
CBD (400 mg; oral; single administration)	Healthy volunteers	SPECT	Anxiolytic	Effects mediated by action on limbic and paralimbic areas	Crippa et al. (2004)

CBD (600 mg; oral; single administration)	SAD patients	Simulated public-speaking	Anxiolytic		Bergamaschi et al. (2011)
CBD (400 mg; oral; single administration)	GAD patients	SPECT	Anxiolytic	Effects mediated by action on limbic and paralimbic areas	Crippa et al. (2011)
CBD (32 mg; oral; single administration)	Healthy volunteers	Fear extinction	↑ Extinction learning		Das et al. (2013)
Rimonabant (90 mg; oral; single administration)	Healthy volunteers	Simulated public-speaking	Anxiogenic		Bergamaschi et al. (2011)

AMY, amygdala; *CS*, conditioned stimulus; *GAD*, generalized anxiety disorder; *HIP*, hippocampus; *PFC*, prefrontal cortex; *SAD*, social anxiety disorder; *SPECT*, single photon emission computed tomography; ↓, decrease/impair; ↑, increase/facilitate.

The eCB system may be involved in the pathophysiology of PTSD. An in vivo imaging study demonstrated higher CB_1 expression in specific brain areas of PTSD patients. Interestingly, increased CB_1 receptor binding was associated with a significant reduction in the peripheral concentrations of AEA and cortisol (Neumeister et al., 2013) and with increased attention bias to threat (Pietrzak et al., 2014). In line with this, PTSD patients with lower peripheral AEA levels exhibited more intrusive symptoms (Hill et al., 2013). It has also been shown that baseline anxiety inversely correlates with peripheral AEA content in healthy individuals (Dlugos, Childs, Stuhr, Hillard, & de Wit, 2012). The involvement of the eCB system in anxiety also emerged from clinical trials of the CB_1 receptor antagonist SR141716A (or Rimonabant). This drug, previously approved for the treatment of obesity and related metabolic disorders, was soon removed from the market due to its psychiatric side effects, mainly anxiety and depression (Moreira & Crippa, 2009). Recently, it was observed that Rimonabant increased public speaking anxiety in healthy humans (Bergamaschi et al., 2014).

Clinical studies also indicated that CBD has anxiolytic properties (Campos, Moreira, Gomes, Del Bel, & Guimaraes, 2012). Following the initial report that it diminishes the anxiogenic effects of high doses of Δ^9-THC (Zuardi, Shirakawa, Finkelfarb, & Karniol, 1982), it was demonstrated that CBD reduces anxiety in healthy volunteers during a neuroimaging study or after a simulated public-speaking procedure (Crippa et al., 2004; Zuardi, Cosme, Graeff, & Guimaraes, 1993). More recently, using the latter procedure, Bergamaschi et al. showed that CBD decreases anxiety in treatment-naive social phobia patients (Bergamaschi et al., 2011). CBD also enhanced consolidation of fear extinction in humans (Das et al., 2013). Neuroimaging studies show that CBD changes the activity of brain regions related to the control of emotional process. It attenuates blood oxygenation in the amygdala and the anterior and posterior cingulate cortex in subjects exposed to fearful faces, impairs the connectivity between the prefrontal and subcortical regions (Fusar-Poli et al., 2010), and decreases the activation of left amygdala–hippocampal complex and left posterior cingulate gyrus (Crippa et al., 2004).

2.1.2 Polymorphisms in the Cannabinoid Receptors/Enzymes Genes

Several studies have investigated associations between single-nucleotide polymorphisms (SNPs) in cannabinoid-related genes and fear states, anxiety and stress-related disorders. For example, in healthy, drug-free, individuals carrying the homozygous allele AA, but not the GG carriers or heterozygotes, of the particular SNP rs2180619, located in the regulatory region of CB_1 gene (6q14-q15), presented failure to extinguish the fear-potentiated eye-blink startle reflex. In addition to that, the SNP rs1049353, located in the coding region of CB_1 gene (CNR1), was associated with normal extinction (Heitland et al., 2012). Failure to extinguish fear can lead to a state of persistent fear and anxiety, potentially resulting in PTSD (Amstadter, Nugent, & Koenen, 2009; Jovanovic et al., 2010). Also supporting possible involvement of CNR1 polymorphisms in the genetic causes of PTSD, a study evaluating the association between SNPs in the CNR1 gene and the presence of attention deficit hyperactivity disorder and PTSD in children and their parents found that SNP haplotypes in the CNR1 gene (CA and CG) were significantly associated with PTSD in some individuals (Lu et al., 2008). Moreover, carriers of the C allele of another polymorphism in the CNR1 gene (rs2023239) present increased CB_1 receptor density in the brain (Schacht, Hutchison, & Filbey, 2012), in addition to increased hippocampal volume. Considering that CB_1 receptors are expressed in the hippocampus, as will be discussed later, and that reduced hippocampal volume has been associated with the pathophysiology of PTSD, this polymorphism in the CNR1 gene could be an important factor in protecting individuals from development of PTSD after traumatic experiences.

In addition to the CB_1 receptors, studies evaluating the relationship between a common SNP in the FAAH enzyme gene (385AA), which converts a conserved proline residue to threonine (rs324420 or Pro129Thr) resulting in reduced FAAH expression in lymphocytes and increased circulating levels of AEA (Chiang, Gerber, Sipe, & Cravatt, 2004; Sipe et al., 2010), found decreased amygdala activation in response to fearful faces and trait anxiety levels (Hariri et al., 2009). Moreover, carriers of this low-expressing FAAH variant also present faster habituation of amygdala reactivity to repeated "threatening faces" (Gunduz-Cinar et al., 2013).

Genetic changes in the cannabinoid system could also interact with other neurotransmitters such as serotonin (5-HT). Although conflicting results do exist, the presence of the short allele 5-HTTLPR, a functional polymorphism in the serotonin transporter (5-HTT) gene (SLCA4) promoter, is associated with reduced 5-HTT expression (Lesch et al., 1996) and the presence of psychiatric disorders, including anxiety (Lesch et al., 1996; Ming et al., 2015; Osher, Hamer, & Benjamin, 2000; Schneck et al., 2016). Considering that CB_1 receptors can inhibit serotonin release and increase their metabolites in the rodent brain (Nakazi, Bauer, Nickel, Kathmann, & Schlicker, 2000; Tzavara et al., 2003), a study evaluated functional polymorphisms in the promoter regions of both SLC6A and CNR1 genes in anxious phenotypes (Lazary et al., 2009). The results showed that SLC6A polymorphism was associated with higher anxiety scores only when a SNP in the promoter region of the CNR1 (rs2180619) was also present. Specifically, individuals with the genotype GG and haplotypes "GTGC" (related to lower expression of CB_1) or "AATT" (related to higher expression of CB_1) of the rs2180619, and also carrying the SS genotype of the 5-HTTLPR gene, presented higher risk for anxiety than the other variants. The authors suggest that these combinations result in higher anxiety as a consequence of deficient inhibition of serotonin release by CB_1 receptors, which in turn leads to extremely high or extremely low levels of serotonin (depending on the CB_1 haplotype), probably by a compensation mechanism, which are both implicated in anxiety disorders. Interestingly, mice-lacking CB_1 receptors particularly in serotonergic neurons have increased anxiety-like behavior independent on previous stress experience (Dubreucq et al., 2012).

In summary, these studies suggest that genetic alterations in the eCB system could predict the occurrence of fear and anxiety disorders. More studies, however, aimed particularly at unveiling the functional impact of these polymorphisms are still needed.

3. eCBs IN ANIMAL MODELS OF FEAR AND ANXIETY

Corroborating the clinical findings, studies with laboratory animals have also indicated the involvement of eCBs in anxiety and fear. Two main categories of tests have been used in these studies: (1) innate anxiety-related tests (File, Lippa, Beer, & Lippa, 2004) and (2) conditioned or learned fear-related tests (LeDoux, 2014). The first takes into consideration immediate reaction to novelty—i.e., novel environment, individual, or object—whereas the second requires learned experience about a harmful stimulus, usually paired with a previous neutral condition—i.e., footshock paired with auditory cue or context.

Considering the scope of this review, we will focus on studies investigating the impact of eCBs modulation in anxiety- and fear-like responses. Table 2, however, will also show results of studies testing the effects of synthetic cannabinoid agonists.

3.1 Effects in Innate Fear Models

The elevated plus-maze (EPM), open field, the light/dark box, and social interaction tests are the most commonly used tests to assess innate fear

Table 2 Effect of Drugs Interfering With Endocannabinoid System in Animal Models

Drug (Dose)	Possible Mechanism	Specie	Paradigm	Effect	References
SR141716A (0.25–10 mg/kg)	CB_1 antagonist/inverse agonist	Mice, rats	EPM/ETM/LDB/SI	Anxiogenic	Akinshola, Chakrabarti, and Onaivi (1999), Arevalo, de Miguel, and Hernandez-Tristan (2001), Balerio, Aso, and Maldonado (2006), Navarro et al. (1997), Patel and Hillardd (2006), Terzian, Micale, and Wotjak (2014), and Thiemann, Watt, Ledent, Molleman, and Hasenohrl (2009)
SR141716A (0.3–3 mg/kg)		Mice, rats	EPM/LDB	Anxiolytic	Akinshola et al. (1999), Griebel, Stemmelin, and Scatton (2005), and Rodgers, Haller, Halasz, and Mikics (2003)
SR141716A (1–10 mg/kg)		Mice, rats	EPM/EZM	No effect	Bortolato et al. (2006), Dubreucq et al. (2012), Kathuria et al. (2003), and Rodgers et al. (2003)
SR141716A (0.3–20 mg/kg)		Rat	FC—context	↓ Acquisition	Varga, Kassai, and Gyertyan (2012)
SR141716A (1–10 mg/kg)		Mice, rats	FC—cued and context	↓ Fear extinction	Finn, Beckett, et al. (2004), Llorente-Berzal et al. (2015), Marsicano et al. (2002), Pamplona, Prediger, Pandolfo, and Takahashi (2006), Plendl and Wotjakk (2010), and Suzuki et al. (2004)

Continued

Table 2 Effect of Drugs Interfering With Endocannabinoid System in Animal Models—cont'd

Drug (Dose)	Possible Mechanism	Specie	Paradigm	Effect	References
SR141716A (1–3 mg/kg)		Mice, rats	FC—cued and context	No effect	Dubreucq et al. (2012), Mazzola, Micale, and Drago (2003), and Pamplona et al. (2006)
SR141716A (1.5–5 mg/kg)		Rat	FS	↓ Fear extinction	Chhatwal, Davis, Maguschak, and Ressler (2005)
AM251 (1–10 mg/kg)		Mice	EPM	Anxiogenic	Patel and Hillardd (2006), Rodgers, Evans, and Murphy (2005), and Thiemann et al. (2009)
AM251 (0.1–5 mg/kg)		Rat	EPM	No effect	Haller et al. (2007)
AM251 (3–5 mg/kg)		Mice, rats	FC—discrete cue	↑ Fear expression No effect on fear extinction	Arenos, Musty, and Bucci (2006) and Reich, Mohammadi, and Alger (2008)
AM251 (0.3–3 mg/kg)		Mice, rats	FC—context	↓ Fear expression	Arenos et al. (2006) and Mikics et al. (2006)
AM281 (1–4 mg/kg)		Mice	LDB/OF	No effect	Rutkowska, Jamontt, and Gliniak (2006)
AM281 (1–4 mg/kg)		Mice	FC—context	↑ Fear expression	Lisboa, Gomes, et al. (2015)

Delta9-THC (0.015–5 mg/kg)	Nonselective CB_1/CB_2 agonist	Mice, rats	LDB/EPM	Anxiolytic	Berrendero and Maldonadoo (2002), Rubino et al. (2007), and Valjent, Mitchell, Besson, Caboche, and Maldonado (2002)
Delta9-THC (0.25–20 mg/kg)		Mice, rats	LDB/EPM	Anxiogenic	Long et al. (2010), O'Brien et al. (2013), Onaivi, Green, and Martin (1990), Patel and Hillardd (2006), and Schramm-Sapyta et al. (2007)
WIN55,212-2 (0.16, 0.32, 0.64, 1.25, 2.5, 5 mg/kg, IP)		Rats	Predator exposure	Anxiolytic	Lisboa, Camargo, da Silva, Resstel, and Guimaraes (2014)
WIN55,212-2 (2.5–5.0 mg/kg)		Mice, rats	FC—context	↑ Fear	Mikics et al. (2006) and Pamplona and Takahashii (2006)
CP-55940 (1–50 μg/kg)	CB_1 agonist	Mice	EPM	Low dose—anxiolytic; high dose—anxiogenic	Genn, Tucci, Marco, Viveros, and File (2004), Marco et al. (2004), and Rey, Purrio, Viveros, and Lutz (2012)
CP-55940 (75–125 μg/kg)		Rats	EPM	Anxiogenic	Arevalo et al. (2001) and Marin et al. (2003)
HU210 (5, 20 or 80 mg/kg, IP)		Rats	Ultrasound-induced aversive response	Antiaversive effects and ↑ plasma corticosterone	Finn, Beckett, et al. (2004)
HU-210 (10–50 μg/kg)		Rats	EPM	Biphasic effect	Hill and Gorzalkaa (2004)
HU-210 (0.1 mg/kg)		Rats	FC	↓ Fear expression	Mackowiak, Chocyk, Dudys, and Wedzony (2009)

Continued

Table 2 Effect of Drugs Interfering With Endocannabinoid System in Animal Models—cont'd

Drug (Dose)	Possible Mechanism	Specie	Paradigm	Effect	References
AM360 (1–3 mg/kg; acute)	CB_2 antagonist	Mice	LDB	Anxiogenic	Garcia-Gutierrez, Garcia-Bueno, Zoppi, Leza, and Manzanares (2012)
AM360 (1–3 mg/kg; chronic)		Mice	LDB/EPM	Anxiolytic	Garcia-Gutierrez et al. (2012)
AM360 (3 mg/kg)		Mice	FC—cued	↑ Fear extinction	Llorente-Berzal et al. (2015)
JWH133 (3–10 mg/kg; acute)	CB_2 agonist	Mice	LDB	No effect	Garcia-Gutierrez et al. (2012)
JWH133 (0.5–2 mg/kg)		Mice	EPM	Anxiolytic	Busquets-Garcia et al. (2011)
JWH133 (0.5–2 mg/kg; chronic)		Mice	EPM/LDB	Anxiogenic	Garcia-Gutierrez et al. (2012)
JWH015 (1–20 mg/kg)		Mice	LDB	Anxiogenic	Onaivi et al. (2008)
SB366791 (0.1–1 mg/kg)	TRPV1 antagonist	Mice	EPM	Anxiolytic	Micale et al. (2009b)
Capsazepine (1.5–10 μg/kg)		Rat	EPM	Anxiolytic	Kasckow, Mulchahey, and Geracioti (2004)
SB366791 (1–3 mg/kg)		Mice	FC—cued	No effect	Llorente-Berzal et al. (2015)
Olvanil (0.1–5 mg/kg)	TRPV1 agonist	Rats	EPM	Anxiogenic	Kasckow et al. (2004)

AM404 (1–4 mg/kg)	Reuptake inhibitor (?)	Mice	LDB	No effect	Rutkowska et al. (2006)
AM404 (0.3–10 mg/kg)		Mice, rat	EPM/MB	Anxiolytic	Bortolato et al. (2006), Braida, Limonta, Malabarba, Zani, and Sala (2007), Gomes, Casarotto, Resstel, and Guimaraes (2011), and Patel and Hillardd (2006)
AM404 (3 mg/kg)		Mice	FC—Cued	↑ Fear extinction	Llorente-Berzal et al. (2015)
AM404 (10 mg/kg)		Rat	FS	↓ Fear reinstatement	Chhatwal et al. (2005)
VDM11 (3 mg/kg)		Mice	FC—cued	No effect	Llorente-Berzal et al. (2015)
PF-3845 (1–10 mg/kg)	FAAH inhibitor	Mice—C57Bl6	MB	Anxiolytic	Kinsey, O'Neal, Long, Cravatt, and Lichtman (2011)
URB597 (0.1–10 mg/kg)		Mice, rats, Syrian hamster	EPM/EZM/LDB/	Anxiolytic	Busquets-Garcia et al. (2011), Hill, Karacabeyli, and Gorzalka (2007), Kathuria et al. (2003), Moise, Eisenstein, Astarita, Piomelli, and Hohmann (2008) Moreira, Kaiser, Monory, and Lutz (2008), and Patel and Hillardd (2006)
URB597 (0.03–10 mg/kg)		Mice, rats, Syrian hamsters	EPM/SD	No effect	Haller et al. (2009), Moise et al. (2008), Naderi et al. (2008), and Naidu et al. (2007)
URB597 (1 mg/kg)		Mice	FC—context	↓ Fear expression	Busquets-Garcia et al. (2011)

Continued

Table 2 Effect of Drugs Interfering With Endocannabinoid System in Animal Models—cont'd

Drug (Dose)	Possible Mechanism	Specie	Paradigm	Effect	References
URB597 (0.3–3 mg/kg)		Mice	FC—context	↑ Fear extinction	Lisboa, Gomes, et al. (2015)
URB597 (0.3 mg/Kg)		Mice	FC—contex 7 days after IPE	Prevented stress-induced fear extinction impairment	Lisboa, Camargo, da Silva, et al. (2014)
AA-5-HT (0.1–5 mg/kg)	FAAH inhibitor/ TRPV1 antagonist	Mice	EPM	Anxiolytic	Micale et al. (2009b)
OMDM198 (1–5 mg/kg)		Mice	FC—cued	No effect	Llorente-Berzal et al. (2015)
JZL184 (1–408 mg/kg)	MAGL inhibitor	Mice, rat	EPM/EZM/ MB	Anxiolytic; ↑ corticosterone	Aliczki et al. (2013), Busquets-Garcia et al. (2011), Kinsey et al. (2011), and Sciolino, Zhou, and Hohmann (2011)
JZL184 (4–8 mg/kg)		Mice	FC—cued	↑ Fear expression	Llorente-Berzal et al. (2015)

EPM, elevated plus-maze; *ETM*, elevated T maze; *EZM*, elevated zero maze; *FC*, fear conditioning; *FS*, fear startle; *HB*, hole board; *IPE*, inescapable predator (rat) exposure; *LDB*, light dark box; *MB*, marble burrying; *NGT*, novelty grooming test; *NOI*, novel object investigation; *OF*, open field; *PA*, passive avoidance; *SD*, social defeat; *SI*, social interaction; ↓, decrease/impair; ↑, increase/facilitate.

and anxiety (Cryan & Holmes, 2005). They are based on the innate conflict between the potentially aversive situation (such as bright open spaces or unknown individuals) and animal's natural drive to explore unknown areas or to contact unknown/known individuals (Cryan & Holmes, 2005).

Since the development of pharmacological tools to investigate the participation of the eCB system in anxiety-like responses, conflicting data have been observed, especially regarding the use of CB_1 antagonists (Griebel et al., 2005; Haller, Bakos, Szirmay, Ledent, & Freund, 2002; Lafenetre, Chaouloff, & Marsicano, 2007; Lan et al., 1999; Patel & Hillard, 2006; Rinaldi-Carmona et al., 1994; Table 2). These contradictory results could be due to a number of factors, such as experimental condition and protocol, treatment dosage, animal strain, or even the antagonist of choice (Lafenetre et al., 2007; Parolaro, Realini, Vigano, Guidali, & Rubino, 2010). For example, although blockade of CB_1 receptor moderately increases excitatory neurotransmission (Slanina & Schweitzer, 2005), indicating a tonic eCB activity (Ruehle, Rey, Remmers, & Lutz, 2012), the two main antagonists used, SR141716A (Rimonabant) and AM251, are in fact inverse agonists (Pertwee, 2005; Sink et al., 2010). Therefore, under certain circumstances they can produce effects per se (Roberto et al., 2010). In addition, pharmacological effects of these drugs have been reported in CB_1-knockout (CB_1 KO) mice, indicating the presence of non-CB_1 action (Haller et al., 2002).

Even if discrepant results exist, possibly due to different background strain and particular experimental conditions (Haller, Varga, Ledent, Barna, & Freund, 2004; Haller, Varga, Ledent, & Freund, 2004; Ledent et al., 1999), CB_1 KO mice are more anxious when tested in anxiety models (Haller, Varga, Ledent, Barna, et al., 2004; Haller, Varga, Ledent, & Freund, 2004; Martin, Ledent, Parmentier, Maldonado, & Valverde, 2002; Table 3). Furthermore, benzodiazepine drugs do not induce anxiolytic-like effects in CB_1 KO mice (Uriguen, Perez-Rial, Ledent, Palomo, & Manzanares, 2004).

Pharmacological (Table 2) and genetic (Table 3) manipulations of the eCBs reuptake/degradation enzymatic processes also indicate their involvement in anxiety modulation. The most common tool used to inhibit eCBs reuptake is AM404 (Beltramo & Piomelli, 2000), which increases the levels of both AEA and 2-AG (Beltramo et al., 1997; Bisogno et al., 2001; Hajos, Kathuria, Dinh, Piomelli, & Freund, 2004). In the EPM, whereas low doses of AM404 showed anxiolytic-like, CB_1-dependent effect (Bortolato et al., 2006; Patel & Hillard, 2006), higher doses tend to increase anxiety (Patel & Hillard, 2006), which is probably related to AM404 targeting other molecules, such as TRPV1 channel (Zygmunt et al., 1999). The use of the FAAH

Table 3 Effect of Genetic Deletion of Endocannabinoid System Components in Animal Models

Genetic Model	Deletion Location	Specie	Paradigm	Effect	References
CB_1 KO	Ubiquitous deletion	Mice	EPM	No alteration	Dubreucq et al. (2012) and Ledent et al. (1999)
			EPM/LDB/OF/SI/NOI	Anxiogenic	Haller, Varga, Ledent, Barna, et al. (2004), Jacob et al. (2009), Maccarrone et al. (2002), and Terzian, Micale, et al. (2014)
			FC—cued and context	↓ Fear extinction	Dubreucq et al. (2012), Jacob, Marsch, Marsicano, Lutz, and Wotjak (2012), Kamprath et al. (2009), and Marsicano et al. (2002)
			FC—context	↓ Fear expression	Mikics et al. (2006)
Glu-CB_1 KO	Deletion on glutamatergic cortical neurons	Mice	EPM/HB	No alteration	Dubreucq et al. (2012) and Rey et al. (2012)
			OF/SI/NOI	Anxiogenic	Haring, Kaiser, Monory, and Lutz (2011), Jacob et al. (2009), and Terzian, Micale, et al. (2014)
			FC—cued	↓ Fear extinction	Dubreucq et al. (2012), Kamprath et al. (2009), and Llorente-Berzal et al. (2015)
GABA-CB_1 KO	Deletion on GABAergic neurons	Mice	EMP/HB	No alteration	Rey et al. (2012)
			EPM/SI/NOI	Anxiolytic	Haring et al. (2011), Dubreucq et al. (2012), Terzian, dos Reis, Guimaraes, Correa, and Resstel (2014), and Terzian, Micale, et al. (2014)
			FC—cued	No alteration	Dubreucq et al. (2012)
			FC—cued	↓ Fear extinction	Llorente-Berzal et al. (2015)

D1-CB_1	Deletion on dopaminergic D1 receptor expression neurons	Mice	SI/NOI	No alteration	Haring et al. (2011)
			EPM/LDB/SI/NOI/NGT	Mild anxiogenic	Terzian, Drago, Wotjak, and Micale (2011)
			FC—cued and context	↓ Fear extiction	Terzian et al. (2011)
5HT-CB_1	Deletion on 5HT neurons	Mice	EPM/LDB/SI	Anxiogenic	Dubreucq et al. (2012) and Haring et al. (2015)
			FC—cued	No alteration	Dubreucq et al. (2012)
CB_2 KO	Ubiquitous deletion	Mice	EPM/LDB	Anxiogenic	Ortega-Alvaro, Aracil-Fernandez, Garcia-Gutierrez, Navarrete, and Manzanares (2011)
FAAH KO	Deletion of FAAH enzyme	Mice	EPM/LDB	Anxiolytic	Moreira et al. (2008)
DGL-alfa	Deletion of DGL-alfa enzyme	Mice—C57Bl6	LDB/EZM/OF	Anxiogenic	Jenniches et al. (2016) and Shonesy et al. (2014)
			FC—cued	↑ Fear expression/Impaired fear extinction	Jenniches et al. (2016)
MAGL	Deletion of MAGL enzyme	Mice—C57Bl6	LDB/MB	Anxiogenic	Imperatore et al. (2015)

EPM, elevated plus-maze; *ETM*, elevated T maze; *EZM*, elevated zero maze; *FC*, fear conditioning; *FS*, fear startle; *HB*, hole board; *LDB*, light dark box; *MB*, marble burrying; *NGT*, novelty grooming test; *NOI*, novel object investigation; *OF*, open field; *PA*, passive avoidance; *SD*, social defeat; *SI*, social interaction; ↓, decrease/impair; ↑, increase/facilitate.

inhibitor URB597 allows us to evaluate the role of AEA, while avoiding 2-AG effects. This drug selectively increases AEA, but not 2-AG, levels in the mouse brain after systemic injection. URB597 also induces anxiolytic-like effects in several tests, such as the EPM and zero maze, suggesting that AEA signaling induces anxiolytic-like activity (Kathuria et al., 2003). These effects are mediated via CB_1 receptors (Busquets-Garcia et al., 2011). Corroborating the pharmacological evidence, genetic deletion of FAAH in mice induced reduction in anxiety levels in different tests (Moreira et al., 2008), suggesting that FAAH inhibition is an important mechanism for the alleviation of anxiety, particularly under high aversive situations (Haller et al., 2009).

More recently, the role of 2-AG in the control of anxiety was also assessed. Inhibition of 2-AG metabolizing enzyme MAGL by JZL184 produced anxiolytic-like effects in the EPM, but also only on highly aversive conditions (Sciolino et al., 2011). Contrasting with URB597 results, however, these effects were CB_2 dependent. Additionally, animals lacking the 2-AG synthesis enzyme, diacylglycerol lipase (DAGLα-KO mice), which exhibits 80% reduction of central, but not circulating, 2-AG levels, presents greater anxiety levels (Shonesy et al., 2014), which are reversed after treatment with JZL184. Nevertheless, genetic deletion of MAGL is also reported to promote anxiogenic-like effects in mice through CB_1 receptors by reducing glutamate levels and altering BLA-medial prefrontal cortex (MPFC) connectivity (Imperatore et al., 2015).

Overall, these studies point out to an important role of eCB system in the modulation of anxiety behavior in models involving innate fear.

3.2 Effects on Induced/Learned Fear

As described for innate fear, the eCB system also plays an important role in learned fear responses. It modulates different phases of fear, from acquisition to expression of fear responses, and to reinstatement of extinct memories. As described earlier, however, contradictory results are also found, depending on the factors, such as experimental conditions, animal strains, and drug selectivity.

The classical fear-conditioning protocol is based on an association between a neutral stimulus (CS) and an aversive stimulus (unconditioned stimulus, US). After one or more CS + US pairings, the presentation of the CS alone elicits fear responses. The usual behavioral parameter evaluated is freezing behavior, characterized by the absence of any movements except those necessary for breathing (LeDoux, 2000, 2014). However, after the CS+US association, the exposure to the CS alone can induce two (opposite) processes: reconsolidation or extinction of the aversive memory, promoting a strengthening or decrease in fear responses, respectively.

Few studies have addressed the participation of the eCB system in *fear acquisition* (Guindon & Hohmann, 2009). Rodents present increased freezing during the conditioning session after administration of the CB_1 antagonist AM251 (Reich et al., 2008), specifically for the paired CS+US presentations. Others studies did not find any impairment in memory acquisition/initial consolidation, as the pretraining administration of another CB_1 antagonist, SR141716A did not affect contextual fear expression (Suzuki et al., 2004). Interestingly, URB597 (FAAH inhibitor) enhanced aversive memory acquisition (Mazzola et al., 2009), although this result was not always observed (Laricchiuta, Centonze, & Petrosini, 2013). Moreover, the enhancing effects of URB597 on memory acquisition seem to be mediated by a bioactive lipid other than AEA, and do not involve cannabinoid receptors (Mazzola et al., 2009). Accordingly, CB_1 KO mice were able to learn a tone-foot shock association (Marsicano et al., 2002). In addition, FAAH inhibition does not affect the aversive memory consolidation phase (Busquets-Garcia et al., 2011).

Fear expression is the response to the first CS exposure after the conditioning session, resulting from acquisition, consolidation and retrieval of the fear memory. There seems that the eCB system plays a different role in cued and contextual learning. One study observed reduced freezing to context and increased freezing to tone after treatment with the CB_1 antagonist AM251 (Arenos et al., 2006), whereas a later work obtained the opposite effect with similar doses of this antagonist (Sink et al., 2010). In agreement with this later study, we recently demonstrated that administration of the CB_1 antagonist AM281 to mice increased fear expression in the contextual fear conditioning, whereas URB597 induced opposite effect (Lisboa, Gomes, et al., 2015). URB597 and the eCBs reuptake inhibitor AM404 also reduced the freezing response in another study (Llorente-Berzal et al., 2015), whereas the MAGL inhibitor JZL184 induced opposite effect depending on CB_1 signaling in GABA neurons (Llorente-Berzal et al., 2015), indicating a differential role for the 2-AG. Therefore, these data contrast to results obtained from innate fear responses, where both AEA and 2-AG attenuate anxiety-like behaviors.

The eCB system is highly involved in *extinction* of fear memory, while not being essential for appetitive memories extinction (Manwell et al., 2009). This reinforces the "on-demand" participation of the eCB system on aversive situations. CB_1 KO mice present impaired short-term (within-session) as well as long-term (between-session) extinction after cued conditioning (Marsicano et al., 2002). Also, treatment with SR141716A before extinction training produced similar results in wild-type mice. In the same study, it was suggested that CB_1 receptors are not involved in consolidation of the extinction memory, as no effect of pharmacological blockade immediately after extinction training was found (Marsicano et al., 2002). The lack of habituation to stimuli suggests that CB_1 receptor signaling critically participates in nonassociative learning processes (Kamprath et al., 2006), which is also observed for the habituation to homotypic stressors (Patel, Roelke, Rademacher, & Hillard, 2005). Furthermore, fear generalization can be observed (Reich et al., 2008), probably related to protocol intensity and due to the fact that the fear-reducing effect of the eCB system greatly depends on the strength of the aversive stimulus presented (Kamprath et al., 2009).

Additionally, several studies have shown that inhibition of eCBs uptake or AEA degradation facilitates extinction (Bitencourt, Pamplona, & Takahashi, 2008; Lisboa, Gomes, et al., 2015; Manwell et al., 2009; Pamplona, Bitencourt, & Takahashi, 2008) that is more resistant to reinstatement (Chhatwal et al., 2005). The effects of FAAH inhibition on extinction depend on CB_1 receptors (Gunduz-Cinar et al., 2013; Laricchiuta et al., 2013). In a genetic mouse model of increased nitric oxide signaling, URB597 was also able to facilitate fear extinction (Lisboa, Gomes, et al., 2015). Furthermore, URB597 treatment counteracted stress-related effects on extinction of aversive memories (Laricchiuta et al., 2013). We recently demonstrated that deficits in contextual fear extinction in mice observed a week after an inescapable predator exposure (a live rat) was attenuated by repeated treatment with URB597 (Lisboa, Camargo, da Silva, et al., 2014). Finally, genetically modified mice presenting 80% reduction of brain 2-AG showed fear extinction deficit (Jenniches et al., 2016).

In summary, these results suggest that eCBs can modify the balance between (re)consolidation and extinction of the original aversive memory. In this way, eCBs may protect the organism from excessive interference induced by aversive stimuli. Consequently, the eCB system can function as a regulatory buffer system for emotional responses, as previously suggested (Ruehle et al., 2012).

4. BRAIN SITES MODULATING CANNABINOID EFFECTS IN ANIMAL MODELS

4.1 Medial Prefrontal Cortex

The PFC is the center for executive functioning, responsible for mediating a range of cognitive, behavioral, and neuroendocrine processes (Rose & Woolsey, 1948; Uylings & van Eden, 1990). The PFC is a structural and functionally heterogeneous brain region. The MPFC in rodents is subdivided into dorsomedial and ventromedial regions (Heidbreder & Groenewegen, 2003). The ventromedial portion of the PFC (vMPFC) comprises the prelimbic (PL), infralimbic (IL), and medial orbital cortices (Heidbreder & Groenewegen, 2003; Uylings, Groenewegen, & Kolb, 2003), which have constantly been implicated in the regulation of mood, emotion, and stress. These subregions present differential modulation and connections to other brain areas (Holmes & Wellman, 2009).

The eCB system is present throughout the PFC of rodents (Marsicano & Lutz, 1999; Tsou, Brown, et al., 1998). CB_1 receptors are highly expressed in GABAergic interneurons, and in less extent in glutamatergic pyramidal neurons (Lafourcade et al., 2007). In addition to CB_1 and CB_2 receptors, eCB synthesis/degradation enzymes are also abundantly expressed in the MPFC (Basavarajappa, 2007; Marsicano & Lutz, 1999; Onaivi et al., 2006; Ramikie & Patel, 2012). Changes in CB_1 receptor expression during a stressful situation or following genetic manipulations can promote morphological changes in the corticoamygdalar pathway, along with marked dendritic atrophy in MPFC pyramidal neurons (Hill, Hillard, & McEwen, 2011). Likewise, signaling alterations in CB_1 receptor might be due to alterations in structural morphology within limbic areas after chronic stress exposure, eventually influencing amygdala-dependent information processing that shifts the emotional balance toward proaversive responses, subsequently developing pathological mood states (McLaughlin, Hill, & Gorzalka, 2014).

Under stressful situations, plastic changes in the eCB system are observed in the vMPFC. For example, AEA signaling is reduced, contributing to anxiety-like outcomes. In addition, overexpression of FAAH by lentivirus injection into the PFC elicits anxiogenic-like response in a stressful environment, whereas local URB597 administration was anxiolytic (Rubino, Realini, et al., 2008; Table 4). Such effects might involve CB_1 receptor-mediated suppression of GABAergic inhibition of PFC projection neurons, increasing PFC-amygdala pathway (McLaughlin et al., 2014). However,

Table 4 Local Effect of Drugs Interfering With Endocannabinoid System in Animal Models

Brain Region	Drug (Dose; Injection Site)	Possible Mechanism	Species	Animal Model	Main Effects	References
MPFC	AM251 (100 nmol); intra-PL + IL	CB_1 antagonist	Rats	CFC	↑ Fear expression (lower shock intensity)	Lisboa, Reis, et al. (2010)
	AM251 (100 nmol); intra-PL + IL	CB_1 antagonist	Rats	CFC	No effect (higher shock intensity)	Lisboa, Reis, et al. (2010)
	NIDA41020 (25–100 pmol); intra-PL + IL	CB_1 antagonist	Rats	CFC	↑ Fear expression	Uliana, Hott, Lisboa, and Resstel (2016)
	mAEA (0.1–10 μg); intra-MPFC	CB_1/TRPV1 agonists	Rats	EPM	Biphasic effect—low dose: anxiolytic; high dose: anxiogenic	Rubino, Realini, et al. (2008)
	AEA (5 pmol); intra-PL + IL	Partial agonist	Rat	CFC	↓ CER via CB_1 receptors	Lisboa, Reis, et al. (2010)
	2-AG (0.01–10 μg); intra-MPFC	CB_1 agonist	Rats	EPM	No effect	Rubino, Realini, et al. (2008)
	THC (2.5–25 μg); intra-CG + PL	CB_1 agonist	Rats	EPM	Anxiolytic (bell-shaped)	Rubino, Guidali, et al. (2008)
	ACEA (0.5–5 pmol); intra-PL AM251	CB_1 agonist CB_1 antagonist	Rats	EPM	Anxiolytic No effect; blocked ACEA effect	Fogaca, Aguiar, et al. (2012) and Fogaca, Lisboa, et al. (2012)

	THC (32 nmol); intra-MPFC	CB_1 agonist	Rat	EPM	Biphasic effect—low dose: anxiolytic; high dose: anxiogenic	Rubino, Guidali, et al. (2008)
	Capsazepine (5 μg); not specified	TRPV1 antagonist	Rats	EPM	No effect; blocked methanandamide effect	Rubino, Realini, et al. (2008)
	Capsazepine (1–60 nmol); intra-PL + IL	TRPV1 antagonist	Rats	EPM/VCT	Anxiolytic	Aguiar, Terzian, Guimaraes, and Moreira (2009) and Uliana et al. (2016)
	Capsazepine (1–60 nmol); intra-PL + IL	TRPV1 antagonist	Rats	CFC	↓ Fear expression	Terzian, Micale, et al. (2014)
	6-Iodonordihydrocapsaicin (3 nmol); intra-PL	TRPV1 antagonist	Rats	EPM/VT	Anxiolytic	Fogaca, Aguiar, et al. (2012) and Fogaca, Lisboa, et al. (2012)
	6-Iodonordihydrocapsaicin (3 nmol); intra-PL + IL	TRPV1 antagonist	Rats	CFC	↓ Fear expression	Terzian, Micale, et al. (2014)

Continued

Table 4 Local Effect of Drugs Interfering With Endocannabinoid System in Animal Models—cont'd

Brain Region	Drug (Dose; Injection Site)	Possible Mechanism	Species	Animal Model	Main Effects	References
	Capsaicin (1–5 μg); intra-MPFC	TRPV1 agonist	Rats	EPM	Anxiogenic	Rubino, Realini, et al. (2008)
	Capsaicin (1 nmol); intra-PL + IL		Rats	CFC	↑ Fear expression	Terzian, Micale, et al. (2014)
	AM404 (50 pmol); intra-PL AM251 (100 pmol)	AEA uptake/metabolism inhibitor CB_1 antagonist	Rats	EPM/VCT	Anxiolytic via CB_1 No effect; blocked AM404 effect	Lisboa, Borges, et al. (2015)
	AM404 (50 pmol); intra-PL + IL AM251 (100 pmol)	AEA uptake/metabolism inhibitor CB_1 antagonist	Rats	CFC	↓ Fear expression via CB_1 No effect; blocked AM404 effect	Lisboa, Reis, et al. (2010)
	URB597 (0.01–0.1 nmol); intra-PL AM251 (100 pmol)	FAAH inhibitor CB_1 antagonist	Rats	EPM/VCT	Anxiolytic via CB_1 No effect; blocked URB effect	Lisboa, Borges, et al. (2015)
	URB597 (0.01–1 μg); intra-MPFC	FAAH inhibitor	Rats	EPM	Anxiolytic	Rubino, Realini, et al. (2008)
	CBD (30 nmol); intra-PL or IL	?	Rat	CFC	↓ (PL) or ↑ (IL) CER	Lemos, Resstel, and Guimaraes (2010)

dHIP	WIN55, 212-2 (1.9–5.6 nmol)	CB_1, CB_2, TRPV1 receptor agonist	Rat	EPM	Anxiogenic via CB_1	Roohbakhsh, Moghaddam, Massoudi, and Zarrindast (2007)
	WIN55, 212-2 (0.48 nmol)		Mice	Hole board	Prevented anxiogenic effect of histamine	Zarrindast, Nasehi, Piri, and Bina (2010)
	AEA (0.17 ng per side)	Partial agonist	Rat	CFC	↑ and ↓, respectively, extinction and reconsolidation	de Oliveira Alvares, Pasqualini Genro, Diehl, Molina, and Quillfeldt (2008)
	AM404 (50 pmol) bilateral	AEA uptake/ metabolism inhibitor	Rat	VCT	Anxiolytic; via CB_1	Lisboa, Borges, et al. (2015)
	URB 597 (0.01 nmol) bilateral	FAAH inhibitor	Rat	VCT/EPM	Anxiolytic; via CB_1	Lisboa, Borges, et al. (2015)
	URB 597 (0.1 and 1 μg) unilateral		Rat	EPM	Anxiolytic	Hakimizadeh, Oryan, Hajizadeh Moghaddam, Shamsizadeh, and Roohbakhsh (2012)

Continued

Table 4 Local Effect of Drugs Interfering With Endocannabinoid System in Animal Models—cont'd

Brain Region	Drug (Dose; Injection Site)	Possible Mechanism	Species	Animal Model	Main Effects	References
	AMG 9810 (unilateral, 0.01 and 0.1 nmol)	TRPV1 antagonist	Rat	EPM	Anxiolytic	Hakimizadeh et al. (2012)
	Capsaicin (1 and 10 pmol unilateral)	TRPV1 agonist	Rat	EPM	Anxiogenic	Hakimizadeh et al. (2012)
	AM251 (0.3 nmol) bilateral	CB_1 receptor antagonist/partial agonist	Rat	CFC	↑ Fear expression, via NMDA-NO pathway	Spiacci, Antero, Reis, Lisboa, and Resstel (2016)
	AM251 (0.6 pmol) bilateral	CB_1 receptor antagonist/partial agonist	Rat	CFC	Prevented the ↓ of CFC retrieval induced by corticosterone	Atsak et al. (2012)
	AM251 (5.5 or 0.27 ng per side)	CB_1 antagonist/partial agonist	Rat	CFC	↓ and ↑, respectively, extinction and reconsolidation	de Oliveira Alvares et al. (2008)
vHIP	Capsazepine (2 nmol)	TRPV1 antagonist	Rat	EPM	Anxiolytic	Santos, Stern, and Bertoglio (2008)
	THC (16–32 nmol)	CB_1 agonist	Rat	EPM	Anxiolytic (bell-shaped)	Rubino, Guidali, et al. (2008)

	URB597 (0.03–3 nmol) AM251 (1.8–180 pmol)	AEA metabolism inhibitor CB_1 antagonist	Rat	EPM	Anxiogenic No effect	Roohbakhsh, Keshavarz, Hasanein, Rezvani, and Moghaddam (2009)
	AM404 (5–50 pmol) AM251 (0.01–1000 pmol)	AEA uptake/and metabolism inhibitor CB_1 antagonist	Rat	EPM, VCT	Anxiogenic (EPM) anxiolytic (EPM postrestraint, VCT) No effect	Campos, Ferreira, Guimaraes, and Lemos (2010)
	WIN55212-2 (11 nmol)	CB_1/CB_2/TRPV1 agonist	Rat	CFC	Prevented SPS-induced ↓ fear extinction (injection immediately after SPS)	Ganon-Elazar and Akiravv (2013)
	2-AG (0.52 nmol)	CB_1/CB_2 agonist	Rat	CFC	↓ Fear expression. Effect disappear under chronic pain	Rea et al. (2014)
dPAG	HU210 (0.1, 1, 5 μg/rat)	CB_1 agonist	Rat	Chemical stimulation DPAG	Antiaversive effect	Finn et al. (2003)

Continued

Table 4 Local Effect of Drugs Interfering With Endocannabinoid System in Animal Models—cont'd

Brain Region	Drug (Dose; Injection Site)	Possible Mechanism	Species	Animal Model	Main Effects	References
	HU210 (5 μg/rat)	CB_1 agonist	Rat	Ultrasound-induced aversive response	Antiaversive effect	Finn, Beckett, et al. (2004)
	URB 602 (0.1 nmol)	MAGL inhibitor; CB_1 activation	Rat	Stress-induced analgesia	Analgesic	Hohmann et al. (2005)
	URB 597 (0.1 nmol)	FAAH inhibitor; via CB_1	Rat	Stress-induced analgesia	Analgesic	Hohmann et al. (2005)
	AEA (5–50 pmol)	Partial agonist; via CB_1	Rat	EPM, VCT, CFC, predator exposure	Anxiolytic	Moreira, Aguiar, and Guimaraes (2007), Lisboa, Resstel, Aguiar, and Guimaraes (2008), Resstel, Lisboa, Aguiar, Correa, and Guimaraes (2008), Lisboa, Camargo, da Silva, et al. (2014), and Lisboa, Camargo, Magesto, Resstel, and Guimaraes (2014)

	AM404 (50 pmol)	AEA uptake/ metabolism inhibitor; via CB_1	Rat	EPM, VCT, CFC	Anxiolytic	Moreira et al. (2007), Lisboa, Resstel, Aguiar, and Guimaraes (2008), and Resstel, Tavares, et al. (2008)
	URB597 (0.01–0.1 nmol)	FAAH inhibitor; via CB_1	Rat	VCT, CFC, chemical stimulation dlPAG	Anxiolytic	Lisboa et al. (2008), Resstel et al. (2008), and Viana, Hott, Resstel, Aguiar, and Moreira (2015)
	ACEA (0.05–5 pmol)	CB_1 activation	Rat	EPM, electrical stimulation dPAG	Anxiolytic	Moreira et al. (2007) and Casarotto et al. (2012)
	2AG (5–500 pmol)	Full agonist; via CB_1 and CB_2	Rat	EPM	Anxiolytic	Almeida-Santos, Gobira et al. (2013)
	URB 602 (30–300 pmol)	MAGL inhibitor; via CB_1 and CB_2	Rat	EPM	Anxiolytic	Almeida-Santos, Gobira, et al. (2013)
BLA	WIN55, 212-2 (9.6 nmol)	CB_1/CB_2/ TRPV1 agonist	Rat	Elevated platform + inhibitory avoidance	↓ Stress effect	Ganon-Elazar and Akirav (2009)
	AEA (5 pmol) bilateral	Partial agonist	Rat	EPM	No effect	Lisboa and Guimarães (2007)

Continued

Table 4 Local Effect of Drugs Interfering With Endocannabinoid System in Animal Models—cont'd

Brain Region	Drug (Dose; Injection Site)	Possible Mechanism	Species	Animal Model	Main Effects	References
	THC (3.2 nmol) unilateral	CB_1 agonist	Rat	EPM	Anxiogenic	Rubino, Guidali, et al. (2008)
	AM251 (1.8 nmol) unilateral	CB_1 antagonist	Rat	EPM	No effect	Rubino, Guidali, et al. (2008)
	AM251 (11 pmol)	CB_1 antagonist		Elevated platform + inhibitory avoidance	↑ Stress effect	Ganon-Elazar and Akirav (2009)
	CBD (7.5–30 nmol) bilateral		Rat	EPM	No effect	Lisboa and Guimarães (2007)
	AM251 (2.5–500 ng/0.5 μL; Sigma)	CB_1 antagonist	Rat	Olfactory fear-conditioning	↓ Acquisition of fear memory	Tan et al. (2011)
	WIN55,212-2 (5–500 ng/0.5 μL)	CB_1/CB_2 antagonist	Rat	Olfactory fear-conditioning	↑ Subthreshold fear-conditioning stimuli	Tan et al. (2011)
	AM404 (5–50 ng/0.5 μL)		Rat	Olfactory fear-conditioning	↑ Subthreshold fear-conditioning stimuli	Tan et al. (2011)

	URB597 (10 ng)	FAAH inhibitor	Rat	EPM	↓ Anxiogenic-like effect induced by CRH injection	Gray et al. (2015)
	AM3506 (0 or 0.1 μg/μL) preextinction	FAAH inhibitor	29S1/Sv1mJ (S1) mice	Cue-fear conditioning	↓ Freezing on retrieval	Gunduz-Cinar et al. (2013)
	Rimonabant (2 μg/μL)	CB_1 antagonist	29S1/Sv1mJ (S1) mice	Cue-fear conditioning	Prevented the ↓ freezing on retrieval induced by AM3506 (1 mg/kg i.p.).	Gunduz-Cinar et al. (2013)
	ACPA (0.4–14.5 pmol)	CB_1 agonist	Rat	EPM	Anxiolytic	Zarrindast et al. (2008)
	THC (320–470 nmol)	CB_1 agonist	Mice	EPM	Anxiogenic	Onaivi, Chakrabarti, Gwebu, and Chaudhuri (1995)
	Rimonabant (0.5, 1.5 μg/side)	CB_1 antagonist	Rat (female)	Abstinence from palatable diet cycling EPM	Anxiogenic	Blasio et al. (2013)
CeA	ACPA (0.4–14.5 pmol)	CB_1 agonist	Rat	EPM	Anxiolytic	Zarrindast et al. (2008)
	THC (320–470 nmol)	CB_1 agonist	Mice	EPM	Anxiogenic	Onaivi et al. (1995)

Continued

Table 4 Local Effect of Drugs Interfering With Endocannabinoid System in Animal Models—cont'd

Brain Region	Drug (Dose; Injection Site)	Possible Mechanism	Species	Animal Model	Main Effects	References
BNST	CBD (30 and 60 nmol) bilateral	5HT1A agonist	Rat	CFC, VCT, EPM	Anxiolytic	Gomes, Resstel, and Guimaraes (2011)
Nucleus accumbens shell	ACEA (0.05 nmol)	CB_1 receptor agonist	Rat	EPM	Anxiogenic	Kochenborger et al. (2014)
VMH	AEA (5 pmol)	CB_1 receptor activation	Rat	Panic-like responses evoked by bicuculline infusion into the VMH	Panicolytic	Dos Anjos-Garcia, Ullah, Falconi-Sobrinho, and Coimbra (in press)

2-AG, 2-arachnoylglycerol; *(m)AEA*, (metha)anandamide; *ACEA*, arachidonyl-2-chloroethylamide; *ACPA*, arachidonyl cyclopropylamide; *CFC*, contextual fear conditioning; *EPM*, elevated plus-maze; *SPS*, single-prolonged stress; *VCT*, Vogel conflict test; *THC*, delta-9-tetrahydrocannabinol; *MPFC*, medial prefrontal cortex; *PL*, prelimbic; *IL*, infralimbic; *dHIP*, dorsal hippocampus; *vHIP*, ventral hippocampus; *dlPAG*, dorsolateral periaqueductal gray; *BLA*, basolateral amygdala; *CeA*, central amygdlaa; *MeA*, medial amygdala; *BNST*, bed nucleus of stria terminalis; *VMH*, ventromedial hypothalamus; ↓, decrease/impair; ↑, increase/facilitate.

adeno-associated virus-mediated overexpression of CB_1 receptors in the MPFC did not change anxiety responses (Klugmann, Goepfrich, Friemel, & Schneider, 2011), possibly as a consequence of overexpression in regions other than PFC, such as the BNST or striatum. Also, CB_1 antagonism in the PFC does not seem to be sufficient to alter anxiety levels under mild aversive conditions (Lisboa, Reis, et al., 2010; Rubino, Realini, et al., 2008; Table 4). Pharmacological blockade of MPFC TRPV1 channels, in contrast, reduced anxiety levels in animals (Aguiar et al., 2009; Fogaca, Aguiar, Moreira, & Guimaraes, 2012; Table 4). Therefore, modulation of anxiety by AEA within the MPFC might be due to a balance between CB_1 and TRPV1 receptors (Moreira, Aguiar, Terzian, Guimaraes, & Wotjak, 2012).

The vMPFC eCB system also plays a role in the expression and extinction of conditioned fear responses. Increase in the eCB tone in this region (Table 4) attenuates fear retrieval (but not acquisition) and facilitates extinction in different paradigms (Kuhnert, Meyer, & Koch, 2013; Lin, Mao, Su, & Gean, 2009; Lisboa, Reis, et al., 2010), whereas CB_1 receptor antagonism impairs fear consolidation and extinction (Kuhnert et al., 2013; Lin et al., 2009). Moreover, CB_1 mRNA increased in the MPFC 24 h after fear conditioning (Lisboa, Reis, et al., 2010). Besides, CB_1, but not CB_2, mRNA levels correlated positively with freezing behavior in a mice strain selectively bred for low and high fear, when submitted to classical fear-conditioning protocols (Choi et al., 2012). Therefore, CB_1 receptors in the MPFC participate in emotional learning of conditioned fear memories, most likely via reciprocal interaction with the basolateral amygdala (BLA) (Laviolette & Grace, 2006). Despite evidence of CB_2 receptor expression in the MPFC (Onaivi et al., 2006), no study addressed its direct involvement in this region on the control of anxiety and fear responses.

Using a proposed PTSD model, the single-prolonged stress (SPS) test, a recent study showed that deficits in contextual fear extinction and increased glucocorticoid receptor (GR) expression in the IL portion of vMPFC observed 1 week after SPS exposure were not attenuated by local or systemic administration of the CB_1/CB_2 agonist WIN55,212-2 after stress (Ganon-Elazar & Akirav, 2013). These data together with the fact that systemic corticosterone administration increases AEA levels in amygdala and hippocampus, but not in the PFC, 10 min after injection (Hill, Karatsoreos, Hillard, & McEwen, 2010), suggest that activation of CB_1 receptors in the IL are not immediately necessary after traumatic experience to modify fear extinction a week later (Ganon-Elazar & Akirav, 2013).

Therefore, eCB signaling in the PFC represents an attractive target for the treatment of neuropsychiatric disorders characterized by broad dysfunction of emotional processing (McLaughlin et al., 2014).

4.1.1 Mechanisms

During aversive situations there was a reduction in AEA (Hill et al., 2008) and an increase in 2-AG levels in the MPFC (Patel, Roelke, et al., 2005). In addition, CB_1 receptor binding and mRNA expression also increase (Hill et al., 2008; Lisboa, Reis, et al., 2010). CB_1 receptor in cortical areas is mainly present on terminals of GABAergic interneurons and, in less extent, on glutamatergic neurons (Monory et al., 2006). Nevertheless, the relative density of CB_1 receptors on GABAergic and glutamatergic neurons is not the only parameter to consider in evaluating the role of these receptors on neurotransmitter release regulation. In fact, at least in hippocampus CB_1 receptors located on glutamatergic terminals show greater capacity to recruit G protein than when they are present on GABAergic terminals (Steindel et al., 2013). It has been suggested that anxiogenic-like responses induced by cannabinoids are mediated by CB_1 receptors located on GABAergic terminal, whereas activation of CB_1 in glutamatergic terminals produces anxiolytic-like effects (Haring et al., 2011; Lafenetre, Chaouloff, & Marsicano, 2009; Metna-Laurent et al., 2012; Terzian, dos Reis, et al., 2014).

Electrophysiological results also suggest that the eCB tone is altered in the MPFC after stress exposure (Rossi et al., 2008). Chronic stress blocks the normal reduction of inhibitory postsynaptic potentials (IPSPs) produced by CB_1 agonist administered directly into mice corticostriatal slices (Rossi et al., 2008). On the other hand, after application of CB_1 agonists, excitatory postsynaptic currents (EPSCs) were strongly depressed, whereas SR141716A reversed these responses. CB_1 antagonists per se increased glutamate release (Auclair, Otani, Soubrie, & Crepel, 2000). Also, WIN55,212-2 (CB_1/CB_2 agonist) strongly favored LTD at the apparent expense of LTP, whereas SR141716A presented the opposite profile (Auclair et al., 2000).

The hypothalamus-pituitary-adrenal (HPA) axis is also controlled by MPFC activity. Corticosterone administration rapidly induced eCB synthesis in the MPFC (Hill et al., 2011). This region, along with other limbic areas (Ulrich-Lai & Herman, 2009), inhibits the HPA axis activity by activating inhibitory projections to PVN neurons (Jankord & Herman, 2008). Furthermore, the MPFC is vastly connected to a great variety of brain areas. For instance, MPFC sends glutamatergic projections to several areas (Del Arco & Mora, 2009; Vazquez-Borsetti, Celada, Cortes, & Artigas,

2011), whereas it receives serotoninergic and noradrenergic afferents from the raphe nuclei (Charnay & Leger, 2010; Fitzgerald, 2011) or *locus coeruleus* (Fitzgerald, 2011), respectively.

Local and systemic administration of the CB1/CB_2 agonists WIN55,212-2 causes an increase in extracellular norepinephrine in the MPFC and anxiogenic-like behavior (Page et al., 2007; Page, Oropeza, & Van Bockstaele, 2008), while acute and chronic treatment with the same agonist blocked the increase in cortical pyramidal cell excitability and input resistance evoked by an alpha2-adrenergic receptor agonist (Cathel et al., 2014), suggesting that the increased cortical norepinephrine efflux may be mediated by cannabinoid receptors-mediated desensitization of α2-adrenergic receptors. Also, activation of CB_1 and CB_2 receptors, or increased eCB tone, upregulates the expression of serotoninergic 5HT2A receptors and elevates serotoninergic tone in the MPFC (Cassano et al., 2011; Franklin & Carrasco, 2012). Therefore, it is suggested that norepinephrine enhancement by eCBs is associated with anxiogenic responses, whereas cannabinoid-mediated stimulation of serotonergic signaling is associated with anxiolytic responses (Ruehle et al., 2012).

Another system involved in the MPFC CB_1 control of emotional responses is the NMDA-nitric oxide (NO) pathway. Glutamate NMDA receptors and the neuronal nitric oxide synthase (nNOS) enzyme are expressed in the MPFC (Brené, Messer, & Nestler, 1998; Vincent & Kimura, 1992). Local inhibition of NMDA receptors or NO signaling prevents the increase in aversive responses induced by CB_1 receptor blockade (Uliana et al., 2016).

4.2 Hippocampus

The septum-hippocampal system has direct and indirect connections to the amygdala and PFC, critical structures involved in defensive responses (Dejean et al., 2015). It provides contextual information associated both with conditioning and innate fear (Canteras et al., 2010; Dejean et al., 2015). In addition, the hippocampus also interferes with defensive responses by other mechanisms related to its mnemonic, neuroendocrine (inhibition of the HPA axis), spatial learning, and stimuli analysis roles (Canteras et al., 2010). Moreover, it was proposed to generate anxiety in response to conflict by interrupting ongoing behavior and increasing the level of arousal and attention, as already discussed (McNaughton & Corr, 2004).

The effects of cannabinoids on hippocampal function and plasticity are complex and depend on the aversiveness of the task (Akirav, 2011). CB_1 receptors, although prominently expressed in GABAergic terminals

(Hajos & Freund, 2002; Marsicano & Lutz, 1999), are also present in glutamatergic, serotonergic, and cholinergic neurons in the hippocampus (Katona & Freund, 2008). Low levels of CB_1 receptors are also present in progenitor cells of the subgranular zone of the dentate gyrus (Hu & Mackie, 2015). Similar to the receptors, the enzymes responsible for the synthesis and metabolism of eCBs are highly expressed in the hippocampus (for review, see Hu & Mackie, 2015).

In classical animal models of anxiety, systemic or direct injection of cannabinoid-related drugs into the hippocampus has produced contradictory results (Tables 2 and 4). The stressful experience of the animal has been proposed as a key factor to explain these conflicting findings (Campos et al., 2010). In less stressful animal models, such as the EPM, intrahippocampal administration of drugs that facilitate eCB signaling, such as AM404, produced anxiogenic effects (Roohbakhsh et al., 2007). However, AM404 was anxiolytic when injected directly into the ventral hippocampus (vHPC) of rats submitted to the Vogel conflict test (VCT), a model that involves pain and a previous stressor experience (water deprivation for 24 h). Moreover, the anxiogenic-like effect observed in the EPM after intra-vHPC AM404 injection turned into anxiolytic in rats previously submitted to an acute restraint stress (Campos et al., 2010). Corroborating this latter finding, Δ^9-THC was also anxiolytic in this brain area only when the animals have been kept isolated in their home cage (Rubino, Guidali, et al., 2008), a stressful situation that increases anxiety in the EPM (Maisonnette, Morato, & Brandao, 1993). Although the vHPC is particularly associated with anxiety (Bannerman et al., 2004; Bertoglio, Joca, & Guimaraes, 2006), contradictory results have been reported for the dorsal hippocampus (dHPC; Table 4). In the EPM, the CB_1/CB_2 agonist WIN55,212-2 induced anxiogenic effect when injected into this brain area. In contrast, enhanced eCB signaling (by AM404 or URB597) was anxiolytic in both the EPM and VCT (Lisboa, Borges, et al., 2015) (Table 4). These results suggest a regional dissociation between the dHPC and vHPC regarding stress interference on eCB modulation of anxiety responses. This possibility, however, needs to be further investigated.

Learning/memory processes involving the hippocampus are modulated by cannabinoids (Clarke et al., 2008). These effects, however, may vary depending on the route of administration, the nature of the task (aversive or not), and whether it involves emotional memory (e.g., conditioned fear and extinction learning) or nonemotional memory formation (e.g., spatial learning) (Akirav, 2011). CB_1 receptor activation, in addition to impairing acquisition (Pamplona & Takahashi, 2006) and facilitating extinction of contextual fear memory (Akirav, 2011), also affects consolidation of this memory. This later effect is associated with a decreased expression of poly-salic acid-neural cell adhesion molecules in the hippocampus, glycoproteins that have been proposed to be important for memory consolidation (Mackowiak et al., 2009). Reinforcing the involvement of the hippocampus in fear-conditioning processes, synthetic Δ^9-THC increases activation of this structure in humans during extinction of an aversive conditioned cue (Rabinak et al., 2014).

Direct drug injections into the hippocampus have also helped to unveil the role of this structure in cannabinoid interference in fear learning. Bilateral infusion of AM251 into the dHPC increases fear expression (Spiacci et al., 2016). Likewise, WIN55,212-2 injected into the ventral subiculum (in the vHIP) impaired retrieval of a contextual fear-conditioning task (Akirav, 2011). Similar effects were reported for 2-AG (Rea et al., 2014). Moreover, while 2-AG blocks fear extinction, AEA facilitates it. On the other hand, AEA blocked fear reconsolidation, whereas AM251 facilitated it. The authors suggest that the opposite effects of AEA in these two processes indicates that the hippocampal eCB system acts as a switching mechanism, deciding which process will take place by either maintaining the original memory (reconsolidation) or promoting a new learning (extinction) (de Oliveira Alvares et al., 2008).

Exposure to a model of PTSD (SPS, mentioned in Section 4.1) impaired fear extinction and increased GR expression in the vHIP a week later. Administration of WIN55,212-2 in this region or systemically attenuates these alterations (Ganon-Elazar & Akirav, 2013). Moreover, the local effect of WIN55,212 was prevented by the GR antagonist RU-486, pointing to an involvement of the HPA axis on cannabinoid effects.

4.2.1 Mechanisms

The mechanisms involved in cannabinoid regulation of emotional response by hippocampus are not clear. Hippocampal CB_1 receptors are widely expressed in cholecystokinin-expressing GABA terminals, but are also found in glutamatergic, serotonergic, and cholinergic presynaptic terminals (for review, see Akirav, 2011; Katona & Freund, 2008). Because CB_1 receptors are present at higher concentration in GABAergic compared to glutamatergic terminals, interference in the former neurotransmission has been related to cannabinoid influence in hippocampal-dependent processes (Akirav, 2011; Hajos & Freund, 2002), such as depolarization-induced suppression of inhibition (DSI); this process facilitates LTP formation by eCB-mediated disinhibition (Wilson, Kunos, & Nicoll, 2001). However, eCBs interference with GABAergic and glutamatergic hippocampal neurotransmission may depend on the rodent species (Haller et al., 2007), brain region, and stress influence (Morena, Patel, Bains, & Hill, 2016).

CB_1 deletion in glutamatergic, but not in GABAergic, neurons impaired the behavior and neuroendocrine responses caused by forced swimming stress (Steiner et al., 2008). Recently, the contribution of these two hippocampal populations to cannabinoid effects was elegantly demonstrated by showing that reconstituting CB_1 receptor function in the dorsal telencephalic glutamatergic neurons partially restored the anxiety-like behavior phenotype of global CB_1 receptor deletion (Ruehle et al., 2013). This effect was associated with CB_1-dependent normalization of hippocampal depolarization-induced suppression of excitation, a CB_1-glutamate associated form of neural plasticity. Together, these results suggest that hippocampal CB_1 receptors located on glutamatergic terminals could be a target for the anxiolytic effects of small doses of systemically injected cannabinoids (Viveros et al., 2005).

Contrasting to CB_1, little is known about the role of hippocampal CB_2 receptors in anxiety modulation. CB_2 overexpression decreases anxiety in the EPM and light–dark tests (Garcia-Gutierrez & Manzanares, 2010). A recent study found that, different from CB_1, short-term CB_2 activation in hippocampal slices does not significantly modify synaptic activity. Long-term exposure to CB_2 agonists, on the other hand, increases synaptic activity and glutamate release. Similar effects were found in vivo after repeated administration (Kim & Li, 2015). Other mechanisms that have been associated with an anxiolytic-like or antistress effect of repeated CB_2 receptor activation are facilitation of hippocampal neurogenesis (Campos et al., 2013; Fogaca, Galve-Roperh, Guimaraes, & Campos, 2013) and antiinflammatory action, which will be further discussed.

4.3 Periaqueductal Gray Matter

The periaqueductal gray matter (PAG) has been initially investigated as an important brain site for the analgesic effects of cannabinoids (Walker, Huang, Strangman, Tsou, & Sanudo-Pena, 1999). Later on, it was proposed as a target for the antiaversive effects of these compounds, with important implications for the modulation of anxiety disorders and panic attacks. Local injection of the synthetic CB_1/CB_2 agonist HU210 in the rat dorsal portion of the PAG (dPAG) inhibits the effects of local stimulation with an excitatory aminoacid, D,L-homocysteic acid (DLH). This compound induces a robust escape response, characterized by running in an arena, which has been proposed as a model of panic attack. HU210 prevents this response, in addition to promoting analgesia (Finn et al., 2003). Moreover, intra-DPAG injection of cannabinoids also inhibited escape responses induced by ultrasound stimulus, further suggesting that cannabinoid receptors modulate aversive responses in this structure (Finn, Jhaveri, et al., 2004). Therefore, cannabinoid mechanisms operate in the PAG to modulate pain and aversive behavior. eCBs are also locally recruited to mediated stress-induced analgesia (Hohmann et al., 2005).

Based on this literature, we have been performing extensive studies to further characterize the role of the eCB system in modulating aversive responses in the PAG, particularly in its dorsolateral portion (dlPAG). The injection of cannabinoids into the dlPAG (Table 4) induces consistent dose-dependent anxiolytic-like effect in rats tested in the EPM. This effect was observed after local AEA injection and mimicked by the selective CB_1 receptor agonist ACEA. AEA effect was prevented by the selective CB_1 receptor agonist, AM251, indicating a role for this receptor in mediating the cannabinoid effects on anxiety in the dlPAG (Moreira et al., 2007). Furthermore, local injection of AM404 and URB597 also induced anxiolytic-like effect in the Vogel test, a model based on the conflict between approaching and avoiding a stimulus, rather than on exploratory activity, supporting the role of the dlPAG in modulating the effects of cannabinoids against different types of aversive responses (Lisboa et al., 2008).

Moreover, dlPAG CB_1 activation in rats by AEA or AM404 also inhibited fear behavior in the contextual fear-conditioning model (Olango, Roche, Ford, Harhen, & Finn, 2012; Resstel et al., 2008), whereas blockade of that receptors increased behavioral and autonomic responses in the same model (Uliana et al., 2016). Similar responses have been reproduced in a more natural setting, in which AEA inhibited fear responses in rats exposed to a live predator (Lisboa, Camargo, da Silva, et al., 2014). Apart from the effects of cannabinoids acting on the CB_1 receptor, the dlPAG also mediate the antiaversive effects of CBD through non-CB_1-mediated mechanisms (Campos & Guimaraes, 2008). Finally, local injection of URB597 inhibits escape response induced by the excitatory aminoacids NMDA or DLH (Batista, Bastos, & Moreira, 2015; Batista, Fogaca, & Guimaraes, 2015; Viana et al., 2015). Thus, the eCB system might be recruited as an endogenous mechanisms protecting against aversive stimuli in the PAG.

2-AG also seems to modulate anxiety effects in the dlPAG. Local injection of an MAGL inhibitor into the dlPAG induced anxiolytic-like effects in the EPM (Almeida-Santos, Gobira, et al., 2013), indicating that both AEA and 2-AG participate in the modulation of this behavior.

4.3.1 Mechanisms

Although several studies have investigated the possible mechanisms involved in the eCB effects modulated by the dlPAG, this is still a matter of debate. Studies showing anxiolytic-like effects of eCB-related compounds administered into the dlPAG have suggested that activation of CB_1 receptors could result in decreased glutamate release, attenuating the aversiveness of the experimental situation (Fogaca, Aguiar, et al., 2012; Fogaca, Lisboa, et al., 2012; Lisboa & Guimaraes, 2012; Lisboa, Magesto, Aguiar, Resstel, & Guimaraes, 2013; Lisboa et al., 2008; Resstel et al., 2008). Accordingly, a higher and ineffective dose of AEA in the dlPAG turned into anxiolytic after local NMDA blockade (Fogaca, Gomes, Moreira, Guimaraes, & Aguiar, 2013). In addition, we previously showed that the panic-like reaction induced by administration of a NO donor into the dlPAG was attenuated by low doses of AEA and URB597 (Lisboa & Guimaraes, 2012). Moreover, a high dose of AEA attenuated the fear behavior evoked by exposing rats to a live cat only after local inhibition of NO synthesis (Lisboa, Camargo, da Silva, et al., 2014; Lisboa, Camargo, Magesto, et al., 2014). In the EPM, combination of ineffective low doses of AEA with NO signaling pathway inhibitors induced anxiolytic-like effect (Lisboa et al., 2013). Corroborating the NMDA-NO involvement in the antiaversive effect of cannabinoids, we recently demonstrated that the increase in contextual conditioned emotional response induced by blocking CB_1 receptors in the dlPAG was prevented by local NMDA antagonist or drugs inhibiting NO signaling (Uliana et al., 2016), suggesting that blockade of CB_1 receptors disinhibits glutamatergic neurotransmission and NO signaling pathway. Overall, CB_1 receptors in the dlPAG seem to decrease mainly glutamatergic/NO neurotransmission to induce anxiolytic and antifear effects, although the involvement of GABAergic mechanisms has also been reported, specially with higher doses of cannabinoids, when unbalanced glutamatergic and GABAergic mechanisms can take place (Lisboa et al., 2013; Uliana et al., 2016).

4.4 Amygdaloid Complex

The amygdala is a heterogeneous complex that modulates neuroendocrine function and is involved in the modulation of complex behaviors such as fear, anxiety, defense, aggression, learning, and memory, including conditioning mechanisms (Adolphs, Tranel, Damasio, & Damasio, 1994; LeDoux, 1994; McNaughton & Corr, 2004; Morris et al., 1996; Phillips & LeDoux, 1992; Schafe et al., 2005; Scott et al., 1997; Sullivan et al., 2004).

The amygdaloid complex has been divided into basolateral (BLA) complex and the centromedial nuclei (central-CeA and medial-MeA amygdala nuclei), identified by cytoarchitectonic, histochemical, and immunocytochemical features (Ramikie & Patel, 2012; Sah, Faber, Lopez De Armentia, & Power, 2003). The different connections of these nuclei can be involved in the different effects promoted by them in a variety of functions, including anxiety. For instance, the amygdala hyperexcitability observed in mood and anxiety disorders has been related to impairments of the inhibitory influence of the hippocampus and PFC (Etkin, Prater, Schatzberg, Menon, & Greicius, 2009; Kim et al., 2011). Moreover, proper synchronization between amygdala, MPFC, and hippocampus is important for fear mechanisms, such as fear retrieval and extinction, and learning and memory (Lesting et al., 2013; Rei et al., 2015; Seidenbecher, Laxmi, Stork, & Pape, 2003). For instance, electrical stimulation of the BLA disrupts hippocampal CA1 LTP (Vouimba & Richter-Levin, 2005), whereas BLA lesion blocks impaired LTP and spatial memory induced by repeated stress (Kim, Koo, Lee, & Han, 2005). Moreover, optogenetic stimulation of BLA projections to the dHIP is sufficient to reproduce the molecular changes and learning and memory deficits induced by repeated stress in mice (Rei et al., 2015). The PL and IL portions of the MPFC differentially modulate CeA neurons, increasing and decreasing, respectively, CeA neurons activity and fear response, through projections to BLA and the intercalated cells nucleus (Sotres-Bayon & Quirk, 2010). Additionally, it was recently demonstrated that optogenetic stimulation and inhibition of IL axons in the BLA facilitates and impairs, respectively, fear extinction acquisition, but not retrieval, demonstrating a causal linking between MPFC and amygdala on modulation of fear memories (Bukalo et al., 2015).

CB_1 receptors are expressed in the amygdala, although at considerably higher densities in the BLA compared to the CeA or MeA (Herkenham et al., 1991; Katona et al., 2001). In addition to the receptors, eCBs are also produced in the amygdala (Bisogno et al., 1999; Herkenham et al., 1991) given that FAAH and MAGL enzymes are present in this brain area (Hu & Mackie, 2015; Tsou, Nogueron, et al., 1998). The levels of the eCBs AEA and 2-AG in different amygdala nuclei are modulated by several stressors (Blasio et al., 2013; Hill et al., 2008, 2009; Patel, Cravatt, & Hillard, 2005;

Patel, Kingsley, Mackie, Marnett, & Winder, 2009; Qin et al., 2015; Rademacher et al., 2008; and others). Local AEA levels can decrease after acute restraint stress, probably due to increased local activity of FAAH (Hill et al., 2009), whereas 2-AG is not altered after a similar procedure (Patel et al., 2009). 2-AG levels, however, increased in the BLA after chronic restraint stress (Patel et al., 2009). In contrast, exposure to a tone previous paired with footshocks increases both AEA and 2-AG release in the BLA (Marsicano et al., 2002). These data indicate that different stressors and their duration, in addition to different animal species, can differently modulate eCB levels in amygdala.

Systemic or central administration of CB_1 receptor agonists induces neuronal activation in the CeA of rats (Arnold, Topple, Mallet, Hunt, & McGregor, 2001; McGregor, Arnold, Weber, Topple, & Hunt, 1998; Patel, Cravatt, et al., 2005; Patel, Moldow, Patel, Wu, & Chang, 1998; Valjent et al., 2002; and others). In contrast, there is also a report that CB_1 agonists only induce neuronal activation in that region when combined with stress exposure (Patel, Cravatt, et al., 2005). Activation of the CeA, MeA or BLA was also observed after systemic administration of CB_1 antagonists (Newsom et al., 2012; Patel, Cravatt, et al., 2005; Patel, Roelke, et al., 2005; Sink et al., 2010).

The anxiolytic-like effects of low doses of systemic Δ^9-THC are associated with decreased amygdala activation and are prevented by the CB_1 antagonist AM251 (Rubino et al., 2007). An anxiogenic dose of AM251, in contrast, increased neuronal activation (Sink et al., 2010). In humans, anxiolytic doses of Δ^9-THC significantly reduce amygdala reactivity to social signals of threat (Phan et al., 2008). However, higher Δ^9-THC doses increased amygdala activation and anxiety feeling in similar situation (Bhattacharyya et al., 2010).

Contradictory results are also observed after local administration of cannabinoid agonists into different amygdala nuclei (Table 4). Administration of low doses of Δ^9-THC into the CeA or BLA induced anxiety-like behavior in rodents (Onaivi et al., 1995; Rubino, Guidali, et al., 2008), whereas administration of arachidonylcyclopropylamide (ACPA), a CB_1 agonist, into the CeA produced an anxiolytic-like effect (Zarrindast et al., 2008). Corroborating this latter effect, the CB_1 antagonist SR141716A in the CeA precipitated anxiety-like behavior in female rats during abstinence from palatable diet cycling (Blasio et al., 2013). In the BLA, administration of the nonselective agonist WIN55,212-2 decreased the enhancing effect of stress on inhibitory avoidance conditioning (Ganon-Elazar & Akirav, 2009), whereas the antagonism of CB_1 receptors impaired extinction of fear conditioning (Roche, O'Connor, Diskin, & Finn, 2007) and acquisition of fear memory (Tan et al., 2011), respectively. Still in the BLA, administration of the FAAH inhibitor AM3506 facilitates fear extinction (Gunduz-Cinar et al., 2013). These later data suggest that reestablishment of an efficient AEA signaling by FAAH inhibition and/or activating cannabinoid receptors signaling could be a potential target to treat disorders associated with impaired fear extinction, such as PTSD. In fact, the fear extinction deficits and anxiety-like behavior observed a week later in a rat model of PTSD (SPS model) were attenuated by acute systemic or intra-BLA administration of WIN55,212-2, 2 min or 24 h after stress, in a CB_1-dependent way (Ganon-Elazar & Akirav, 2012). Overall, the results reported in the literature suggest that the eCB system exerts an inhibitory role in amygdala activity related to threatening situations, which is similar to what was observed with individuals carrying FAAH mutations that confer decreased FAAH, as discussed earlier in this chapter. In contrast, at high doses CB_1 agonists could facilitate anxiety-like behaviors. Studies evaluating the involvement of amygdala CB_2 receptors on modulation of anxiety-like behavior, however, are still missing.

Although administration of CBD (7.5–30 nmol) into the BLA of rats did not induce any behavioral alteration in the EPM (Lisboa & Guimarães, 2007), evidence from brain image studies in healthy humans showed that this drug did not attenuate left amygdala activity and decrease anxiety, which implicates this brain region in the effects of CBD (Crippa et al., 2004; Fusar-Poli et al., 2009). In addition, a recent study showed that although CBD in a single dose was not able to attenuate anxiety-like behavior in mice by itself, it decreased neuronal activation of the CeA and attenuated the anxiogenic-like effect induced by Δ^9-THC (Todd & Arnold, 2016).

4.4.1 Mechanisms

The mechanisms involved in the opposite effects of eCB system in amygdala, as in other brain regions, seem to involve modulation of glutamatergic and GABAergic neurotransmissions by CB_1 receptors. For instance, activation of these receptors within the BLA decreases both GABAergic and glutamatergic neurotransmissions, interfering with the plastic mechanisms of LTD and LTP, respectively (Azad et al., 2003, 2008; Katona et al., 2001). Moreover, superfusion of WIN55,212-2 onto CeA neurons decreased evoked $GABA_A$ receptor-mediated IPSPs in a CB_1-dependent way (Roberto et al., 2010). Both SR141716A and AM251 (CB_1 antagonists) increased inhibitory responses, suggesting the presence of a tonic eCB mechanism that decreases inhibitory transmission in CeA (Roberto et al., 2010).

As commented earlier, plastic alterations within limbic regions after chronic stress exposure can result in signaling alterations in CB_1 receptors, shifting the emotional balance toward proaversive responses (McLaughlin et al., 2014). Considering that spontaneous activity of pyramidal BLA neurons is maintained at low levels by high local GABAergic inhibitory tone, stress-induced impairments in GABAergic transmission within the BLA may contribute to anxiety behavior (Quirk & Gehlert, 2003), an effect that could be modulate by eCBs. For instance, repeated stress increases 2-AG-mediated short-term synaptic suppression of GABAergic transmission in BLA slices, suggesting that 2-AG signaling is involved with adaptations induced by repeated stress (Patel et al., 2009). This effect could depend on eCB release driven, in a Ca^{+2}-independent way, by postsynaptic metabotropic glutamate receptor type 5 (mGluR5), as has already been suggested for DSI in the BLA (Zhu & Lovinger, 2005).

A recent work investigating the mechanisms by which chronic stress in the amygdala affects mGluR5 signaling and eCB synthesis resulting in molecular and behavioral alterations revealed a crucial role for the tyrosine phosphatase PTP1B (Qin et al., 2015). Stress-induced PTP1B activity in the amygdala or the genetic deletion of an endogenous inhibitor of this enzyme (LMO4) was associated with decreased mGluR5 phosphorylation, along with decreased AEA and 2-AG levels in the amygdala. These changes were associated with the presence of anxiety-like behavior and impaired fear extinction, and depended on GRs (Qin et al., 2015). Therefore, proper phosphorylation of mGluR5 is crucial for normal eCB signaling in the amygdala. Increased levels of glucocorticoids during stressful situations can disrupt this mechanism, contributing to the development of anxiety.

As will be discussed later, several studies have shown a close relationship between glucocorticoids and eCBs. CB_1 receptors in the BLA, but not in the CeA or MeA, are important modulators of the HPA axis. Their inhibition enhances HPA axis activity, increasing corticosterone levels (Ganon-Elazar & Akirav, 2009; Hill et al., 2009). Stress-induced corticosterone elevation is attenuated by intra-BLA administration of CB_1 agonists or URB597 (Ganon-Elazar & Akirav, 2009; Hill et al., 2009). These data suggest that increased FAAH activity could contribute to the decrease in BLA eCB levels, driven HPA axis activation. Corroborating this hypothesis, stress exposure mobilizes FAAH in the BLA, depleting AEA levels and increasing neuronal excitability. In contrast, pharmacological inhibition of FAAH by AM3506 prevents the reduction of AEA levels in the BLA, in addition to preventing stress-induced dendritic hypertrophy and anxiety-like behavior (Gunduz-Cinar et al., 2013).

eCBs could also modulate anxiety behavior, at least in the BLA, due to an interaction with the corticotrophin release factor (CRF). BLA CRF1 receptors are involved in stress- and anxiety-like behavior (Jochman, Newman, Kalin, & Bakshi, 2005; Roozendaal, Brunson, Holloway, McGaugh, & Baram, 2002; Sztainberg, Kuperman, Tsoory, Lebow, & Chen, 2010). Acute restraint stress in rats induces a quick increase in FAAH activity in the amygdala pyramidal neurons (mainly BLA), an effect prevented by systemic treatment with a CRF1 antagonist. Moreover, the anxiogenic-like behavior induced by intra-BLA administration of CRF, as well as increased circulating levels of corticosterone, was attenuated by previous local administration of URB597 (Gray et al., 2015). Considering that glucocorticoids increase AEA levels in BLA (Hill et al., 2010), it was suggested that the rapid initial AEA signaling disruption mediated by CRF interfering with FAAH activity would be later counteracted by production and release of glucocorticoids as part of a negative feedback mechanism.

Overall, these results suggest that the eCB system in the amygdala, particularly in the BLA, plays an inhibitory role during threatening situations, decreasing the magnitude of HPA axis and the behavioral consequences of stress, such as anxiety and fear extinction.

4.5 Bed Nucleus of Stria Terminalis

The BNST, which is part of what is called the extended amygdala, is a highly heterogeneous and complex limbic structure associated with autonomic, neuroendocrine and behavioral functions (Crestani et al., 2013). It has reciprocal connections with the MeA and CeA (Dong, Petrovich, & Swanson, 2001; Shammah-Lagnado et al., 2000) and receives projections from the hippocampus, BLA, and MPFC (Dong, Petrovich, & Swanson, 2001; Dong, Petrovich, Watts, & Swanson, 2001; Vertes, 2006). Several studies have shown that the BNST plays a key role in expression of anxiety-like responses (Davis, Walker, Miles, & Grillon, 2010; Walker, Toufexis, & Davis, 2003).

CB_1 receptors are present in glutamatergic and GABAergic terminals in the BNST (Matsuda, Bonner, & Lolait, 1993; Puente et al., 2010; Tsou, Brown, et al., 1998). In vitro electrophysiology studies demonstrated that

CB_1 receptor activation inhibits both excitatory and inhibitory synaptic transmission in this region, which could be important for the regulation of aversive responses (Massi et al., 2008; Puente et al., 2010). Additionally, the activation of CB_1 receptors within the BNST alters the excitability of VTA dopamine neurons evoked by the stimulation of the IL portion of the MPFC and, based on those findings, it was suggested that disturbances in this circuitry might underlie abnormal emotional processing observed in disorders such as addiction and pathological stress (Massi et al., 2008). Despite these pieces of evidence, no study so far has investigated the effects of direct injections of CB_1 receptor agonists or antagonists into the BNST of animals submitted to aversive situations. However, recent studies from our group provide evidence that the BNST would be involved in the antiaversive effects of CBD.

Systemic CBD treatment attenuated the increase in c-Fos protein expression, a marker of neuronal activation, in the BNST induced by the reexposure to an aversive context previously paired with footshock (Lemos et al., 2010), suggesting a possible involvement of this structure in the effects of this drug. Afterward, we showed that intra-BNST administration of CBD attenuated freezing behavioral and the increase in heart rate and arterial pressure induced by contextual fear conditioning (Gomes et al., 2012). It should be noted that CBD did not induce any significant change in baseline values of these cardiovascular parameters. Therefore, it is unlikely that the attenuation of the cardiovascular responses to the aversive context depends on direct cardiovascular effects, but rather on an attenuation of the emotional response. In fact, intra-BNST CBD administration also induced anxiolytic-like effects in animals submitted to the EPM and VCT (Gomes, Resstel, et al., 2011). However, a direct influence of CBD in cardiovascular function could not be ruled out. Accordingly, intra-BNST administration of CBD facilitates bradycardiac baroreflex response (Alves et al., 2010). In addition, an enhanced tachycardiac response evoked by acute exposure to restraint stress, without affecting blood pressure changes, was also observed after BNST treatment with CBD (Gomes et al., 2013). These data are in contrast with those obtained after systemic or intracisternal administration of CBD, showing an attenuation of cardiovascular changes induced by restraint stress (Granjeiro, Gomes, Guimaraes, Correa, & Resstel, 2011; Resstel et al., 2009). Additionally, although the facilitation of restraint-evoked cardiovascular responses observed after intra-BNST CBD administration seems in contrast to its antiaversive effects, it suggests that CBD effects in behavioral and autonomic responses to aversive situations could involve different neurobiological mechanisms. We cannot also exclude the possibility that neurobiological actions of CBD within BNST may depend on the type of stress.

Curiously, all the effects induced by CBD administration into the BNST were blocked by the pretreatment with the 5-HT1A receptor antagonist WAY100635, indicating that CBD acts through these receptors (Alves et al., 2010; Gomes et al., 2013, 2012; Gomes, Resstel, et al., 2011).

4.6 Other Brain Sites

The effects of cannabinoids in other brain areas that express CB_1 receptors and have been linked to anxiety, such as nucleus accumbens and hypothalamic nuclei, are less well known (Moreira, Aguiar, Resstel, et al., 2012). Onaivi and colleagues observed that administration of Δ^9-THC into the nucleus accumbens did not induce anxiety-like behavior (Onaivi et al., 1995). However, it was recently found that administration of ACEA into the shell region of the nucleus accumbens induced an anxiogenic-like effect in the EPM (Kochenborger et al., 2014). Regarding hypothalamic nuclei, several lines of evidence suggest a role for the eCBs as a negative modulator of the HPA axis. Therefore, changes in the eCB tone might be linked to stress-related diseases (Morena et al., 2016). The dorsomedial and ventromedial hypothalamic nuclei also play a role in the modulation of anxiety-like behaviors (McNaughton & Corr, 2004). Infusion of AEA into the ventromedial hypothalamus (VMH) induced a panicolytic-like effect through CB_1 receptors activation (Dos Anjos-Garcia et al., in press). Likewise, preliminary data from our group have indicated that intra-dorsomedial hypothalamus administration of drugs targeting the eCBs induces antiaversive effects, but this remains under investigation. Recently, it was showed that CB_1 receptor presynaptically expressed on medial habenula control the expression of aversive memories by selectively modulating cholinergic transmission at synapses in the interpeduncular nucleus (Soria-Gomez et al., 2015).

5. ADDITIONAL MECHANISMS INVOLVED IN THE eCB EFFECTS ON ANXIETY

5.1 Role of Specific eCBs

An important topic that remains to be further explored is the role of specific eCBs in modulating anxiety and aversive responses. AEA and 2-AG have different molecular targets in the brain and might interfere with emotional responses in particular ways. Both eCBs seem to be involved in anxiety, as revealed by studies with compounds that selectively inhibit the enzymes reponsible for terminating their actions. As previously mentioned, systemic treatment with FAAH inhibitors, which increase mainly the levels of AEA, induces CB_1-dependent anxiolytic-like effects in experimental animals (Kathuria et al., 2003), whereas the effects of MAGL inhibition, preventing 2-AG hydrolysis, may depend on CB_2 receptor (Busquets-Garcia et al., 2011). Local injection of an MAGL inhibitor into the dlPAG also demonstrates that the anxiolytic-like effects of this compound may depend on both receptors (Almeida-Santos, Gobira, et al., 2013). Other enzymes responsible for eCB metabolism, such as the cyclo-oxygenase-2 (COX-2), have also been implicated in anxiety responses. For instance, site-especific inhibition of COX-2 induces anxiolytic-like effects in rats (Hermanson et al., 2013).

Other substances that have been proposed as part of the eCB system might also represent interesting targets. Hemopressin is an endogenous peptide that also binds to the CB_1 receptor, although as an inverse agonist, rather than an agonist (Heimann et al., 2007). In line with this mechanism, hemopressin administration into the dlPAG induces anxiogenic-like effects in rats (Fogaca et al., 2015).

5.2 Different Receptors and Modulation of Different Neurotransmitters

The CB_1 receptor is largely expressed throughout several brain regions related to anxiety and aversive responses (Herkenham et al., 1991), in different neuronal subpopulations (Marsicano & Lutz, 1999) and also in glia cells (Han et al., 2012; Navarrete & Araque, 2010), as already discussed. Its expression in presynaptic terminals of a variety of neuronal populations allows the modulation of several neurotransmitters released in a dose-dependent manner (Castillo, Younts, Chavez, & Hashimotodani, 2012). CB_1 are localized in GABAergic, cholinergic, glutamatergic, noradrenergic, serotonergic, among others, neuron terminals and are differentially involved in the responses to acute and chronic stress in mice (Haring, Guggenhuber, & Lutz, 2012; Lutz, Marsicano, Maldonado, & Hillard, 2015; Marsicano & Lutz, 1999; Monory et al., 2006).

CB_1 receptor activation modulates both glutamatergic and GABAergic neural subpopulations (Hoffman, Laaris, Kawamura, Masino, & Lupica, 2010; Laaris, Good, & Lupica, 2010; Monory et al., 2007) and it is broadly accepted that CB_1 localization on cortical glutamatergic and GABAergic neurons play opposing roles on the control of anxiety and fear-related behaviors, as discussed earlier. It also has been proposed that CB_2 receptor is expressed in glia and neurons in various regions (Onaivi et al., 2006). CB_2 might modulate behavior relevant to neuropsychiatric disorders, including mood and reward (Onaivi et al., 2008), although its role in modulating anxiety-related responses has remained uncertain. This topic, therefore, warrants further investigation. Since its effects have been mostly related to the immune system, this will be addressed further.

The effects of low doses of cannabinoids are accepted to depend on the presence of CB_1 receptor on cortical glutamatergic neurons, whereas the anxiogenic effect of higher doses is mediated by CB_1 receptors on forebrain GABAergic neurons (Lutz et al., 2015). Animals with specific CB_1 deletion on glutamatergic neurons showed increased anxiety-like responses in highly aversive or arousing situations (Haring et al., 2011; Jacob et al., 2009; Lafenetre et al., 2009; Rey et al., 2012), in addition to deficits in fear learning and active-passive copying strategies (Metna-Laurent et al., 2012). Specific rescue of CB_1 receptors gene on cortical glutamatergic neurons restored anxiety-like, but not fear-related behavior (Ruehle et al., 2013). In contrast, deletion of CB_1 receptors in forebrain GABAergic neurons reduced anxiety-like or neophobic behavior in mild aversive or arousing conditions (Haring et al., 2011; Lafenetre et al., 2009; Rey et al., 2012). Also, although CB_1 receptors in GABAergic neurons do not seem to have an essential role in fear alleviation (Dubreucq et al., 2012), other studies reported decreased freezing in conditional mutant mice (Llorente-Berzal et al., 2015), as well as a more efficient switch from passive to active avoidance strategies (Metna-Laurent et al., 2012).

Another possible step in the eCB action mechanism involves NO signaling, as already commented. NO is involved in several stress-related disorders (Guimaraes, Beijamini, Moreira, Aguiar, & de Lucca, 2005) and interacts with the eCB system in the dlPAG (Batista, Bastos, et al., 2015; Batista, Fogaca, et al., 2015; Lisboa & Guimaraes, 2012; Lisboa et al., 2013; Uliana et al., 2016) and in the dHIP (Spiacci et al., 2016) on the control of emotions. Moreover, pharmacological and genetic inhibition of nitric oxide synthase (neuronal, nNOS; or inducible, iNOS) induce changes in anxiety- and fear-like behaviors (Buskila et al., 2007; Kelley, Anderson, & Itzhak, 2010; Kelley, Balda, Anderson, & Itzhak, 2009; Lisboa, Gomes, et al., 2015; Wultsch et al., 2007). Recently, it was showed that the fear extinction disruption observed in mice-lacking iNOS was attenuated by the FAAH inhibition (Lisboa, Gomes, et al., 2015), suggesting that increased eCB tone could balance NO signaling toward fear and anxiety relief.

Several studies demonstrated that the eCB system could also modulate monoaminergic systems (Haj-Dahmane & Shen, 2011; Laviolette & Grace, 2006; Micale et al., 2009a; Nakazi et al., 2000; Terzian et al., 2011). CB_1 agonists, for example, stimulate the release of norepinephrine and the expression of different adrenergic receptors in the rodent PFC, as previously mentioned (Carvalho, Mackie, & Van Bockstaele, 2010; Oropeza, Mackie, & Van Bockstaele, 2007; Oropeza, Page, & Van Bockstaele, 2005). Under acute stress exposure, an ineffective dose of URB597 increased norepinephrine levels in PFC and BLA of rats in a CB_1-dependent way (Bedse et al., 2015), suggesting a possible noradrenergic mechanism by which eCBs modulate stress responses. These effects could be due to increased firing of noradrenergic neurons, since URB597 enhances this mechanism, in addition to inducing antidepressant-like effect in mice (Gobbi et al., 2005). In relation to serotonin, mice-lacking CB_1 receptors in serotonergic neurons are more anxious and less sociable than control littermates, but only under stressful situations (Haring et al., 2015). Added to that, increases of eCB tone produces anxiolytic-like responses via serotoninergic 5-HT1A and 5-HT2A receptors (Bambico et al., 2010), also enhancing serotonin neurons firing activity (Gobbi et al., 2005). In addition, eCBs could downregulate serotonin release (Haring, Marsicano, Lutz, & Monory, 2007). There is also evidence for the serotonin involvement in the biphasic effects of cannabinoid drugs, since the anxiogenic-like effect of a higher dose of the CB_1 agonist CP55,940, but not the anxiolytic effect of a lower dose, was prevented by a 5-HT1A antagonist (Marco et al., 2004). CB_1 and eCBs also interact with the dopaminergic system, since deletion of dopamine D3 receptor, for example, promotes anxiolytic-like behaviors accompanied by changes in AEA and 2-AG levels in several brain areas (Micale et al., 2009a). Moreover, CB_1 deletion on dopaminergic D1- expressing neurons induces anxiety- and fear-like behaviors in mice (Terzian et al., 2011).

Apart from the cannabinoid receptors, other molecular targets for AEA expressed in brain regions related to anxiety might be involved in eCB effects on anxiety. The expression of the TRPV1 channel, which is activated by AEA, as already discussed, has been reported in various brain regions (Toth et al., 2005). This topic, however, has been controversial. Recently, its expression was detected at significant levels only in specific brain regions, including the PAG (Cavanaugh et al., 2011). In contrast, other studies have not only detected TRPV1 expression in the brain but also observed a significant colocalization with CB_1 in various brain regions (Cristino et al., 2006), including the dlPAG (Casarotto et al., 2012). These results are in line with pharmacological evidence for a reciprocal interaction between CB_1 and TRPV1 in mediating the effects of AEA. At low doses, this eCB may bind predominantly to CB_1 receptor and exert anxiolytic effects. At higher levels, however, it may activate TRPV1 at the same synapse and promote anxiogenic and proaversive responses (Almeida-Santos, Moreira, Guimaraes, & Aguiar, 2013; Casarotto et al., 2012; Fogaca, Aguiar, et al., 2012; Fogaca, Lisboa, et al., 2012; Lisboa & Guimaraes, 2012; Terzian, Aguiar, Guimaraes, & Moreira, 2009).

Therefore, considering eCBs role in emotional processing, the fine-tune control of neurotransmitters release and different receptors involved assures the necessary balance in neuronal signaling, especially under stressful situations, as suggested elsewhere (Kano, Ohno-Shosaku, Hashimotodani, Uchigashima, & Watanabe, 2009; Fogaca, Lisboa, et al., 2012). Together, these results could help to explain why the resultant effect on anxiety behavior of drugs interfering with eCB signaling depend on several factors, such as the location and type of cannabinoid receptors and the stressors characteristic, including their duration.

5.3 Modulation of the HPA Axis and the Immune System

Exposure to acute and/or repeated stressors increases anxiety in rodents and is associated with the onset of several psychiatric disorders in humans. Cannabinoid effects in animal models of anxiety are also influenced by stress, being usually more marked in highly stressful situations (Morena et al., 2016). Several plastic mechanisms seem to be responsible for this influence. Activation of CB_1 receptors modulates several mechanisms of synaptic plasticity such as LTP, LTD, and depolarization-induced suppression of inhibition/excitation (DSI/E) that are influenced by stress (for review, see Morena et al., 2016). Stress, on the other hand, changes eCB signaling (Morena et al., 2016). For example, chronic unpredictable stress (CUS) decreases CB_1 receptors and reduces AEA levels, and eCB-mediated synaptic plasticity in brain areas associated with anxiety such as MPFC, amygdala, and hippocampus (Gorzalka & Hill, 2011; Hill et al., 2008), as discussed previously in this chapter. In adult rats, CUS decreases CB_1 receptors in the hippocampus of male rats by still unknown mechanisms. This reduction seems to involve GABAergic terminals, with the less inhibition resulting in enhanced local excitatory transmission (Reich, Mihalik, Iskander, Seckler, & Weiss, 2013). In this way, considering that stressful situations decrease AEA levels and drive AEA to act preferentially in CB receptors located in GABAergic terminals in the hippocampus, we could argue that administration of an AEA metabolism/uptake inhibitor into the ventral hippocampus turns the anxiogenic effect of this drug in nonstressed rats into anxiolytic effect after restraint stress since it is normalizing the local neurotransmission (Campos et al., 2010).

Since the effects of chronic stressors on eCBs are similar to those produced by repeated corticosterone administration, activation of the HPA axis has been proposed as the causal factor relating stress and changes in eCB signaling (Gorzalka & Hill, 2011). Glucocorticoid exogenous administration, similar to stress, preferentially modulates eCB signaling in glutamatergic or GABAergic terminals, depending on the region (Morena et al., 2016). The eCB system also seems to be very sensitive to early stress in a gender-dependent manner. Maternal deprivation, for example, increases expression of most eCB-related genes (Marco et al., 2014).

Although there are some inconsistent data, stress exposure or corticosterone administration treatment increases 2-AG, but decreases AEA levels, in the hippocampus and other brain regions (Morena et al., 2016). The decline in AEA levels induced by stress seems to depend on increased expression/activity of FAAH caused by a rise in CRF signaling. This initial AEA decrease has been proposed to mediate the initiation of the stress response. The subsequent corticosterone-dependent increase in 2-AG levels would facilitate the termination of this response (Morena et al., 2016). Furthermore, CB_1 KO mice exhibit increased CRF mRNA levels in the PVN, suggesting impaired HPA fast feedback mechanism, showed by sustained HPA axis activity (Steiner et al., 2008).

These results suggest that the hippocampus, together with other brain regions such as the MPFC and BLA, could be one of the structures where the HPA and eCB systems interact to modulate stress-induced changes in anxiety behavior (Lee & Gorzalka, 2012; Morena et al., 2016). For example, blockade of dorsal hippocampal CB_1 receptors with AM251 prevented the impairment of contextual fear retrieval induced by systemic injection of corticosterone in rats (Atsak et al., 2012). Moreover, WIN55,212-2 administered before each daily restraint stress episode for 2 weeks prevented the long-lasting anxiogenic effects, impaired LTP in the *vHIP-accumbens* pathway and hippocampal-dependent learning performance, and reduced GR expression in the hippocampus (Abush & Akirav, 2013). In addition, exposure to stressors such as SPS upregulates GR in the hippocampus, in the BLA and frontal cortex, an effect prevented by WIN55,212-2 systemic administration (Ganon-Elazar & Akirav, 2013).

The attenuation of stress consequences by cannabinoids seems to depend on GR activation, since they can be prevented by the GR antagonist RU-486. In the BLA, however, the stress-induced increase in glutamate currents depends on mineralocorticoid receptors (Karst, Berger, Erdmann, Schutz, & Joels, 2010). Beta-adrenergic receptors could also be involved, since propranolol injected into the hippocampus blocked the memory retrieval impairment induced by WIN55,212-2. These results suggest that stressful-arousing situations increase GCs levels in the hippocampus that would interact, by nongenomic mechanisms, with the noradrenergic system to impair emotional memory retrieval (Hill et al., 2010).

Glucocorticoids, similar to noradrenergic signaling, are well known to modulate immune response induced by stressors (Carrillo-de Sauvage et al., 2013), including microglia activation (Frank, Thompson, Watkins, & Maier, 2012). They can also suppress hippocampal and amygdala eCB signaling (Bowles et al., 2012). These alterations have been implicated in anxiety-like behavior induced by stressful situations involving the immune system activation (Reader et al., 2015). Microglia, the resident immune cells in the brain, peripheral macrophages and other immune cells express cannabinoid receptors (Atwood, Huffman, Straiker, & Mackie, 2010; Ehrhart et al., 2005; Maresz, Carrier, Ponomarev, Hillard, & Dittel, 2005; Martin-Moreno et al., 2011; Nunez et al., 2004; and others) and functional eCB degradation enzymes (Tham, Whitaker, Luo, & Webb, 2007). Under pathological conditions, expression of CB_2 and production of eCBs are increased (Cabral & Marciano-Cabral, 2005; Stella, 2009). It has been suggested that such increase may contribute to defense mechanisms through accumulation of antiinflammatory microglia phenotype (Stella, 2009).

Cannabinoids and eCBs can interfere on microglia activation and the production of inflammatory mediators (Henry, Kerr, Finn, & Roche, 2016; Lisboa, Gomes, Guimaraes, & Campos, 2016; Mecha, Carrillo-Salinas, Feliú, Mestres, & Guaza, 2016). Over the past years, increasing amount of evidence points to an important modulatory role of immune system in anxiety- and stress-related disorders (Baker, Nievergelt, & O'Connor, 2012; Capuron & Miller, 2011; Haroon, Raison, & Miller, 2012; Maes, Leonard, Myint, Kubera, & Verkerk, 2011; O'Donovan et al., 2015; Yirmiya, Rimmerman, & Reshef, 2015; and others). In contrast, microglia cells also contribute to the CNS homeostasis and plasticity, removing redundant synapses and eliminating dying neurons, in addition to modulating neurotransmitter release and neurogenesis (Beumer et al., 2012; Jones & Lynch, 2015; Kettenmann, Kirchhoff, & Verkhratsky, 2013; Ziv et al., 2006).

There is a substantial body of evidence suggesting that cannabinoid receptors, mostly CB_2 on microglia, may be potential targets for cannabinoid drugs that modulate immune system associated disorders, including psychiatry disorders (Lisboa, Gomes, et al., 2016). In vitro and in vivo studies have revealed that activation of those receptors suppresses the release of proinflammatory cytokines, enhances the release of antiinflammatory cytokines and reduces microglia activation and proliferation (Ehrhart et al., 2005; Fernandez-Ruiz, Pazos, Garcia-Arencibia, Sagredo, & Ramos, 2008; Romero-Sandoval, Horvath, Landry, & DeLeo, 2009; Walter et al., 2003; and others). Additionally, an extensive literature shows that cannabinoid drugs modulate stress-induced immune system alterations along with behavioral effects. For instance, the increased proinflammatory profile in the frontal cortex induced by subchronic stress in mice was attenuated by pharmacological CB_2 activation or by genetic overexpression of this receptor, whereas CB_2 deletion induced effects similar to stress (Zoppi et al., 2014). Similar results were reported with CB_1 activation and CB_1 KO mice (Zoppi et al., 2011), suggesting that both receptors exert an inhibitory role in stress-induced neuroinflammation and its behavioral consequences. We recently provided the first evidence linking microglia cells, cannabinoid receptors, and anxiety by showing that administration of the nonselective cannabinoid receptors agonist WIN55,212-2 daily before social defeat stress in mice attenuates stress-induced immune system activation, including brain microglia activation and anxiety-like behavior (Lisboa, Niraula, et al., 2016).

In addition to stress exposure, other evidence for interaction between cannabinoid and immune systems on anxiety modulation came from studies with IL-1β signaling. Central administration of IL-1β, a proinflammatory cytokine that has been related to learning and memory, HPA axis modulation, and neurogenesis, induces anxiety-like behavior similar to a social defeat stress paradigm in mice. The stress effect is dependent on the IL-1β receptor signaling (Rossi, Sacchetti, et al., 2012). Moreover, both IL-1β and stress (Rossi et al., 2008) alter the sensitivity of $CB_{1/GABA}$ receptors in striatum in a cholesterol- and TRPV1 receptors-dependent fashion (Maccarrone et al., 2008; Rossi et al., 2008; Rossi, Sacchetti, et al., 2012). Meanwhile, it was also showed that IL-1β increases the frequency of glutamatergic sEPSCs in the striatum via TRPV1 channels (Rossi, Furlan, et al., 2012). Different eCBs (2-AG and AEA, respectively) and intracellular mechanisms, however, are related to modulation of $CB_{1/GABA}$ and $CB_{1/Glutamate}$ (De Chiara et al., 2013; Rossi, Sacchetti, et al., 2012). These data suggest that inflammatory mediators, such as IL-1β, induce anxiety by interfering with eCB signaling and alteration of GABA and glutamate neurotransmissions. In contrast, activation of CB_1 and CB_2 receptors could attenuate anxiety disorders by inducing IL-1ra release, an endogenous antagonist of IL-1β, by glial cells (Molina-Holgado et al., 2003). In addition to the effects on immune cells and directly modulation of the HPA axis, as aforementioned, considering the relationship between cannabinoids and IL-1β signaling and that this signaling pathway modulates the HPA axis activity and, consequently, glucocorticoid release (Goshen, Yirmiya, Iverfeldt, & Weidenfeld, 2003), cannabinoids could also indirectly interfere with the HPA axis through IL-1β signaling. Overall, the studies published so far indicate that eCB signaling could modulate anxiety behavior by interfering with the HPA axis and immune system activation, dampening the overactivation of the immune system.

5.4 Plastic Mechanisms

Two cellular mechanisms have been particularly associated with the behavioral consequences of stress exposure: decreased hippocampal neurogenesis and dendritic remodeling (Castren & Hen, 2013). In adult mice, repeated treatment with drugs that facilitate eCB signaling share similarities with antidepressants. Both activate intracellular pathways that regulate cell proliferation and neural cell survival such as mitogen-activated protein kinase and phosphoinositide 3-kinase/Akt or PKB signaling (Fogaca, Galve-Roperh, et al., 2013). Nonselective CB_1 and CB_2 agonists, or drugs that facilitate eCB signaling, could enhance adult hippocampal neurogenesis (Fogaca, Galve-Roperh, et al., 2013). This effect is similar to that observed with antidepressants and has been associated with the anxiolytic/antistress effects of repeated treatment with these drugs (Campos et al., 2013; Jiang et al., 2005). The effects of cannabinoid agonists on neurogenesis, however, could also be deleterious. For example, adolescent exposure to CB_1 agonists could interfere with hippocampal function by decreasing neurogenesis (Abboussi, Tazi, Paizanis, & El Ganouni, 2014) and Δ^9-THC can reduce hippocampal cell proliferation and impair spatial memory (Wolf et al., 2010).

CB_1 receptor maintenance of functional hippocampal homeostasis is proposed to depend on the structural synaptic plasticity of pyramidal neurons (Monory, Polack, Remus, Lutz, & Korte, 2015). However, this role depends on their synaptic localization. Deletion of CB_1 receptors located in glutamatergic or GABAergic terminals in the hippocampus increase and decrease, respectively, the density of dendrite spines in the dentate gyrus. These effects were associated with similar changes in long-term potentiation (Monory et al., 2015).

Exposure to cannabinoid agonists can decrease dendritic spine density in the dentate gyrus, a finding that was associated with memory impairment (Candelaria-Cook & Hamilton, 2014). A direct interference by cannabinoids on stress-induced synaptic remodeling in the hippocampus, however, has not been investigated.

6. CLOSING REMARKS AND FUTURE DIRECTIONS

It has become increasingly clear by both preclinical and clinical studies that cannabinoids interfere with anxiety responses. The final outcome, however, depends on several factors such as the drugs and doses used, the aversiveness of the task, and the animal species. These cannabinoid drugs exert these multifaceted effects by interfering with components of the eCB system. Numerous pieces of evidence now suggest that this system regulates and buffers exaggerated responses to several insults, including threatening and aversive stimuli. Several studies have also suggested that eCB dysfunctions play a role in the pathophysiology of anxiety- and stress-related disorders. Consequently, their components are potential targets for the development of new therapeutic approaches for these disorders. Along this review, several of these targets have been discussed. However, paralleling the increase in the number of studies investigating the modulatory effects of cannabinoids on emotions, our knowledge about the complexity of the eCB system and the neurobiology of anxiety is also exponentially growing. The challenge for future studies and the development of new therapeutic targets is to integrate these recent findings in an intelligible picture.

ACKNOWLEDGMENTS

We thank the financial support from FAPESP (2011/22523-0, 2012/17626-7, 2014/21226-0, 2014/24828-0), Capes, and CNPq.

REFERENCES

Abboussi, O., Tazi, A., Paizanis, E., & El Ganouni, S. (2014). Chronic exposure to WIN55,212-2 affects more potently spatial learning and memory in adolescents than in adult rats via a negative action on dorsal hippocampal neurogenesis. *Pharmacology, Biochemistry, and Behavior*, *120*, 95–102. http://dx.doi.org/10.1016/j.pbb.2014.02.014. S0091-3057(14)00066-5 [pii].

Abush, H., & Akirav, I. (2013). Cannabinoids ameliorate impairments induced by chronic stress to synaptic plasticity and short-term memory. *Neuropsychopharmacology*, *38*(8), 1521–1534. http://dx.doi.org/10.1038/npp.2013.51. npp201351 [pii].

Adolphs, R., Tranel, D., Damasio, H., & Damasio, A. (1994). Impaired recognition of emotion in facial expressions following bilateral damage to the human amygdala. *Nature*, *372*(6507), 669–672. http://dx.doi.org/10.1038/372669a0.

Aguiar, D. C., Terzian, A. L., Guimaraes, F. S., & Moreira, F. A. (2009). Anxiolytic-like effects induced by blockade of transient receptor potential vanilloid type 1 (TRPV1) channels in the medial prefrontal cortex of rats. *Psychopharmacology*, *205*(2), 217–225. http://dx.doi.org/10.1007/s00213-009-1532-5.

Akinshola, B. E., Chakrabarti, A., & Onaivi, E. S. (1999). In-vitro and in-vivo action of cannabinoids. *Neurochemical Research*, *24*(10), 1233–1240.

Akirav, I. (2011). The role of cannabinoids in modulating emotional and non-emotional memory processes in the hippocampus. *Frontiers in Behavioral Neuroscience*, *5*, 34. http://dx.doi.org/10.3389/fnbeh.2011.00034.

Aliczki, M., Zelena, D., Mikics, E., Varga, Z. K., Pinter, O., Bakos, N. V., & Haller, J. (2013). Monoacylglycerol lipase inhibition-induced changes in plasma corticosterone levels, anxiety and locomotor activity in male CD1 mice. *Hormones and Behavior*, *63*(5), 752–758. http://dx.doi.org/10.1016/j.yhbeh.2013.03.017.

Almeida-Santos, A. F., Gobira, P. H., Rosa, L. C., Guimaraes, F. S., Moreira, F. A., & Aguiar, D. C. (2013). Modulation of anxiety-like behavior by the endocannabinoid 2-arachidonoylglycerol (2-AG) in the dorsolateral periaqueductal gray. *Behavioural Brain Research*, *252*, 10–17. http://dx.doi.org/10.1016/j.bbr.2013.05.027.

Almeida-Santos, A. F., Moreira, F. A., Guimaraes, F. S., & Aguiar, D. C. (2013). Role of TRPV1 receptors on panic-like behaviors mediated by the dorsolateral periaqueductal gray in rats. *Pharmacology, Biochemistry, and Behavior*, *105*, 166–172. http://dx.doi.org/10.1016/j.pbb.2013.02.012.

Alves, F. H., Crestani, C. C., Gomes, F. V., Guimaraes, F. S., Correa, F. M., & Resstel, L. B. (2010). Cannabidiol injected into the bed nucleus of the stria terminalis modulates baroreflex activity through 5-HT1A receptors. *Pharmacological Research*, *62*(3), 228–236. http://dx.doi.org/10.1016/j.phrs.2010.05.003. S1043-6618(10)00119-2 [pii].

Amstadter, A. B., Nugent, N. R., & Koenen, K. C. (2009). Genetics of PTSD: Fear conditioning as a model for future research. *Psychiatric Annals*, *39*(6), 358–367. http://dx.doi.org/10.3928/00485713-20090526-01.

Arenos, J. D., Musty, R. E., & Bucci, D. J. (2006). Blockade of cannabinoid CB1 receptors alters contextual learning and memory. *European Journal of Pharmacology*, *539*(3), 177–183. http://dx.doi.org/10.1016/j.ejphar.2006.04.017.

Arevalo, C., de Miguel, R., & Hernandez-Tristan, R. (2001). Cannabinoid effects on anxiety-related behaviours and hypothalamic neurotransmitters. *Pharmacology, Biochemistry, and Behavior*, *70*(1), 123–131.

Arnold, J. C., Topple, A. N., Mallet, P. E., Hunt, G. E., & McGregor, I. S. (2001). The distribution of cannabinoid-induced Fos expression in rat brain: Differences between the Lewis and Wistar strain. *Brain Research*, *921*(1–2), 240–255. S0006-8993(01)03127-4 [pii].

Atsak, P., Hauer, D., Campolongo, P., Schelling, G., McGaugh, J. L., & Roozendaal, B. (2012). Glucocorticoids interact with the hippocampal endocannabinoid system in impairing retrieval of contextual fear memory. *Proceedings of the National Academy of Sciences of the United States of America*, *109*(9), 3504–3509. http://dx.doi.org/10.1073/pnas.1200742109. 1200742109 [pii].

Atwood, B. K., Huffman, J., Straiker, A., & Mackie, K. (2010). JWH018, a common constituent of 'Spice' herbal blends, is a potent and efficacious cannabinoid CB receptor agonist. *British Journal of Pharmacology*, *160*(3), 585–593. http://dx.doi.org/10.1111/j.1476-5381.2009.00582.x.

Auclair, N., Otani, S., Soubrie, P., & Crepel, F. (2000). Cannabinoids modulate synaptic strength and plasticity at glutamatergic synapses of rat prefrontal cortex pyramidal neurons. *Journal of Neurophysiology*, *83*(6), 3287–3293.

Azad, S. C., Eder, M., Marsicano, G., Lutz, B., Zieglgansberger, W., & Rammes, G. (2003). Activation of the cannabinoid receptor type 1 decreases glutamatergic and GABAergic synaptic transmission in the lateral amygdala of the mouse. *Learning & Memory*, *10*(2), 116–128. http://dx.doi.org/10.1101/lm.53303.

Azad, S. C., Kurz, J., Marsicano, G., Lutz, B., Zieglgansberger, W., & Rammes, G. (2008). Activation of CB1 specifically located on GABAergic interneurons inhibits LTD in the lateral amygdala. *Learning & Memory*, *15*(3), 143–152. http://dx.doi.org/10.1101/lm.741908. 15/3/143 [pii].

Baker, D. G., Nievergelt, C. M., & O'Connor, D. T. (2012). Biomarkers of PTSD: Neuropeptides and immune signaling. *Neuropharmacology*, *62*(2), 663–673. http://dx.doi.org/10.1016/j.neuropharm.2011.02.027.

Balerio, G. N., Aso, E., & Maldonado, R. (2006). Role of the cannabinoid system in the effects induced by nicotine on anxiety-like behaviour in mice. *Psychopharmacology*, *184*(3–4), 504–513. http://dx.doi.org/10.1007/s00213-005-0251-9.

Bambico, F. R., Cassano, T., Dominguez-Lopez, S., Katz, N., Walker, C. D., Piomelli, D., & Gobbi, G. (2010). Genetic deletion of fatty acid amide hydrolase alters emotional behavior and serotonergic transmission in the dorsal raphe, prefrontal cortex, and hippocampus. *Neuropsychopharmacology*, *35*(10), 2083–2100. http://dx.doi.org/10.1038/npp.2010.80.

Bannerman, D. M., Rawlins, J. N., McHugh, S. B., Deacon, R. M., Yee, B. K., Bast, T., & Feldon, J. (2004). Regional dissociations within the hippocampus—Memory and anxiety. *Neuroscience and Biobehavioral Reviews*, *28*(3), 273–283. http://dx.doi.org/10.1016/j.neubiorev.2004.03.004. S0149763404000314 [pii].

Basavarajappa, B. S. (2007). Neuropharmacology of the endocannabinoid signaling system-molecular mechanisms, biological actions and synaptic plasticity. *Current Neuropharmacology*, *5*(2), 81–97.

Batista, L. A., Bastos, J. R., & Moreira, F. A. (2015). Role of endocannabinoid signalling in the dorsolateral periaqueductal grey in the modulation of distinct panic-like responses. *Journal of Psychopharmacology*, *29*(3), 335–343. http://dx.doi.org/10.1177/0269881114566259.

Batista, P. A., Fogaca, M. V., & Guimaraes, F. S. (2015). The endocannabinoid, endovanilloid and nitrergic systems could interact in the rat dorsolateral periaqueductal gray matter to control anxiety-like behaviors. *Behavioural Brain Research*, *293*, 182–188. http://dx.doi.org/10.1016/j.bbr.2015.07.019.

Bedse, G., Romano, A., Tempesta, B., Lavecchia, M. A., Pace, L., Bellomo, A., & Gaetani, S. (2015). Inhibition of anandamide hydrolysis enhances noradrenergic and GABAergic transmission in the prefrontal cortex and basolateral amygdala of rats subjected to acute swim stress. *Journal of Neuroscience Research*, *93*(5), 777–787. http://dx.doi.org/10.1002/jnr.23539.

Beltramo, M., & Piomelli, D. (2000). Carrier-mediated transport and enzymatic hydrolysis of the endogenous cannabinoid 2-arachidonylglycerol. *Neuroreport*, *11*(6), 1231–1235.

Beltramo, M., Stella, N., Calignano, A., Lin, S. Y., Makriyannis, A., & Piomelli, D. (1997). Functional role of high-affinity anandamide transport, as revealed by selective inhibition. *Science*, *277*(5329), 1094–1097.

Bergamaschi, M. M., Queiroz, R. H., Chagas, M. H., de Oliveira, D. C., De Martinis, B. S., Kapczinski, F., … Crippa, J. A. (2011). Cannabidiol reduces the anxiety induced by simulated public speaking in treatment-naive social phobia patients. *Neuropsychopharmacology*, *36*(6), 1219–1226. http://dx.doi.org/10.1038/npp.2011.6. npp20116 [pii].

Bergamaschi, M. M., Queiroz, R. H., Chagas, M. H., Linares, I. M., Arrais, K. C., de Oliveira, D. C., … Crippa, J. A. (2014). Rimonabant effects on anxiety induced by simulated public speaking in healthy humans: A preliminary report. *Human Psychopharmacology*, *29*(1), 94–99. http://dx.doi.org/10.1002/hup.2374.

Berrendero, F., & Maldonado, R. (2002). Involvement of the opioid system in the anxiolytic-like effects induced by delta(9)-tetrahydrocannabinol. *Psychopharmacology, 163*(1), 111–117. http://dx.doi.org/10.1007/s00213-002-1144-9.

Bertoglio, L. J., Joca, S. R., & Guimaraes, F. S. (2006). Further evidence that anxiety and memory are regionally dissociated within the hippocampus. *Behavioural Brain Research, 175*(1), 183–188. http://dx.doi.org/10.1016/j.bbr.2006.08.021. S0166-4328(06)00479-7 [pii].

Beumer, W., Gibney, S. M., Drexhage, R. C., Pont-Lezica, L., Doorduin, J., Klein, H. C., & Drexhage, H. A. (2012). The immune theory of psychiatric diseases: A key role for activated microglia and circulating monocytes. *Journal of Leukocyte Biology, 92*(5), 959–975. http://dx.doi.org/10.1189/jlb.0212100.

Bhattacharyya, S., Morrison, P. D., Fusar-Poli, P., Martin-Santos, R., Borgwardt, S., Winton-Brown, T., & McGuire, P. K. (2010). Opposite effects of delta-9-tetrahydrocannabinol and cannabidiol on human brain function and psychopathology. *Neuropsychopharmacology, 35*(3), 764–774. http://dx.doi.org/10.1038/npp.2009.184. npp2009184 [pii].

Bishop, S. J. (2007). Neurocognitive mechanisms of anxiety: An integrative account. *Trends in Cognitive Sciences, 11*(7), 307–316.

Bisogno, T., Berrendero, F., Ambrosino, G., Cebeira, M., Ramos, J. A., Fernandez-Ruiz, J. J., & Di Marzo, V. (1999). Brain regional distribution of endocannabinoids: Implications for their biosynthesis and biological function. *Biochemical and Biophysical Research Communications, 256*(2), 377–380. http://dx.doi.org/10.1006/bbrc.1999.0254. S0006-291X(99)90254-1 [pii].

Bisogno, T., Hanus, L., De Petrocellis, L., Tchilibon, S., Ponde, D. E., Brandi, I., & Di Marzo, V. (2001). Molecular targets for cannabidiol and its synthetic analogues: Effect on vanilloid VR1 receptors and on the cellular uptake and enzymatic hydrolysis of anandamide. *British Journal of Pharmacology, 134*(4), 845–852. http://dx.doi.org/10.1038/sj.bjp.0704327.

Bitencourt, R. M., Pamplona, F. A., & Takahashi, R. N. (2008). Facilitation of contextual fear memory extinction and anti-anxiogenic effects of AM404 and cannabidiol in conditioned rats. *European Neuropsychopharmacology, 18*(12), 849–859. http://dx.doi.org/10.1016/j.euroneuro.2008.07.001.

Blanchard, R. J., Blanchard, D. C., Griebel, G., & Nutt, D. J. (2008). Introduction to the handbook on fear and anxiety. In R. J. Blanchard, D. C. Blanchard, G. Griebel, & D. J. Nutt (Eds.), *Handbook of anxiety and fear.* Oxford: Elsevier.

Blasio, A., Iemolo, A., Sabino, V., Petrosino, S., Steardo, L., Rice, K. C., & Cottone, P. (2013). Rimonabant precipitates anxiety in rats withdrawn from palatable food: Role of the central amygdala. *Neuropsychopharmacology, 38*(12), 2498–2507. http://dx.doi.org/10.1038/npp.2013.153.

Bortolato, M., Campolongo, P., Mangieri, R. A., Scattoni, M. L., Frau, R., Trezza, V., & Piomelli, D. (2006). Anxiolytic-like properties of the anandamide transport inhibitor AM404. *Neuropsychopharmacology, 31*(12), 2652–2659. http://dx.doi.org/10.1038/sj.npp.1301061.

Bowles, N. P., Hill, M. N., Bhagat, S. M., Karatsoreos, I. N., Hillard, C. J., & McEwen, B. S. (2012). Chronic, noninvasive glucocorticoid administration suppresses limbic endocannabinoid signaling in mice. *Neuroscience, 204*, 83–89. http://dx.doi.org/10.1016/j.neuroscience.2011.08.048.

Braida, D., Limonta, V., Malabarba, L., Zani, A., & Sala, M. (2007). 5-HT1A receptors are involved in the anxiolytic effect of Delta9-tetrahydrocannabinol and AM 404, the anandamide transport inhibitor, in Sprague-Dawley rats. *European Journal of Pharmacology, 555*(2–3), 156–163. http://dx.doi.org/10.1016/j.ejphar.2006.10.038.

Brené, S., Messer, C., & Nestler, E. J. (1998). Expression of messenger RNAs encoding ionotropic glutamate receptors in rat brain: Regulation by haloperidol. *Neuroscience*, *84*(3), 813–823.

Bukalo, O., Pinard, C. R., Silverstein, S., Brehm, C., Hartley, N. D., Whittle, N., & Holmes, A. (2015). Prefrontal inputs to the amygdala instruct fear extinction memory formation. *Science Advances*, *1*(6), 1–8. http://dx.doi.org/10.1126/sciadv.1500251. e1500251 [pii].

Buskila, Y., Abu-Ghanem, Y., Levi, Y., Moran, A., Grauer, E., & Amitai, Y. (2007). Enhanced astrocytic nitric oxide production and neuronal modifications in the neocortex of a NOS2 mutant mouse. *PloS One*, *2*(9), e843. http://dx.doi.org/10.1371/journal.pone.0000843.

Busquets-Garcia, A., Puighermanal, E., Pastor, A., de la Torre, R., Maldonado, R., & Ozaita, A. (2011). Differential role of anandamide and 2-arachidonoylglycerol in memory and anxiety-like responses. *Biological Psychiatry*, *70*(5), 479–486. http://dx.doi.org/10.1016/j.biopsych.2011.04.022.

Cabral, G. A., & Marciano-Cabral, F. (2005). Cannabinoid receptors in microglia of the central nervous system: Immune functional relevance. *Journal of Leukocyte Biology*, *78*(6), 1192–1197. http://dx.doi.org/10.1189/jlb.0405216. jlb.0405216 [pii].

Cameron, C., Watson, D., & Robinson, J. (2014). Use of a synthetic cannabinoid in a correctional population for posttraumatic stress disorder-related insomnia and nightmares, chronic pain, harm reduction, and other indications: A retrospective evaluation. *Journal of Clinical Psychopharmacology*, *34*(5), 559–564. http://dx.doi.org/10.1097/JCP.0000000000000180.

Campos, A. C., Ferreira, F. R., Guimaraes, F. S., & Lemos, J. I. (2010). Facilitation of endocannabinoid effects in the ventral hippocampus modulates anxiety-like behaviors depending on previous stress experience. *Neuroscience*, *167*(2), 238–246. http://dx.doi.org/10.1016/j.neuroscience.2010.01.062. S0306-4522(10)00154-5 [pii].

Campos, A. C., & Guimaraes, F. S. (2008). Involvement of 5HT1A receptors in the anxiolytic-like effects of cannabidiol injected into the dorsolateral periaqueductal gray of rats. *Psychopharmacology*, *199*(2), 223–230. http://dx.doi.org/10.1007/s00213-008-1168-x.

Campos, A. C., Moreira, F. A., Gomes, F. V., Del Bel, E. A., & Guimaraes, F. S. (2012). Multiple mechanisms involved in the large-spectrum therapeutic potential of cannabidiol in psychiatric disorders. *Philosophical Transactions of the Royal Society of London. Series B, Biological Sciences*, *367*, 3364–3378. http://dx.doi.org/10.1098/rstb.2011.0389.

Campos, A. C., Ortega, Z., Palazuelos, J., Fogaca, M. V., Aguiar, D. C., Diaz-Alonso, J., & Guimaraes, F. S. (2013). The anxiolytic effect of cannabidiol on chronically stressed mice depends on hippocampal neurogenesis: Involvement of the endocannabinoid system. *International Journal of Neuropsychopharmacology*, *16*(6), 1407–1419. http://dx.doi.org/10.1017/S1461145712001502.

Candelaria-Cook, F. T., & Hamilton, D. A. (2014). Chronic cannabinoid agonist (WIN 55,212-2) exposure alters hippocampal dentate gyrus spine density in adult rats. *Brain Research*, *1542*, 104–110. http://dx.doi.org/10.1016/j.brainres.2013.10.039. S0006-8993(13)01430-3 [pii].

Canteras, N. S., Resstel, L. B., Bertoglio, L. J., Carobrez, A. P., & Guimaraes, F. S. (2010). Neuroanatomy of anxiety. *Current Topics in Behavioral Neurosciences*, *2*, 77–96.

Capuron, L., & Miller, A. H. (2011). Immune system to brain signaling: Neuropsychopharmacological implications. *Pharmacology & Therapeutics*, *130*(2), 226–238. http://dx.doi.org/10.1016/j.pharmthera.2011.01.014.

Carrillo-de Sauvage, M. A., Maatouk, L., Arnoux, I., Pasco, M., Sanz Diez, A., Delahaye, M., & Vyas, S. (2013). Potent and multiple regulatory actions of microglial glucocorticoid receptors during CNS inflammation. *Cell Death and Differentiation, 20*(11), 1546–1557. http://dx.doi.org/10.1038/cdd.2013.108.

Carvalho, A. F., Mackie, K., & Van Bockstaele, E. J. (2010). Cannabinoid modulation of limbic forebrain noradrenergic circuitry. *European Journal of Neuroscience, 31*(2), 286–301. http://dx.doi.org/10.1111/j.1460-9568.2009.07054.x.

Casarotto, P. C., Terzian, A. L., Aguiar, D. C., Zangrossi, H., Guimaraes, F. S., Wotjak, C. T., & Moreira, F. A. (2012). Opposing roles for cannabinoid receptor type-1 (CB(1)) and transient receptor potential vanilloid type-1 channel (TRPV1) on the modulation of panic-like responses in rats. *Neuropsychopharmacology, 37*(2), 478–486. http://dx.doi.org/10.1038/npp.2011.207.

Cassano, T., Gaetani, S., Macheda, T., Laconca, L., Romano, A., Morgese, M. G., & Piomelli, D. (2011). Evaluation of the emotional phenotype and serotonergic neurotransmission of fatty acid amide hydrolase-deficient mice. *Psychopharmacology, 214*(2), 465–476. http://dx.doi.org/10.1007/s00213-010-2051-0.

Castillo, P. E., Younts, T. J., Chavez, A. E., & Hashimotodani, Y. (2012). Endocannabinoid signaling and synaptic function. *Neuron, 76*(1), 70–81. http://dx.doi.org/10.1016/j.neuron.2012.09.020.

Castren, E., & Hen, R. (2013). Neuronal plasticity and antidepressant actions. *Trends in Neurosciences, 36*(5), 259–267. http://dx.doi.org/10.1016/j.tins.2012.12.010. S0166-2236(13)00002-7 [pii].

Cathel, A. M., Reyes, B. A., Wang, Q., Palma, J., Mackie, K., Van Bockstaele, E. J., & Kirby, L. G. (2014). Cannabinoid modulation of alpha2 adrenergic receptor function in rodent medial prefrontal cortex. *European Journal of Neuroscience, 40*(8), 3202–3214. http://dx.doi.org/10.1111/ejn.12690.

Cavanaugh, D. J., Chesler, A. T., Jackson, A. C., Sigal, Y. M., Yamanaka, H., Grant, R., & Basbaum, A. I. (2011). Trpv1 reporter mice reveal highly restricted brain distribution and functional expression in arteriolar smooth muscle cells. *Journal of Neuroscience, 31*(13), 5067–5077. http://dx.doi.org/10.1523/JNEUROSCI.6451-10.2011.

Charnay, Y., & Leger, L. (2010). Brain serotonergic circuitries. *Dialogues in Clinical Neuroscience, 12*(4), 471–487.

Chhatwal, J. P., Davis, M., Maguschak, K. A., & Ressler, K. J. (2005). Enhancing cannabinoid neurotransmission augments the extinction of conditioned fear. *Neuropsychopharmacology, 30*(3), 516–524. http://dx.doi.org/10.1038/sj.npp.1300655.

Chiang, K. P., Gerber, A. L., Sipe, J. C., & Cravatt, B. F. (2004). Reduced cellular expression and activity of the P129T mutant of human fatty acid amide hydrolase: Evidence for a link between defects in the endocannabinoid system and problem drug use. *Human Molecular Genetics, 13*(18), 2113–2119. http://dx.doi.org/10.1093/hmg/ddh216.

Choi, K., Le, T., McGuire, J., Xing, G., Zhang, L., Li, H., & Ursano, R. J. (2012). Expression pattern of the cannabinoid receptor genes in the frontal cortex of mood disorder patients and mice selectively bred for high and low fear. *Journal of Psychiatric Research, 46*(7), 882–889. http://dx.doi.org/10.1016/j.jpsychires.2012.03.021.

Clarke, J. R., Rossato, J. I., Monteiro, S., Bevilaqua, L. R., Izquierdo, I., & Cammarota, M. (2008). Posttraining activation of CB1 cannabinoid receptors in the CA1 region of the dorsal hippocampus impairs object recognition long-term memory. *Neurobiology of Learning and Memory, 90*(2), 374–381. http://dx.doi.org/10.1016/j.nlm.2008.04.009. S1074-7427(08)00069-5 [pii].

Crestani, C. C., Alves, F. H., Gomes, F. V., Resstel, L. B., Correa, F. M., & Herman, J. P. (2013). Mechanisms in the bed nucleus of the stria terminalis involved in control of autonomic and neuroendocrine functions: A review. *Current Neuropharmacology, 11*(2), 141–159. http://dx.doi.org/10.2174/1570159X11311020002. CN-11-141 [pii].

Crippa, J. A., Derenusson, G. N., Ferrari, T. B., Wichert-Ana, L., Duran, F. L., Martin-Santos, R., & Hallak, J. E. (2011). Neural basis of anxiolytic effects of cannabidiol (CBD) in generalized social anxiety disorder: A preliminary report. *Journal of Psychopharmacology*, *25*(1), 121–130. http://dx.doi.org/10.1177/0269881110379283.

Crippa, J. A., Zuardi, A. W., Garrido, G. E., Wichert-Ana, L., Guarnieri, R., Ferrari, L., & Filho Busatto, G. (2004). Effects of cannabidiol (CBD) on regional cerebral blood flow. *Neuropsychopharmacology*, *29*(2), 417–426. http://dx.doi.org/10.1038/sj.npp.1300340. 1300340 [pii].

Cristino, L., de Petrocellis, L., Pryce, G., Baker, D., Guglielmotti, V., & Di Marzo, V. (2006). Immunohistochemical localization of cannabinoid type 1 and vanilloid transient receptor potential vanilloid type 1 receptors in the mouse brain. *Neuroscience*, *139*(4), 1405–1415. http://dx.doi.org/10.1016/j.neuroscience.2006.02.074. S0306-4522(06)00240-5 [pii].

Cryan, J. F., & Holmes, A. (2005). The ascent of mouse: Advances in modelling human depression and anxiety. *Nature Reviews. Drug Discovery*, *4*(9), 775–790. http://dx.doi.org/10.1038/nrd1825.

Das, R. K., Kamboj, S. K., Ramadas, M., Yogan, K., Gupta, V., Redman, E., & Morgan, C. J. (2013). Cannabidiol enhances consolidation of explicit fear extinction in humans. *Psychopharmacology*, *226*(4), 781–792. http://dx.doi.org/10.1007/s00213-012-2955-y.

Davis, M., Walker, D. L., Miles, L., & Grillon, C. (2010). Phasic vs sustained fear in rats and humans: Role of the extended amygdala in fear vs anxiety. *Neuropsychopharmacology*, *35*(1), 105–135. http://dx.doi.org/10.1038/npp.2009.109. npp2009109 [pii].

De Chiara, V., Motta, C., Rossi, S., Studer, V., Barbieri, F., Lauro, D., & Centonze, D. (2013). Interleukin-1beta alters the sensitivity of cannabinoid CB1 receptors controlling glutamate transmission in the striatum. *Neuroscience*, *250*, 232–239. http://dx.doi.org/10.1016/j.neuroscience.2013.06.069.

de Oliveira Alvares, L., Pasqualini Genro, B., Diehl, F., Molina, V. A., & Quillfeldt, J. A. (2008). Opposite action of hippocampal CB1 receptors in memory reconsolidation and extinction. *Neuroscience*, *154*(4), 1648–1655. http://dx.doi.org/10.1016/j.neuroscience.2008.05.005. S0306-4522(08)00750-1 [pii].

Dejean, C., Courtin, J., Rozeske, R. R., Bonnet, M. C., Dousset, V., Michelet, T., & Herry, C. (2015). Neuronal circuits for fear expression and recovery: Recent advances and potential therapeutic strategies. *Biological Psychiatry*, *78*(5), 298–306. http://dx.doi.org/10.1016/j.biopsych.2015.03.017. S0006-3223(15)00255-3 [pii].

Del Arco, A., & Mora, F. (2009). Neurotransmitters and prefrontal cortex-limbic system interactions: Implications for plasticity and psychiatric disorders. *Journal of Neural Transmission (Vienna)*, *116*(8), 941–952. http://dx.doi.org/10.1007/s00702-009-0243-8.

Devane, W. A., Dysarz, F. A., 3rd, Johnson, M. R., Melvin, L. S., & Howlett, A. C. (1988). Determination and characterization of a cannabinoid receptor in rat brain. *Molecular Pharmacology*, *34*(5), 605–613.

Devane, W. A., Hanus, L., Breuer, A., Pertwee, R. G., Stevenson, L. A., Griffin, G., & Mechoulam, R. (1992). Isolation and structure of a brain constituent that binds to the cannabinoid receptor. *Science*, *258*(5090), 1946–1949.

Di Marzo, V., & De Petrocellis, L. (2012). Why do cannabinoid receptors have more than one endogenous ligand? *Philosophical Transactions of the Royal Society of London Series B, Biological Sciences*, *367*(1607), 3216–3228. http://dx.doi.org/10.1098/rstb.2011.0382.

Dlugos, A., Childs, E., Stuhr, K. L., Hillard, C. J., & de Wit, H. (2012). Acute stress increases circulating anandamide and other N-acylethanolamines in healthy humans. *Neuropsychopharmacology*, *37*(11), 2416–2427. http://dx.doi.org/10.1038/npp.2012.100. npp2012100 [pii].

Dong, H. W., Petrovich, G. D., & Swanson, L. W. (2001a). Topography of projections from amygdala to bed nuclei of the stria terminalis. *Brain Research. Brain Research Reviews*, *38*(1–2), 192–246. S0165017301000790 [pii].

Dong, H. W., Petrovich, G. D., Watts, A. G., & Swanson, L. W. (2001b). Basic organization of projections from the oval and fusiform nuclei of the bed nuclei of the stria terminalis in adult rat brain. *Journal of Comparative Neurology*, *436*(4), 430–455.

Dos Anjos-Garcia T., Ullah F., Falconi-Sobrinho L.L. and Coimbra N.C., CB1 cannabinoid receptor-mediated anandamide signalling reduces the defensive behaviour evoked through GABAA receptor blockade in the dorsomedial division of the ventromedial hypothalamus, *Neuropharmacology,* in press. http://dx.doi.org/10.1016/j.neuropharm.2016.04.003. S0028-3908(16)30139-3 [pii].

Dubreucq, S., Matias, I., Cardinal, P., Haring, M., Lutz, B., Marsicano, G., & Chaouloff, F. (2012). Genetic dissection of the role of cannabinoid type-1 receptors in the emotional consequences of repeated social stress in mice. *Neuropsychopharmacology*, *37*(8), 1885–1900. http://dx.doi.org/10.1038/npp.2012.36.

Ehrhart, J., Obregon, D., Mori, T., Hou, H., Sun, N., Bai, Y., & Shytle, R. D. (2005). Stimulation of cannabinoid receptor 2 (CB2) suppresses microglial activation. *Journal of Neuroinflammation*, *2*, 29. http://dx.doi.org/10.1186/1742-2094-2-29.

Etkin, A., Prater, K. E., Schatzberg, A. F., Menon, V., & Greicius, M. D. (2009). Disrupted amygdalar subregion functional connectivity and evidence of a compensatory network in generalized anxiety disorder. *Archives of General Psychiatry*, *66*(12), 1361–1372. http://dx.doi.org/10.1001/archgenpsychiatry.2009.104.

Fabre, L. F., & McLendon, D. (1981). The efficacy and safety of nabilone (a synthetic cannabinoid) in the treatment of anxiety. *Journal of Clinical Pharmacology*, *21*(8–9 Suppl.), 377S–382S.

Fernandez-Ruiz, J., Pazos, M. R., Garcia-Arencibia, M., Sagredo, O., & Ramos, J. A. (2008). Role of CB2 receptors in neuroprotective effects of cannabinoids. *Molecular and Cellular Endocrinology*, *286*(1–2 Suppl. 1), S91–S96. http://dx.doi.org/10.1016/j.mce.2008.01.001.

File, S. E., Lippa, A. S., Beer, B., & Lippa, M. T. (2004). Animal tests of anxiety. *Current Protocols in Neuroscience*. http://dx.doi.org/10.1002/0471142301.ns0803s26. Chapter 8, Unit 8 3.

Finn, D. P., Beckett, S. R., Richardson, D., Kendall, D. A., Marsden, C. A., & Chapman, V. (2004). Evidence for differential modulation of conditioned aversion and fear-conditioned analgesia by CB1 receptors. *European Journal of Neuroscience*, *20*(3), 848–852. http://dx.doi.org/10.1111/j.1460-9568.2004.03509.x.

Finn, D. P., Jhaveri, M. D., Beckett, S. R., Kendall, D. A., Marsden, C. A., & Chapman, V. (2004). Cannabinoids modulate ultrasound-induced aversive responses in rats. *Psychopharmacology*, *172*(1), 41–51. http://dx.doi.org/10.1007/s00213-003-1629-1.

Finn, D. P., Jhaveri, M. D., Beckett, S. R., Roe, C. H., Kendall, D. A., Marsden, C. A., & Chapman, V. (2003). Effects of direct periaqueductal grey administration of a cannabinoid receptor agonist on nociceptive and aversive responses in rats. *Neuropharmacology*, *45*(5), 594–604.

Fitzgerald, P. J. (2011). A neurochemical yin and yang: Does serotonin activate and norepinephrine deactivate the prefrontal cortex? *Psychopharmacology*, *213*(2–3), 171–182. http://dx.doi.org/10.1007/s00213-010-1856-1.

Fogaca, M. V., Aguiar, D. C., Moreira, F. A., & Guimaraes, F. S. (2012). The endocannabinoid and endovanilloid systems interact in the rat prelimbic medial prefrontal cortex to control anxiety-like behavior. *Neuropharmacology*, *63*(2), 202–210. http://dx.doi.org/10.1016/j.neuropharm.2012.03.007.

Fogaca, M. V., Galve-Roperh, I., Guimaraes, F. S., & Campos, A. C. (2013). Cannabinoids, neurogenesis and antidepressant drugs: Is there a link? *Current Neuropharmacology*, *11*(3), 263–275. http://dx.doi.org/10.2174/1570159X11311030003. CN-11-263 [pii].

Fogaca, M. V., Gomes, F. V., Moreira, F. A., Guimaraes, F. S., & Aguiar, D. C. (2013). Effects of glutamate NMDA and TRPV1 receptor antagonists on the biphasic responses to anandamide injected into the dorsolateral periaqueductal grey of Wistar rats. *Psychopharmacology*, *226*(3), 579–587. http://dx.doi.org/10.1007/s00213-012-2927-2.

Fogaca, M. V., Lisboa, S. F., Aguair, D. C., Moreira, F. A., Gomes, F. V., Casarotto, P. C., & Guimaraes, F. S. (2012). Fine-tuning of defensive behaviors in the dorsal periaqueductal gray by atypical neurotransmitters. *Brazilian Journal of Medical and Biological Research*, *45*(4), 357–365. http://dx.doi.org/10.1590/S0100-879X2012007500029.

Fogaca, M. V., Sonego, A. B., Rioli, V., Gozzo, F. C., Dale, C. S., Ferro, E. S., & Guimaraes, F. S. (2015). Anxiogenic-like effects induced by hemopressin in rats. *Pharmacology, Biochemistry, and Behavior*, *129*, 7–13. http://dx.doi.org/10.1016/j.pbb.2014.11.013.

Frank, M. G., Thompson, B. M., Watkins, L. R., & Maier, S. F. (2012). Glucocorticoids mediate stress-induced priming of microglial pro-inflammatory responses. *Brain, Behavior, and Immunity*, *26*(2), 337–345. http://dx.doi.org/10.1016/j.bbi.2011.10.005.

Franklin, J. M., & Carrasco, G. A. (2012). Cannabinoid-induced enhanced interaction and protein levels of serotonin 5-HT(2A) and dopamine D(2) receptors in rat prefrontal cortex. *Journal of Psychopharmacology*, *26*(10), 1333–1347. http://dx.doi.org/10.1177/0269881112450786.

Fraser, G. A. (2009). The use of a synthetic cannabinoid in the management of treatment-resistant nightmares in posttraumatic stress disorder (PTSD). *CNS Neuroscience & Therapeutics*, *15*(1), 84–88. http://dx.doi.org/10.1111/j.1755-5949.2008.00071.x. CNS071 [pii].

Fusar-Poli, P., Allen, P., Bhattacharyya, S., Crippa, J. A., Mechelli, A., Borgwardt, S., & McGuire, P. (2010). Modulation of effective connectivity during emotional processing by delta 9-tetrahydrocannabinol and cannabidiol. *International Journal of Neuropsychopharmacology*, *13*(4), 421–432. http://dx.doi.org/10.1017/S1461145709990617. S1461145709990617 [pii].

Fusar-Poli, P., Crippa, J. A., Bhattacharyya, S., Borgwardt, S. J., Allen, P., Martin-Santos, R., & McGuire, P. K. (2009). Distinct effects of {delta}9-tetrahydrocannabinol and cannabidiol on neural activation during emotional processing. *Archives of General Psychiatry*, *66*(1), 95–105. http://dx.doi.org/10.1001/archgenpsychiatry.2008.519. 66/1/95 [pii].

Ganon-Elazar, E., & Akirav, I. (2009). Cannabinoid receptor activation in the basolateral amygdala blocks the effects of stress on the conditioning and extinction of inhibitory avoidance. *Journal of Neuroscience*, *29*(36), 11078–11088. http://dx.doi.org/10.1523/JNEUROSCI.1223-09.2009. 29/36/11078 [pii].

Ganon-Elazar, E., & Akirav, I. (2012). Cannabinoids prevent the development of behavioral and endocrine alterations in a rat model of intense stress. *Neuropsychopharmacology*, *37*(2), 456–466. http://dx.doi.org/10.1038/npp.2011.204.

Ganon-Elazar, E., & Akirav, I. (2013). Cannabinoids and traumatic stress modulation of contextual fear extinction and GR expression in the amygdala-hippocampal-prefrontal circuit. *Psychoneuroendocrinology*, *38*(9), 1675–1687. http://dx.doi.org/10.1016/j.psyneuen.2013.01.014. S0306-4530(13)00019-X [pii].

Garcia-Gutierrez, M. S., Garcia-Bueno, B., Zoppi, S., Leza, J. C., & Manzanares, J. (2012). Chronic blockade of cannabinoid CB2 receptors induces anxiolytic-like actions associated with alterations in GABA(A) receptors. *British Journal of Pharmacology*, *165*(4), 951–964. http://dx.doi.org/10.1111/j.1476-5381.2011.01625.x.

Garcia-Gutierrez, M. S., & Manzanares, J. (2010). The cannabinoid CB1 receptor is involved in the anxiolytic, sedative and amnesic actions of benzodiazepines. *Journal of Psychopharmacology*, *24*(5), 757–765. http://dx.doi.org/10.1177/0269881109106910.

Genn, R. F., Tucci, S., Marco, E. M., Viveros, M. P., & File, S. E. (2004). Unconditioned and conditioned anxiogenic effects of the cannabinoid receptor agonist CP 55,940 in the social interaction test. *Pharmacology, Biochemistry and Behavior*, 77(3), 567–573.

Gobbi, G., Bambico, F. R., Mangieri, R., Bortolato, M., Campolongo, P., Solinas, M., & Piomelli, D. (2005). Antidepressant-like activity and modulation of brain monoaminergic transmission by blockade of anandamide hydrolysis. *Proceedings of the National Academy of Sciences of the United States of America*, *102*(51), 18620–18625. http://dx.doi.org/10.1073/pnas.0509591102.

Gomes, F. V., Alves, F. H., Guimaraes, F. S., Correa, F. M., Resstel, L. B., & Crestani, C. C. (2013). Cannabidiol administration into the bed nucleus of the stria terminalis alters cardiovascular responses induced by acute restraint stress through 5-HT(1)A receptor. *European Neuropsychopharmacology*, *23*(9), 1096–1104. http://dx.doi.org/10.1016/j.euroneuro.2012.09.007. S0924-977X(12)00271-4 [pii].

Gomes, F. V., Casarotto, P. C., Resstel, L. B., & Guimaraes, F. S. (2011). Facilitation of CB1 receptor-mediated neurotransmission decreases marble burying behavior in mice. *Progress in Neuro-Psychopharmacology & Biological Psychiatry*, *35*(2), 434–438. http://dx.doi.org/10.1016/j.pnpbp.2010.11.027.

Gomes, F. V., Reis, D. G., Alves, F. H., Correa, F. M., Guimaraes, F. S., & Resstel, L. B. (2012). Cannabidiol injected into the bed nucleus of the stria terminalis reduces the expression of contextual fear conditioning via 5-HT1A receptors. *Journal of Psychopharmacology*, *26*(1), 104–113. http://dx.doi.org/10.1177/0269881110389095. 0269881110389095 [pii].

Gomes, F. V., Resstel, L. B., & Guimaraes, F. S. (2011). The anxiolytic-like effects of cannabidiol injected into the bed nucleus of the stria terminalis are mediated by 5-HT1A receptors. *Psychopharmacology*, *213*(2–3), 465–473. http://dx.doi.org/10.1007/s00213-010-2036-z.

Gorka, S. M., Fitzgerald, D. A., de Wit, H., & Phan, K. L. (2015). Cannabinoid modulation of amygdala subregion functional connectivity to social signals of threat. *International Journal of Neuropsychopharmacology*, *18*(3). http://dx.doi.org/10.1093/ijnp/pyu104. pyu104 [pii].

Gorzalka, B. B., & Hill, M. N. (2011). Putative role of endocannabinoid signaling in the etiology of depression and actions of antidepressants. *Progress in Neuro-Psychopharmacology & Biological Psychiatry*, *35*(7), 1575–1585. http://dx.doi.org/10.1016/j.pnpbp.2010.11.021. S0278-5846(10)00438-0 [pii].

Goshen, I., Yirmiya, R., Iverfeldt, K., & Weidenfeld, J. (2003). The role of endogenous interleukin-1 in stress-induced adrenal activation and adrenalectomy-induced adrenocorticotropic hormone hypersecretion. *Endocrinology*, *144*(10), 4453–4458.

Granjeiro, E. M., Gomes, F. V., Guimaraes, F. S., Correa, F. M., & Resstel, L. B. (2011). Effects of intracisternal administration of cannabidiol on the cardiovascular and behavioral responses to acute restraint stress. *Pharmacology, Biochemistry, and Behavior*, *99*(4), 743–748. http://dx.doi.org/10.1016/j.pbb.2011.06.027. S0091-3057(11)00231-0 [pii].

Gray, J., & McNaughton, N. (Eds.), (2000). *The neuropsychology of anxiety*. (2nd ed.). Oxford: Oxford Medical Publications.

Gray, J. M., Vecchiarelli, H. A., Morena, M., Lee, T. T., Hermanson, D. J., Kim, A. B., & Hill, M. N. (2015). Corticotropin-releasing hormone drives anandamide hydrolysis in the amygdala to promote anxiety. *Journal of Neuroscience*, *35*(9), 3879–3892. http://dx.doi.org/10.1523/JNEUROSCI.2737-14.2015.

Griebel, G., Stemmelin, J., & Scatton, B. (2005). Effects of the cannabinoid CB1 receptor antagonist rimonabant in models of emotional reactivity in rodents. *Biological Psychiatry*, *57*(3), 261–267. http://dx.doi.org/10.1016/j.biopsych.2004.10.032.

Gross, C. T., & Canteras, N. S. (2012). The many paths to fear. *Nature Reviews. Neuroscience, 13*(9), 651–658. http://dx.doi.org/10.1038/nrn3301. nrn3301 [pii].

Guimaraes, F. S., Beijamini, V., Moreira, F. A., Aguiar, D. C., & de Lucca, A. C. (2005). Role of nitric oxide in brain regions related to defensive reactions. *Neuroscience and Biobehavioral Reviews, 29*(8), 1313–1322. http://dx.doi.org/10.1016/j.neubiorev.2005.03.026.

Guindon, J., & Hohmann, A. G. (2009). The endocannabinoid system and pain. *CNS & Neurological Disorders Drug Targets, 8*(6), 403–421.

Gunduz-Cinar, O., MacPherson, K. P., Cinar, R., Gamble-George, J., Sugden, K., Williams, B., & Holmes, A. (2013). Convergent translational evidence of a role for anandamide in amygdala-mediated fear extinction, threat processing and stress-reactivity. *Molecular Psychiatry, 18*(7), 813–823. http://dx.doi.org/10.1038/mp.2012.72.

Haj-Dahmane, S., & Shen, R. Y. (2011). Modulation of the serotonin system by endocannabinoid signaling. *Neuropharmacology, 61*(3), 414–420. http://dx.doi.org/10.1016/j.neuropharm.2011.02.016.

Hajos, N., & Freund, T. F. (2002). Distinct cannabinoid sensitive receptors regulate hippocampal excitation and inhibition. *Chemistry and Physics of Lipids, 121*(1–2), 73–82. S0009308402001494 [pii].

Hajos, N., Kathuria, S., Dinh, T., Piomelli, D., & Freund, T. F. (2004). Endocannabinoid transport tightly controls 2-arachidonoyl glycerol actions in the hippocampus: Effects of low temperature and the transport inhibitor AM404. *European Journal of Neuroscience, 19*(11), 2991–2996. http://dx.doi.org/10.1111/j.0953-816X.2004.03433.x.

Hakimizadeh, E., Oryan, S., Hajizadeh Moghaddam, A., Shamsizadeh, A., & Roohbakhsh, A. (2012). Endocannabinoid system and TRPV1 receptors in the dorsal hippocampus of the rats modulate anxiety-like behaviors. *Iranian Journal of Basic Medical Sciences, 15*(3), 795–802.

Hall, W., & Solowij, N. (1998). Adverse effects of cannabis. *Lancet, 352*(9140), 1611–1616. http://dx.doi.org/10.1016/S0140-6736(98)05021-1. S0140-6736(98)05021-1 [pii].

Haller, J., Bakos, N., Szirmay, M., Ledent, C., & Freund, T. F. (2002). The effects of genetic and pharmacological blockade of the CB1 cannabinoid receptor on anxiety. *European Journal of Neuroscience, 16*(7), 1395–1398.

Haller, J., Barna, I., Barsvari, B., Gyimesi Pelczer, K., Yasar, S., Panlilio, L. V., & Goldberg, S. (2009). Interactions between environmental aversiveness and the anxiolytic effects of enhanced cannabinoid signaling by FAAH inhibition in rats. *Psychopharmacology, 204*(4), 607–616. http://dx.doi.org/10.1007/s00213-009-1494-7.

Haller, J., Matyas, F., Soproni, K., Varga, B., Barsy, B., Nemeth, B., & Hajos, N. (2007). Correlated species differences in the effects of cannabinoid ligands on anxiety and on GABAergic and glutamatergic synaptic transmission. *European Journal of Neuroscience, 25*(8), 2445–2456. http://dx.doi.org/10.1111/j.1460-9568.2007.05476.x. EJN5476 [pii].

Haller, J., Varga, B., Ledent, C., Barna, I., & Freund, T. F. (2004a). Context-dependent effects of CB1 cannabinoid gene disruption on anxiety-like and social behaviour in mice. *European Journal of Neuroscience, 19*(7), 1906–1912. http://dx.doi.org/10.1111/j.1460-9568.2004.03293.x.

Haller, J., Varga, B., Ledent, C., & Freund, T. F. (2004b). CB1 cannabinoid receptors mediate anxiolytic effects: Convergent genetic and pharmacological evidence with CB1-specific agents. *Behavioural Pharmacology, 15*(4), 299–304.

Han, J., Kesner, P., Metna-Laurent, M., Duan, T., Xu, L., Georges, F., & Zhang, X. (2012). Acute cannabinoids impair working memory through astroglial CB1 receptor modulation of hippocampal LTD. *Cell, 148*(5), 1039–1050. http://dx.doi.org/10.1016/j.cell.2012.01.037.

Haring, M., Enk, V., Aparisi Rey, A., Loch, S., Ruiz de Azua, I., Weber, T., & Lutz, B. (2015). Cannabinoid type-1 receptor signaling in central serotonergic neurons regulates anxiety-like behavior and sociability. *Frontiers in Behavioral Neuroscience*, *9*, 235. http://dx.doi.org/10.3389/fnbeh.2015.00235.

Haring, M., Guggenhuber, S., & Lutz, B. (2012). Neuronal populations mediating the effects of endocannabinoids on stress and emotionality. *Neuroscience*, *204*, 145–158. http://dx.doi.org/10.1016/j.neuroscience.2011.12.035.

Haring, M., Kaiser, N., Monory, K., & Lutz, B. (2011). Circuit specific functions of cannabinoid CB1 receptor in the balance of investigatory drive and exploration. *PloS One*, *6*(11), e26617. http://dx.doi.org/10.1371/journal.pone.0026617.

Haring, M., Marsicano, G., Lutz, B., & Monory, K. (2007). Identification of the cannabinoid receptor type 1 in serotonergic cells of raphe nuclei in mice. *Neuroscience*, *146*(3), 1212–1219. http://dx.doi.org/10.1016/j.neuroscience.2007.02.021.

Hariri, A. R., Gorka, A., Hyde, L. W., Kimak, M., Halder, I., Ducci, F., & Manuck, S. B. (2009). Divergent effects of genetic variation in endocannabinoid signaling on human threat- and reward-related brain function. *Biological Psychiatry*, *66*(1), 9–16. http://dx.doi.org/10.1016/j.biopsych.2008.10.047.

Haroon, E., Raison, C. L., & Miller, A. H. (2012). Psychoneuroimmunology meets neuropsychopharmacology: Translational implications of the impact of inflammation on behavior. *Neuropsychopharmacology*, *37*(1), 137–162. http://dx.doi.org/10.1038/npp.2011.205.

Heidbreder, C. A., & Groenewegen, H. J. (2003). The medial prefrontal cortex in the rat: Evidence for a dorso-ventral distinction based upon functional and anatomical characteristics. *Neuroscience and Biobehavioral Reviews*, *27*(6), 555–579.

Heimann, A. S., Gomes, I., Dale, C. S., Pagano, R. L., Gupta, A., de Souza, L. L., & Devi, L. A. (2007). Hemopressin is an inverse agonist of CB1 cannabinoid receptors. *Proceedings of the National Academy of Sciences of the United States of America*, *104*(51), 20588–20593. http://dx.doi.org/10.1073/pnas.0706980105.

Heitland, I., Klumpers, F., Oosting, R. S., Evers, D. J., Leon Kenemans, J., & Baas, J. M. (2012). Failure to extinguish fear and genetic variability in the human cannabinoid receptor 1. *Translational Psychiatry*, *2*, e162. http://dx.doi.org/10.1038/tp.2012.90.

Henry, R. J., Kerr, D. M., Finn, D. P., & Roche, M. (2016). For whom the endocannabinoid tolls: Modulation of innate immune function and implications for psychiatric disorders. *Progress in Neuro-Psychopharmacology & Biological Psychiatry*, *64*, 167–180. http://dx.doi.org/10.1016/j.pnpbp.2015.03.006.

Herkenham, M., Lynn, A. B., Johnson, M. R., Melvin, L. S., de Costa, B. R., & Rice, K. C. (1991). Characterization and localization of cannabinoid receptors in rat brain: A quantitative in vitro autoradiographic study. *Journal of Neuroscience*, *11*(2), 563–583.

Hermanson, D. J., Hartley, N. D., Gamble-George, J., Brown, N., Shonesy, B. C., Kingsley, P. J., & Patel, S. (2013). Substrate-selective COX-2 inhibition decreases anxiety via endocannabinoid activation. *Nature Neuroscience*, *16*(9), 1291–1298. http://dx.doi.org/10.1038/nn.3480.

Hill, M. N., Bierer, L. M., Makotkine, I., Golier, J. A., Galea, S., McEwen, B. S., … Yehuda, R. (2013). Reductions in circulating endocannabinoid levels in individuals with post-traumatic stress disorder following exposure to the World Trade Center attacks. *Psychoneuroendocrinology*, *38*(12), 2952–2961. http://dx.doi.org/10.1016/j.psyneuen.2013.08.004. S0306-4530(13)00292-8 [pii].

Hill, M. N., Carrier, E. J., McLaughlin, R. J., Morrish, A. C., Meier, S. E., Hillard, C. J., & Gorzalka, B. B. (2008). Regional alterations in the endocannabinoid system in an animal model of depression: Effects of concurrent antidepressant treatment. *Journal of Neurochemistry*, *106*(6), 2322–2336. http://dx.doi.org/10.1111/j.1471-4159.2008.05567.x.

Hill, M. N., & Gorzalka, B. B. (2004). Enhancement of anxiety-like responsiveness to the cannabinoid CB(1) receptor agonist HU-210 following chronic stress. *European Journal of Pharmacology*, *499*(3), 291–295. http://dx.doi.org/10.1016/j.ejphar.2004.06.069.

Hill, M. N., Hillard, C. J., & McEwen, B. S. (2011). Alterations in corticolimbic dendritic morphology and emotional behavior in cannabinoid CB1 receptor-deficient mice parallel the effects of chronic stress. *Cerebral Cortex*, *21*(9), 2056–2064. http://dx.doi.org/10.1093/cercor/bhq280.

Hill, M. N., Karacabeyli, E. S., & Gorzalka, B. B. (2007). Estrogen recruits the endocannabinoid system to modulate emotionality. *Psychoneuroendocrinology*, *32*(4), 350–357. http://dx.doi.org/10.1016/j.psyneuen.2007.02.003.

Hill, M. N., Karatsoreos, I. N., Hillard, C. J., & McEwen, B. S. (2010). Rapid elevations in limbic endocannabinoid content by glucocorticoid hormones in vivo. *Psychoneuroendocrinology*, *35*(9), 1333–1338. http://dx.doi.org/10.1016/j.psyneuen.2010.03.005. S0306-4530(10)00083-1 [pii].

Hill, M. N., McLaughlin, R. J., Morrish, A. C., Viau, V., Floresco, S. B., Hillard, C. J., & Gorzalka, B. B. (2009). Suppression of amygdalar endocannabinoid signaling by stress contributes to activation of the hypothalamic-pituitary-adrenal axis. *Neuropsychopharmacology*, *34*(13), 2733–2745. http://dx.doi.org/10.1038/npp.2009.114.

Hoffman, A. F., Laaris, N., Kawamura, M., Masino, S. A., & Lupica, C. R. (2010). Control of cannabinoid CB1 receptor function on glutamate axon terminals by endogenous adenosine acting at A1 receptors. *Journal of Neuroscience*, *30*(2), 545–555. http://dx.doi.org/10.1523/JNEUROSCI.4920-09.2010.

Hohmann, A. G., Suplita, R. L., Bolton, N. M., Neely, M. H., Fegley, D., Mangieri, R., & Piomelli, D. (2005). An endocannabinoid mechanism for stress-induced analgesia. *Nature*, *435*(7045), 1108–1112.

Holmes, A., & Wellman, C. L. (2009). Stress-induced prefrontal reorganization and executive dysfunction in rodents. *Neuroscience and Biobehavioral Reviews*, *33*(6), 773–783. http://dx.doi.org/10.1016/j.neubiorev.2008.11.005.

Howlett, A. C., Barth, F., Bonner, T. I., Cabral, G., Casellas, P., Devane, W. A., & Pertwee, R. G. (2002). International union of pharmacology. XXVII. Classification of cannabinoid receptors. *Pharmacological Reviews*, *54*(2), 161–202.

Hu, S. S., & Mackie, K. (2015). Distribution of the endocannabinoid system in the central nervous system. *Handbook of Experimental Pharmacology*, *231*, 59–93. http://dx.doi.org/10.1007/978-3-319-20825-1_3.

Imperatore, R., Morello, G., Luongo, L., Taschler, U., Romano, R., De Gregorio, D., & Cristino, L. (2015). Genetic deletion of monoacylglycerol lipase leads to impaired cannabinoid receptor CB1 R signaling and anxiety-like behavior. *Journal of Neurochemistry*, *135*(4), 799–813. http://dx.doi.org/10.1111/jnc.13267.

Jacob, W., Marsch, R., Marsicano, G., Lutz, B., & Wotjak, C. T. (2012). Cannabinoid CB1 receptor deficiency increases contextual fear memory under highly aversive conditions and long-term potentiation in vivo. *Neurobiology of Learning and Memory*, *98*(1), 47–55. http://dx.doi.org/10.1016/j.nlm.2012.04.008.

Jacob, W., Yassouridis, A., Marsicano, G., Monory, K., Lutz, B., & Wotjak, C. T. (2009). Endocannabinoids render exploratory behaviour largely independent of the test aversiveness: Role of glutamatergic transmission. *Genes, Brain, and Behavior*, *8*(7), 685–698. http://dx.doi.org/10.1111/j.1601-183X.2009.00512.x.

Jankord, R., & Herman, J. P. (2008). Limbic regulation of hypothalamo-pituitary-adrenocortical function during acute and chronic stress. *Annals of the New York Academy of Sciences*, *1148*, 64–73. http://dx.doi.org/10.1196/annals.1410.012.

Jenniches, I., Ternes, S., Albayram, O., Otte, D. M., Bach, K., Bindila, L., … Zimmer, A. (2016). Anxiety, stress, and fear response in mice with reduced endocannabinoid levels. *Biological Psychiatry*, *79*(10), 858–868. http://dx.doi.org/10.1016/j.biopsych.2015.03.033.

Jetly, R., Heber, A., Fraser, G., & Boisvert, D. (2015). The efficacy of nabilone, a synthetic cannabinoid, in the treatment of PTSD-associated nightmares: A preliminary randomized, double-blind, placebo-controlled cross-over design study. *Psychoneuroendocrinology*, *51*, 585–588. http://dx.doi.org/10.1016/j.psyneuen.2014.11.002. S0306-4530(14)00413-2 [pii].

Jiang, W., Zhang, Y., Xiao, L., Van Cleemput, J., Ji, S. P., Bai, G., & Zhang, X. (2005). Cannabinoids promote embryonic and adult hippocampus neurogenesis and produce anxiolytic- and antidepressant-like effects. *Journal of Clinical Investigation, 115*(11), 3104–3116. http://dx.doi.org/10.1172/JCI25509.

Jochman, K. A., Newman, S. M., Kalin, N. H., & Bakshi, V. P. (2005). Corticotropin-releasing factor-1 receptors in the basolateral amygdala mediate stress-induced anorexia. *Behavioral Neuroscience, 119*(6), 1448–1458. http://dx.doi.org/10.1037/0735-7044.119.6.1448.

Jones, R. S., & Lynch, M. A. (2015). How dependent is synaptic plasticity on microglial phenotype? *Neuropharmacology, 96*(Pt A), 3–10. http://dx.doi.org/10.1016/j.neuropharm.2014.08.012.

Jovanovic, T., Norrholm, S. D., Blanding, N. Q., Davis, M., Duncan, E., Bradley, B., & Ressler, K. J. (2010). Impaired fear inhibition is a biomarker of PTSD but not depression. *Depression and Anxiety, 27*(3), 244–251. http://dx.doi.org/10.1002/da.20663.

Kamprath, K., Marsicano, G., Tang, J., Monory, K., Bisogno, T., Di Marzo, V., & Wotjak, C. T. (2006). Cannabinoid CB1 receptor mediates fear extinction via habituation-like processes. *Journal of Neuroscience, 26*(25), 6677–6686. http://dx.doi.org/10.1523/JNEUROSCI.0153-06.2006. 26/25/6677 [pii].

Kamprath, K., Plendl, W., Marsicano, G., Deussing, J. M., Wurst, W., Lutz, B., & Wotjak, C. T. (2009). Endocannabinoids mediate acute fear adaptation via glutamatergic neurons independently of corticotropin-releasing hormone signaling. *Genes, Brain, and Behavior, 8*(2), 203–211. http://dx.doi.org/10.1111/j.1601-183X.2008.00463.x.

Kano, M., Ohno-Shosaku, T., Hashimotodani, Y., Uchigashima, M., & Watanabe, M. (2009). Endocannabinoid-mediated control of synaptic transmission. *Physiological Reviews, 89*(1), 309–380. http://dx.doi.org/10.1152/physrev.00019.2008.

Karst, H., Berger, S., Erdmann, G., Schutz, G., & Joels, M. (2010). Metaplasticity of amygdalar responses to the stress hormone corticosterone. *Proceedings of the National Academy of Sciences of the United States of America, 107*(32), 14449–14454. http://dx.doi.org/10.1073/pnas.0914381107. 0914381107 [pii].

Kasckow, J. W., Mulchahey, J. J., & Geracioti, T. D., Jr. (2004). Effects of the vanilloid agonist olvanil and antagonist capsazepine on rat behaviors. *Progress in Neuro-Psychopharmacology & Biological Psychiatry, 28*(2), 291–295. http://dx.doi.org/10.1016/j.pnpbp.2003.10.007.

Kathuria, S., Gaetani, S., Fegley, D., Valino, F., Duranti, A., Tontini, A., & Piomelli, D. (2003). Modulation of anxiety through blockade of anandamide hydrolysis. *Nature Medicine, 9*(1), 76–81. http://dx.doi.org/10.1038/nm803.

Katona, I., & Freund, T. F. (2008). Endocannabinoid signaling as a synaptic circuit breaker in neurological disease. *Nature Medicine, 14*(9), 923–930. http://dx.doi.org/10.1038/nm.f.1869. nm.f.1869 [pii].

Katona, I., Rancz, E. A., Acsady, L., Ledent, C., Mackie, K., Hajos, N., & Freund, T. F. (2001). Distribution of CB1 cannabinoid receptors in the amygdala and their role in the control of GABAergic transmission. *Journal of Neuroscience, 21*(23), 9506–9518. 21/23/9506 [pii].

Kelley, J. B., Anderson, K. L., & Itzhak, Y. (2010). Pharmacological modulators of nitric oxide signaling and contextual fear conditioning in mice. *Psychopharmacology, 210*(1), 65–74. http://dx.doi.org/10.1007/s00213-010-1817-8.

Kelley, J. B., Balda, M. A., Anderson, K. L., & Itzhak, Y. (2009). Impairments in fear conditioning in mice lacking the nNOS gene. *Learning & Memory, 16*(6), 371–378. http://dx.doi.org/10.1101/lm.1329209.

Kettenmann, H., Kirchhoff, F., & Verkhratsky, A. (2013). Microglia: New roles for the synaptic stripper. *Neuron, 77*(1), 10–18. http://dx.doi.org/10.1016/j.neuron.2012.12.023.

Kim, J. J., Koo, J. W., Lee, H. J., & Han, J. S. (2005). Amygdalar inactivation blocks stress-induced impairments in hippocampal long-term potentiation and spatial memory. *Journal of Neuroscience*, *25*(6), 1532–1539. http://dx.doi.org/10.1523/JNEUROSCI.4623-04.2005.

Kim, J., & Li, Y. (2015). Chronic activation of CB2 cannabinoid receptors in the hippocampus increases excitatory synaptic transmission. *The Journal of Physiology*, *593*(4), 871–886. http://dx.doi.org/10.1113/jphysiol.2014.286633. jphysiol.2014.286633 [pii].

Kim, M. J., Loucks, R. A., Palmer, A. L., Brown, A. C., Solomon, K. M., Marchante, A. N., & Whalen, P. J. (2011). The structural and functional connectivity of the amygdala: From normal emotion to pathological anxiety. *Behavioural Brain Research*, *223*(2), 403–410. http://dx.doi.org/10.1016/j.bbr.2011.04.025.

Kinsey, S. G., O'Neal, S. T., Long, J. Z., Cravatt, B. F., & Lichtman, A. H. (2011). Inhibition of endocannabinoid catabolic enzymes elicits anxiolytic-like effects in the marble burying assay. *Pharmacology, Biochemistry, and Behavior*, *98*(1), 21–27. http://dx.doi.org/10.1016/j.pbb.2010.12.002.

Klugmann, M., Goepfrich, A., Friemel, C. M., & Schneider, M. (2011). AAV-mediated overexpression of the CB1 receptor in the mPFC of adult rats alters cognitive flexibility, social behavior, and emotional reactivity. *Frontiers in Behavioral Neuroscience*, *5*, 37. http://dx.doi.org/10.3389/fnbeh.2011.00037.

Kochenborger, L., Levone, B. R., da Silva, E. S., Taschetto, A. P., Terenzi, M. G., Paschoalini, M. A., & Faria, M. S. (2014). The microinjection of a cannabinoid agonist into the accumbens shell induces anxiogenesis in the elevated plus-maze. *Pharmacology, Biochemistry, and Behavior*, *124*, 160–166. http://dx.doi.org/10.1016/j.pbb.2014.05.017. S0091-3057(14)00160-9 [pii].

Kuhnert, S., Meyer, C., & Koch, M. (2013). Involvement of cannabinoid receptors in the amygdala and prefrontal cortex of rats in fear learning, consolidation, retrieval and extinction. *Behavioural Brain Research*, *250*, 274–284. http://dx.doi.org/10.1016/j.bbr.2013.05.002.

Laaris, N., Good, C. H., & Lupica, C. R. (2010). Delta9-tetrahydrocannabinol is a full agonist at CB1 receptors on GABA neuron axon terminals in the hippocampus. *Neuropharmacology*, *59*(1–2), 121–127. http://dx.doi.org/10.1016/j.neuropharm.2010.04.013.

Lafenetre, P., Chaouloff, F., & Marsicano, G. (2007). The endocannabinoid system in the processing of anxiety and fear and how CB1 receptors may modulate fear extinction. *Pharmacological Research*, *56*(5), 367–381. http://dx.doi.org/10.1016/j.phrs.2007.09.006.

Lafenetre, P., Chaouloff, F., & Marsicano, G. (2009). Bidirectional regulation of novelty-induced behavioral inhibition by the endocannabinoid system. *Neuropharmacology*, *57*(7–8), 715–721. http://dx.doi.org/10.1016/j.neuropharm.2009.07.014.

Lafourcade, M., Elezgarai, I., Mato, S., Bakiri, Y., Grandes, P., & Manzoni, O. J. (2007). Molecular components and functions of the endocannabinoid system in mouse prefrontal cortex. *PloS One*, *2*(8), e709. http://dx.doi.org/10.1371/journal.pone.0000709.

Lan, R., Liu, Q., Fan, P., Lin, S., Fernando, S. R., McCallion, D., & Makriyannis, A. (1999). Structure-activity relationships of pyrazole derivatives as cannabinoid receptor antagonists. *Journal of Medicinal Chemistry*, *42*(4), 769–776. http://dx.doi.org/10.1021/jm980363y.

Laricchiuta, D., Centonze, D., & Petrosini, L. (2013). Effects of endocannabinoid and endovanilloid systems on aversive memory extinction. *Behavioural Brain Research*, *256*, 101–107. http://dx.doi.org/10.1016/j.bbr.2013.08.010.

Laviolette, S. R., & Grace, A. A. (2006). Cannabinoids potentiate emotional learning plasticity in neurons of the medial prefrontal cortex through basolateral amygdala inputs. *Journal of Neuroscience*, *26*(24), 6458–6468. http://dx.doi.org/10.1523/JNEUROSCI.0707-06.2006.

Lazary, J., Lazary, A., Gonda, X., Benko, A., Molnar, E., Hunyady, L., & Bagdy, G. (2009). Promoter variants of the cannabinoid receptor 1 gene (CNR1) in interaction with

5-HTTLPR affect the anxious phenotype. *American Journal of Medical Genetics Part B, Neuropsychiatric Genetics*, *150B*(8), 1118–1127. http://dx.doi.org/10.1002/ajmg.b.31024.

Ledent, C., Valverde, O., Cossu, G., Petitet, F., Aubert, J. F., Beslot, F., & Parmentier, M. (1999). Unresponsiveness to cannabinoids and reduced addictive effects of opiates in CB1 receptor knockout mice. *Science*, *283*(5400), 401–404.

LeDoux, J. E. (1994). Emotion, memory and the brain. *Scientific American*, *270*(6), 50–57.

LeDoux, J. E. (2000). Emotion circuits in the brain. *Annual Review of Neuroscience*, *23*, 155–184. http://dx.doi.org/10.1146/annurev.neuro.23.1.155.

LeDoux, J. E. (2014). Coming to terms with fear. *Proceedings of the National Academy of Sciences of the United States of America*, *111*(8), 2871–2878. http://dx.doi.org/10.1073/pnas.1400335111.

Lee, T. T., & Gorzalka, B. B. (2012). Timing is everything: Evidence for a role of corticolimbic endocannabinoids in modulating hypothalamic-pituitary-adrenal axis activity across developmental periods. *Neuroscience*, *204*, 17–30. http://dx.doi.org/10.1016/j.neuroscience.2011.10.006. S0306-4522(11)01180-8 [pii].

Lemos, J. I., Resstel, L. B., & Guimaraes, F. S. (2010). Involvement of the prelimbic prefrontal cortex on cannabidiol-induced attenuation of contextual conditioned fear in rats. *Behavioural Brain Research*, *207*(1), 105–111. http://dx.doi.org/10.1016/j.bbr.2009.09.045. S0166-4328(09)00589-0 [pii].

Lesch, K. P., Bengel, D., Heils, A., Sabol, S. Z., Greenberg, B. D., Petri, S., & Murphy, D. L. (1996). Association of anxiety-related traits with a polymorphism in the serotonin transporter gene regulatory region. *Science*, *274*(5292), 1527–1531.

Lesting, J., Daldrup, T., Narayanan, V., Himpe, C., Seidenbecher, T., & Pape, H. C. (2013). Directional theta coherence in prefrontal cortical to amygdalo-hippocampal pathways signals fear extinction. *PloS One*, *8*(10), e77707. http://dx.doi.org/10.1371/journal.pone.0077707.

Lin, H. C., Mao, S. C., Su, C. L., & Gean, P. W. (2009). The role of prefrontal cortex CB1 receptors in the modulation of fear memory. *Cerebral Cortex*, *19*(1), 165–175. http://dx.doi.org/10.1093/cercor/bhn075.

Lisboa, S. F., Borges, A. A., Nejo, P., Fassini, A., Guimaraes, F. S., & Resstel, L. B. (2015). Cannabinoid CB1 receptors in the dorsal hippocampus and prelimbic medial prefrontal cortex modulate anxiety-like behavior in rats: Additional evidence. *Progress in Neuro-Psychopharmacology & Biological Psychiatry*, *59*, 76–83. http://dx.doi.org/10.1016/j.pnpbp.2015.01.005. S0278-5846(15)00006-8 [pii].

Lisboa, S. F., Camargo, L. H. A., da Silva, A. L., Resstel, L. B. M., & Guimaraes, F. S. (2014a). P.4.a.014 Repeated treatment with an anandamide metabolism inhibitor attenuates long-lasting consequences in a mouse model of post-traumatic stress disorder (PTSD). *European Neuropsychopharmacology*, *24*, S584–S585.

Lisboa, S. F., Camargo, L. H., Magesto, A. C., Resstel, L. B., & Guimaraes, F. S. (2014b). Cannabinoid modulation of predator fear: Involvement of the dorsolateral periaqueductal gray. *International Journal of Neuropsychopharmacology*, *17*(8), 1193–1206. http://dx.doi.org/10.1017/S1461145713001788.

Lisboa, S. F., Gomes, F. V., Guimaraes, F. S., & Campos, A. C. (2016). Microglial cells as a link between cannabinoids and the immune hypothesis of psychiatric disorders. *Frontiers in Neurology*, 7(5). http://dx.doi.org/10.3389/fneur.2016.00005.

Lisboa, S. F., Gomes, F. V., Silva, A. L., Uliana, D. L., Camargo, L. H., Guimaraes, F. S., & Resstel, L. B. (2015). Increased contextual fear conditioning in iNOS knockout mice: Additional evidence for the involvement of nitric oxide in stress-related disorders and contribution of the endocannabinoid system. *International Journal of Neuropsychopharmacology*, *18*(8), 1–8, pyv005. http://dx.doi.org/10.1093/ijnp/pyv005.

Lisboa, S. F., & Guimarães, F. S. (2007). Cannabidiol into basolateral amygdaloid complex does not modify anxiety-like behavior in the elevated plus maze. Program No. 171.14/FF15. *Neuroscience Meeting Planner*. San Diego. Society for Neuroscience, Online.

Lisboa, S. F., & Guimaraes, F. S. (2012). Differential role of CB1 and TRPV1 receptors on anandamide modulation of defensive responses induced by nitric oxide in the dorsolateral periaqueductal gray. *Neuropharmacology*, *62*(8), 2455–2462. http://dx.doi.org/10.1016/j.neuropharm.2012.02.008.

Lisboa, S. F., Magesto, A. C., Aguiar, J. C., Resstel, L. B. M., & Guimaraes, F. S. (2013). Complex interaction between anandamide and the nitrergic system in the dorsolateral periaqueductal gray to modulate anxiety-like behavior in rats. *Neuropharmacology*, *75*, 86–94. http://dx.doi.org/10.1016/j.neuropharm.2013.07.008.

Lisboa, S. F., Niraula, A., Shea, D., Resstel, L. B., Guimaraes, F. S., Godbout, J. P., & Sheridan, J. F. (2016). Abstract # 1831 Repeated social defeat stress-induced neuroinflammation contributes to later fear sensitization and impaired extinction recall of fear conditioning in mice: Involvement of the endocannabinoid system. *Brain, Behavior, and Immunity*, *57*(Supplement), e35. PsychoNeuroImmunology Research Society's 23rd Annual Scientific Meeting.

Lisboa, S. F., Reis, D. G., da Silva, A. L., Correa, F. M., Guimaraes, F. S., & Resstel, L. B. (2010). Cannabinoid CB1 receptors in the medial prefrontal cortex modulate the expression of contextual fear conditioning. *International Journal of Neuropsychopharmacology*, *13*(9), 1163–1173. http://dx.doi.org/10.1017/S1461145710000684.

Lisboa, S. F., Resstel, L. B., Aguiar, D. C., & Guimaraes, F. S. (2008). Activation of cannabinoid CB1 receptors in the dorsolateral periaqueductal gray induces anxiolytic effects in rats submitted to the Vogel conflict test. *European Journal of Pharmacology*, *593*(1–3), 73–78. http://dx.doi.org/10.1016/j.ejphar.2008.07.032. S0014-2999(08)00770-X [pii].

Llorente-Berzal, A., Terzian, A. L., di Marzo, V., Micale, V., Viveros, M. P., & Wotjak, C. T. (2015). 2-AG promotes the expression of conditioned fear via cannabinoid receptor type 1 on GABAergic neurons. *Psychopharmacology*, *232*(15), 2811–2825. http://dx.doi.org/10.1007/s00213-015-3917-y.

Long, L. E., Chesworth, R., Huang, X. F., McGregor, I. S., Arnold, J. C., & Karl, T. (2010). A behavioural comparison of acute and chronic delta9-tetrahydrocannabinol and cannabidiol in C57BL/6JArc mice. *International Journal of Neuropsychopharmacology*, *13*(7), 861–876. http://dx.doi.org/10.1017/S1461145709990605.

Lu, A. T., Ogdie, M. N., Jarvelin, M. R., Moilanen, I. K., Loo, S. K., McCracken, J. T., & Smalley, S. L. (2008). Association of the cannabinoid receptor gene (CNR1) with ADHD and post-traumatic stress disorder. *American Journal of Medical Genetics. Part B, Neuropsychiatric Genetics*, *147B*(8), 1488–1494. http://dx.doi.org/10.1002/ajmg.b.30693.

Lutz, B., Marsicano, G., Maldonado, R., & Hillard, C. J. (2015). The endocannabinoid system in guarding against fear, anxiety and stress. *Nature Reviews. Neuroscience*, *16*(12), 705–718. http://dx.doi.org/10.1038/nrn4036.

Maccarrone, M., Rossi, S., Bari, M., De Chiara, V., Fezza, F., Musella, A., & Centonze, D. (2008). Anandamide inhibits metabolism and physiological actions of 2-arachidonoylglycerol in the striatum. *Nature Neuroscience*, *11*(2), 152–159. http://dx.doi.org/10.1038/nn2042.

Maccarrone, M., Valverde, O., Barbaccia, M. L., Castane, A., Maldonado, R., Ledent, C., & Finazzi-Agro, A. (2002). Age-related changes of anandamide metabolism in CB1 cannabinoid receptor knockout mice: Correlation with behaviour. *European Journal of Neuroscience*, *15*(7), 1178–1186.

Mackowiak, M., Chocyk, A., Dudys, D., & Wedzony, K. (2009). Activation of CB1 cannabinoid receptors impairs memory consolidation and hippocampal polysialylated neural

cell adhesion molecule expression in contextual fear conditioning. *Neuroscience, 158*(4), 1708–1716. http://dx.doi.org/10.1016/j.neuroscience.2008.11.037. S0306-4522(08) 01740-5 [pii].
Maes, M., Leonard, B. E., Myint, A. M., Kubera, M., & Verkerk, R. (2011). The new '5-HT' hypothesis of depression: Cell-mediated immune activation induces indoleamine 2,3-dioxygenase, which leads to lower plasma tryptophan and an increased synthesis of detrimental tryptophan catabolites (TRYCATs), both of which contribute to the onset of depression. *Progress in Neuro-Psychopharmacology & Biological Psychiatry, 35*(3), 702–721. http://dx.doi.org/10.1016/j.pnpbp.2010.12.017.
Maisonnette, S., Morato, S., & Brandao, M. L. (1993). Role of resocialization and of 5-HT1A receptor activation on the anxiogenic effects induced by isolation in the elevated plus-maze test. *Physiology & Behavior, 54*(4), 753–758. 0031-9384(93)90087-V [pii].
Manwell, L. A., Satvat, E., Lang, S. T., Allen, C. P., Leri, F., & Parker, L. A. (2009). FAAH inhibitor, URB-597, promotes extinction and CB(1) antagonist, SR141716, inhibits extinction of conditioned aversion produced by naloxone-precipitated morphine withdrawal, but not extinction of conditioned preference produced by morphine in rats. *Pharmacology, Biochemistry, and Behavior, 94*(1), 154–162. http://dx.doi.org/10.1016/j.pbb.2009.08.002.
Marco, E. M., Echeverry-Alzate, V., Lopez-Moreno, J. A., Gine, E., Penasco, S., & Viveros, M. P. (2014). Consequences of early life stress on the expression of endocannabinoid-related genes in the rat brain. *Behavioural Pharmacology, 25*(5–6), 547–556. http://dx.doi.org/10.1097/FBP.0000000000000068. 00008877-201409000-00018 [pii].
Marco, E. M., Perez-Alvarez, L., Borcel, E., Rubio, M., Guaza, C., Ambrosio, E., & Viveros, M. P. (2004). Involvement of 5-HT1A receptors in behavioural effects of the cannabinoid receptor agonist CP 55,940 in male rats. *Behavioural Pharmacology, 15*(1), 21–27.
Maresz, K., Carrier, E. J., Ponomarev, E. D., Hillard, C. J., & Dittel, B. N. (2005). Modulation of the cannabinoid CB2 receptor in microglial cells in response to inflammatory stimuli. *Journal of Neurochemistry, 95*(2), 437–445. http://dx.doi.org/10.1111/j.1471-4159.2005.03380.x.
Marin, S., Marco, E., Biscaia, M., Fernandez, B., Rubio, M., Guaza, C., ... Viveros, M. P. (2003). Involvement of the kappa-opioid receptor in the anxiogenic-like effect of CP 55,940 in male rats. *Pharmacology, Biochemistry, and Behavior, 74*(3), 649–656.
Marsicano, G., & Lutz, B. (1999). Expression of the cannabinoid receptor CB1 in distinct neuronal subpopulations in the adult mouse forebrain. *European Journal of Neuroscience, 11*(12), 4213–4225. ejn847 [pii].
Marsicano, G., Wotjak, C. T., Azad, S. C., Bisogno, T., Rammes, G., Cascio, M. G., & Lutz, B. (2002). The endogenous cannabinoid system controls extinction of aversive memories. *Nature, 418*(6897), 530–534. http://dx.doi.org/10.1038/nature00839.
Martin, B. R., Compton, D. R., Thomas, B. F., Prescott, W. R., Little, P. J., Razdan, R. K., & Ward, S. J. (1991). Behavioral, biochemical, and molecular modeling evaluations of cannabinoid analogs. *Pharmacology, Biochemistry, and Behavior, 40*(3), 471–478.
Martin, M., Ledent, C., Parmentier, M., Maldonado, R., & Valverde, O. (2002). Involvement of CB1 cannabinoid receptors in emotional behaviour. *Psychopharmacology, 159*(4), 379–387. http://dx.doi.org/10.1007/s00213-001-0946-5.
Martin-Moreno, A. M., Reigada, D., Ramirez, B. G., Mechoulam, R., Innamorato, N., Cuadrado, A., & de Ceballos, M. L. (2011). Cannabidiol and other cannabinoids reduce microglial activation in vitro and in vivo: Relevance to Alzheimer's disease. *Molecular Pharmacology, 79*(6), 964–973. http://dx.doi.org/10.1124/mol.111.071290.
Massi, L., Elezgarai, I., Puente, N., Reguero, L., Grandes, P., Manzoni, O. J., & Georges, F. (2008). Cannabinoid receptors in the bed nucleus of the stria terminalis control cortical excitation of midbrain dopamine cells in vivo. *Journal of*

Neuroscience, *28*(42), 10496–10508. http://dx.doi.org/10.1523/JNEUROSCI.2291-08.2008. 28/42/10496 [pii].

Matsuda, L. A., Bonner, T. I., & Lolait, S. J. (1993). Localization of cannabinoid receptor mRNA in rat brain. *Journal of Comparative Neurology*, *327*(4), 535–550. http://dx.doi.org/10.1002/cne.903270406.

Matsuda, L. A., Lolait, S. J., Brownstein, M. J., Young, A. C., & Bonner, T. I. (1990). Structure of a cannabinoid receptor and functional expression of the cloned cDNA. *Nature*, *346*(6284), 561–564. http://dx.doi.org/10.1038/346561a0.

Mazzola, C., Medalie, J., Scherma, M., Panlilio, L. V., Solinas, M., Tanda, G., & Yasar, S. (2009). Fatty acid amide hydrolase (FAAH) inhibition enhances memory acquisition through activation of PPAR-alpha nuclear receptors. *Learning & Memory*, *16*(5), 332–337. http://dx.doi.org/10.1101/lm.1145209.

Mazzola, C., Micale, V., & Drago, F. (2003). Amnesia induced by beta-amyloid fragments is counteracted by cannabinoid CB1 receptor blockade. *European Journal of Pharmacology*, *477*(3), 219–225.

McGregor, I. S., Arnold, J. C., Weber, M. F., Topple, A. N., & Hunt, G. E. (1998). A comparison of delta 9-THC and anandamide induced c-fos expression in the rat forebrain. *Brain Research*, *802*(1–2), 19–26. S0006-8993(98)00549-6 [pii].

McLaughlin, R. J., Hill, M. N., & Gorzalka, B. B. (2014). A critical role for prefrontocortical endocannabinoid signaling in the regulation of stress and emotional behavior. *Neuroscience and Biobehavioral Reviews*, *42*, 116–131. http://dx.doi.org/10.1016/j.neubiorev.2014.02.006.

McNaughton, N., & Corr, P. J. (2004). A two-dimensional neuropsychology of defense: Fear/anxiety and defensive distance. *Neuroscience and Biobehavioral Reviews*, *28*(3), 285–305.

McNaughton, N., & Zangrossi, H., Jr. (2008). Theoretical approaches to the modeling of anxiety in animals. I. In R. J. Blanchard, D. C. Blanchard, G. Griebel, & D. Nutt (Eds.), *Handbook of anxiety and fear* (pp. 11–27). Oxford: Elsevier.

Mecha, M., Carrillo-Salinas, F. J., Feliú, A., Mestres, L., & Guaza, C. (2016). Microglia activation states and cannabinoid system: Therapeutic implications. *Pharmacology & Therapeutics*, *166*, 40–55. S0163-7258(16)30101-2 [pii].

Mechoulam, R. (1970). Marihuana chemistry. *Science*, *168*(3936), 1159–1166.

Mechoulam, R., Ben-Shabat, S., Hanus, L., Ligumsky, M., Kaminski, N. E., Schatz, A. R., … Vogel, Z. (1995). Identification of an endogenous 2-monoglyceride, present in canine gut, that binds to cannabinoid receptors. *Biochemical Pharmacology*, *50*(1), 83–90.

Metna-Laurent, M., Soria-Gomez, E., Verrier, D., Conforzi, M., Jego, P., Lafenetre, P., & Marsicano, G. (2012). Bimodal control of fear-coping strategies by CB(1) cannabinoid receptors. *Journal of Neuroscience*, *32*(21), 7109–7118. http://dx.doi.org/10.1523/JNEUROSCI.1054-12.2012.

Micale, V., Cristino, L., Tamburella, A., Petrosino, S., Leggio, G. M., Drago, F., & Di Marzo, V. (2009a). Altered responses of dopamine D3 receptor null mice to excitotoxic or anxiogenic stimuli: Possible involvement of the endocannabinoid and endovanilloid systems. *Neurobiology of Disease*, *36*(1), 70–80. http://dx.doi.org/10.1016/j.nbd.2009.06.015.

Micale, V., Cristino, L., Tamburella, A., Petrosino, S., Leggio, G. M., Drago, F., & Di Marzo, V. (2009b). Anxiolytic effects in mice of a dual blocker of fatty acid amide hydrolase and transient receptor potential vanilloid type-1 channels. *Neuropsychopharmacology*, *34*(3), 593–606. http://dx.doi.org/10.1038/npp.2008.98.

Mikics, E., Dombi, T., Barsvari, B., Varga, B., Ledent, C., Freund, T. F., & Haller, J. (2006). The effects of cannabinoids on contextual conditioned fear in CB1 knockout and CD1 mice. *Behavioural Pharmacology*, *17*(3), 223–230.

Ming, Q., Zhang, Y., Yi, J., Wang, X., Zhu, X., & Yao, S. (2015). Serotonin transporter gene polymorphism (5-HTTLPR) L allele interacts with stress to increase anxiety symptoms in Chinese adolescents: A multiwave longitudinal study. *BMC Psychiatry*, *15*, 248. http://dx.doi.org/10.1186/s12888-015-0639-y.

Moise, A. M., Eisenstein, S. A., Astarita, G., Piomelli, D., & Hohmann, A. G. (2008). An endocannabinoid signaling system modulates anxiety-like behavior in male Syrian hamsters. *Psychopharmacology*, *200*(3), 333–346. http://dx.doi.org/10.1007/s00213-008-1209-5.

Molina-Holgado, F., Pinteaux, E., Moore, J. D., Molina-Holgado, E., Guaza, C., Gibson, R. M., & Rothwell, N. J. (2003). Endogenous interleukin-1 receptor antagonist mediates anti-inflammatory and neuroprotective actions of cannabinoids in neurons and glia. *Journal of Neuroscience*, *23*(16), 6470–6474.

Monory, K., Blaudzun, H., Massa, F., Kaiser, N., Lemberger, T., Schutz, G., & Marsicano, G. (2007). Genetic dissection of behavioural and autonomic effects of delta(9)-tetrahydrocannabinol in mice. *PLoS Biology*, *5*(10), e269. http://dx.doi.org/10.1371/journal.pbio.0050269. 07-PLBI-RA-0386 [pii].

Monory, K., Massa, F., Egertova, M., Eder, M., Blaudzun, H., Westenbroek, R., & Lutz, B. (2006). The endocannabinoid system controls key epileptogenic circuits in the hippocampus. *Neuron*, *51*(4), 455–466. http://dx.doi.org/10.1016/j.neuron.2006.07.006.

Monory, K., Polack, M., Remus, A., Lutz, B., & Korte, M. (2015). Cannabinoid CB1 receptor calibrates excitatory synaptic balance in the mouse hippocampus. *Journal of Neuroscience*, *35*(9), 3842–3850. http://dx.doi.org/10.1523/JNEUROSCI.3167-14.2015. 35/9/3842 [pii].

Moreira, F. A., Aguiar, D. C., & Guimaraes, F. S. (2007). Anxiolytic-like effect of cannabinoids injected into the rat dorsolateral periaqueductal gray. *Neuropharmacology*, *52*(3), 958–965.

Moreira, F. A., Aguiar, D. C., Resstel, L. B., Lisboa, S. F., Campos, A. C., Gomes, F. V., & Guimaraes, F. S. (2012a). Neuroanatomical substrates involved in cannabinoid modulation of defensive responses. *Journal of Psychopharmacology*, *26*(1), 40–55. http://dx.doi.org/10.1177/0269881111400651. 0269881111400651 [pii].

Moreira, F. A., Aguiar, D. C., Terzian, A. L., Guimaraes, F. S., & Wotjak, C. T. (2012b). Cannabinoid type 1 receptors and transient receptor potential vanilloid type 1 channels in fear and anxiety-two sides of one coin? *Neuroscience*, *204*, 186–192. http://dx.doi.org/10.1016/j.neuroscience.2011.08.046.

Moreira, F. A., & Crippa, J. A. (2009). The psychiatric side-effects of rimonabant. *Revista Brasileira de Psiquiatria*, *31*(2), 145–153. S1516-44462009000200012 [pii].

Moreira, F. A., Kaiser, N., Monory, K., & Lutz, B. (2008). Reduced anxiety-like behaviour induced by genetic and pharmacological inhibition of the endocannabinoid-degrading enzyme fatty acid amide hydrolase (FAAH) is mediated by CB1 receptors. *Neuropharmacology*, *54*(1), 141–150. http://dx.doi.org/10.1016/j.neuropharm. 2007.07.005.

Morena, M., Patel, S., Bains, J. S., & Hill, M. N. (2016). Neurobiological interactions between stress and the endocannabinoid system. *Neuropsychopharmacology*, *41*(1), 80–102. http://dx.doi.org/10.1038/npp.2015.166. npp2015166 [pii].

Morris, J. S., Frith, C. D., Perrett, D. I., Rowland, D., Young, A. W., Calder, A. J., & Dolan, R. J. (1996). A differential neural response in the human amygdala to fearful and happy facial expressions. *Nature*, *383*(6603), 812–815. http://dx.doi.org/10.1038/383812a0.

Munro, S., Thomas, K. L., & Abu-Shaar, M. (1993). Molecular characterization of a peripheral receptor for cannabinoids. *Nature*, *365*(6441), 61–65. http://dx.doi.org/10.1038/365061a0.

Naderi, N., Haghparast, A., Saber-Tehrani, A., Rezaii, N., Alizadeh, A. M., Khani, A., & Motamedi, F. (2008). Interaction between cannabinoid compounds and diazepam on anxiety-like behaviour of mice. *Pharmacology, Biochemistry, and Behavior, 89*(1), 64–75. http://dx.doi.org/10.1016/j.pbb.2007.11.001.

Naidu, P. S., Varvel, S. A., Ahn, K., Cravatt, B. F., Martin, B. R., & Lichtman, A. H. (2007). Evaluation of fatty acid amide hydrolase inhibition in murine models of emotionality. *Psychopharmacology, 192*(1), 61–70. http://dx.doi.org/10.1007/s00213-006-0689-4.

Nakazi, M., Bauer, U., Nickel, T., Kathmann, M., & Schlicker, E. (2000). Inhibition of serotonin release in the mouse brain via presynaptic cannabinoid CB1 receptors. *Naunyn-Schmiedeberg's Archives of Pharmacology, 361*(1), 19–24.

Navarrete, M., & Araque, A. (2010). Endocannabinoids potentiate synaptic transmission through stimulation of astrocytes. *Neuron, 68*(1), 113–126. http://dx.doi.org/10.1016/j.neuron.2010.08.043.

Navarro, M., Hernandez, E., Munoz, R. M., del Arco, I., Villanua, M. A., Carrera, M. R., & Rodriguez de Fonseca, F. (1997). Acute administration of the CB1 cannabinoid receptor antagonist SR 141716A induces anxiety-like responses in the rat. *Neuroreport, 8*(2), 491–496.

Neumeister, A., Normandin, M. D., Pietrzak, R. H., Piomelli, D., Zheng, M. Q., Gujarro-Anton, A., & Huang, Y. (2013). Elevated brain cannabinoid CB1 receptor availability in post-traumatic stress disorder: A positron emission tomography study. *Molecular Psychiatry, 18*(9), 1034–1040. http://dx.doi.org/10.1038/mp.2013.61. mp201361 [pii].

Newsom, R. J., Osterlund, C., Masini, C. V., Day, H. E., Spencer, R. L., & Campeau, S. (2012). Cannabinoid receptor type 1 antagonism significantly modulates basal and loud noise induced neural and hypothalamic-pituitary-adrenal axis responses in male Sprague-Dawley rats. *Neuroscience, 204*, 64–73. http://dx.doi.org/10.1016/j.neuroscience.2011.11.043.

Nunez, E., Benito, C., Pazos, M. R., Barbachano, A., Fajardo, O., Gonzalez, S., & Romero, J. (2004). Cannabinoid CB2 receptors are expressed by perivascular microglial cells in the human brain: An immunohistochemical study. *Synapse, 53*(4), 208–213. http://dx.doi.org/10.1002/syn.20050.

O'Brien, L. D., Wills, K. L., Segsworth, B., Dashney, B., Rock, E. M., Limebeer, C. L., & Parker, L. A. (2013). Effect of chronic exposure to rimonabant and phytocannabinoids on anxiety-like behavior and saccharin palatability. *Pharmacology, Biochemistry, and Behavior, 103*(3), 597–602. http://dx.doi.org/10.1016/j.pbb.2012.10.008.

O'Donovan, A., Cohen, B. E., Seal, K. H., Bertenthal, D., Margaretten, M., Nishimi, K., & Neylan, T. C. (2015). Elevated risk for autoimmune disorders in Iraq and Afghanistan veterans with posttraumatic stress disorder. *Biological Psychiatry*, 77(4), 365–374. http://dx.doi.org/10.1016/j.biopsych.2014.06.015.

Oehrn, C. R., Baumann, C., Fell, J., Lee, H., Kessler, H., Habel, U., & Axmacher, N. (2015). Human hippocampal dynamics during response conflict. *Current Biology, 25*(17), 2307–2313. http://dx.doi.org/10.1016/j.cub.2015.07.032. S0960-9822(15)00866-0 [pii].

Olango, W. M., Roche, M., Ford, G. K., Harhen, B., & Finn, D. P. (2012). The endocannabinoid system in the rat dorsolateral periaqueductal grey mediates fear-conditioned analgesia and controls fear expression in the presence of nociceptive tone. *British Journal of Pharmacology, 165*(8), 2549–2560. http://dx.doi.org/10.1111/j.1476-5381.2011.01478.x.

Onaivi, E. S., Chakrabarti, A., Gwebu, E. T., & Chaudhuri, G. (1995). Neurobehavioral effects of delta 9-THC and cannabinoid (CB1) receptor gene expression in mice. *Behavioural Brain Research, 72*(1–2), 115–125.

Onaivi, E. S., Green, M. R., & Martin, B. R. (1990). Pharmacological characterization of cannabinoids in the elevated plus maze. *Journal of Pharmacology and Experimental Therapeutics, 253*(3), 1002–1009.

Onaivi, E. S., Ishiguro, H., Gong, J. P., Patel, S., Meozzi, P. A., Myers, L., & Uhl, G. R. (2008). Brain neuronal CB2 cannabinoid receptors in drug abuse and depression: From mice to human subjects. *PloS One*, *3*(2), e1640. http://dx.doi.org/10.1371/journal.pone.0001640.

Onaivi, E. S., Ishiguro, H., Gong, J. P., Patel, S., Perchuk, A., Meozzi, P. A., & Uhl, G. R. (2006). Discovery of the presence and functional expression of cannabinoid CB2 receptors in brain. *Annals of the New York Academy of Sciences*, *1074*, 514–536. http://dx.doi.org/10.1196/annals.1369.052.

Oropeza, V. C., Mackie, K., & Van Bockstaele, E. J. (2007). Cannabinoid receptors are localized to noradrenergic axon terminals in the rat frontal cortex. *Brain Research*, *1127*(1), 36–44. http://dx.doi.org/10.1016/j.brainres.2006.09.110.

Oropeza, V. C., Page, M. E., & Van Bockstaele, E. J. (2005). Systemic administration of WIN 55,212-2 increases norepinephrine release in the rat frontal cortex. *Brain Research*, *1046*(1–2), 45–54. http://dx.doi.org/10.1016/j.brainres.2005.03.036.

Ortega-Alvaro, A., Aracil-Fernandez, A., Garcia-Gutierrez, M. S., Navarrete, F., & Manzanares, J. (2011). Deletion of CB2 cannabinoid receptor induces schizophrenia-related behaviors in mice. *Neuropsychopharmacology*, *36*(7), 1489–1504. http://dx.doi.org/10.1038/npp.2011.34.

Osher, Y., Hamer, D., & Benjamin, J. (2000). Association and linkage of anxiety-related traits with a functional polymorphism of the serotonin transporter gene regulatory region in Israeli sibling pairs. *Molecular Psychiatry*, *5*(2), 216–219.

Page, M. E., Oropeza, V. C., Sparks, S. E., Qian, Y., Menko, A. S., & Van Bockstaele, E. J. (2007). Repeated cannabinoid administration increases indices of noradrenergic activity in rats. *Pharmacology, Biochemistry, and Behavior*, *86*(1), 162–168. http://dx.doi.org/10.1016/j.pbb.2006.12.020.

Page, M. E., Oropeza, V. C., & Van Bockstaele, E. J. (2008). Local administration of a cannabinoid agonist alters norepinephrine efflux in the rat frontal cortex. *Neuroscience Letters*, *431*(1), 1–5. http://dx.doi.org/10.1016/j.neulet.2007.11.009.

Pamplona, F. A., Bitencourt, R. M., & Takahashi, R. N. (2008). Short- and long-term effects of cannabinoids on the extinction of contextual fear memory in rats. *Neurobiology of Learning and Memory*, *90*(1), 290–293. http://dx.doi.org/10.1016/j.nlm.2008.04.003.

Pamplona, F. A., Prediger, R. D., Pandolfo, P., & Takahashi, R. N. (2006). The cannabinoid receptor agonist WIN 55,212-2 facilitates the extinction of contextual fear memory and spatial memory in rats. *Psychopharmacology*, *188*(4), 641–649. http://dx.doi.org/10.1007/s00213-006-0514-0.

Pamplona, F. A., & Takahashi, R. N. (2006). WIN 55212-2 impairs contextual fear conditioning through the activation of CB1 cannabinoid receptors. *Neuroscience Letters*, *397*(1–2), 88–92. http://dx.doi.org/10.1016/j.neulet.2005.12.026. S0304-3940(05)01390-X [pii].

Parolaro, D., Realini, N., Vigano, D., Guidali, C., & Rubino, T. (2010). The endocannabinoid system and psychiatric disorders. *Experimental Neurology*, *224*(1), 3–14. http://dx.doi.org/10.1016/j.expneurol.2010.03.018.

Patel, S., Cravatt, B. F., & Hillard, C. J. (2005). Synergistic interactions between cannabinoids and environmental stress in the activation of the central amygdala. *Neuropsychopharmacology*, *30*(3), 497–507. http://dx.doi.org/10.1038/sj.npp.1300535. 1300535 [pii].

Patel, S., & Hillard, C. J. (2006). Pharmacological evaluation of cannabinoid receptor ligands in a mouse model of anxiety: Further evidence for an anxiolytic role for endogenous cannabinoid signaling. *Journal of Pharmacology and Experimental Therapeutics*, *318*(1), 304–311. http://dx.doi.org/10.1124/jpet.106.101287.

Patel, S., Kingsley, P. J., Mackie, K., Marnett, L. J., & Winder, D. G. (2009). Repeated homotypic stress elevates 2-arachidonoylglycerol levels and enhances short-term endocannabinoid signaling at inhibitory synapses in basolateral amygdala. *Neuropsychopharmacology*, *34*(13), 2699–2709. http://dx.doi.org/10.1038/npp.2009.101.

Patel, N. A., Moldow, R. L., Patel, J. A., Wu, G., & Chang, S. L. (1998). Arachidonylethanolamide (AEA) activation of FOS proto-oncogene protein immunoreactivity in the rat brain. *Brain Research*, *797*(2), 225–233. S0006899398003643 [pii].

Patel, S., Roelke, C. T., Rademacher, D. J., & Hillard, C. J. (2005). Inhibition of restraint stress-induced neural and behavioural activation by endogenous cannabinoid signalling. *European Journal of Neuroscience*, *21*(4), 1057–1069. http://dx.doi.org/10.1111/j.1460-9568.2005.03916.x.

Pertwee, R. G. (2005). Inverse agonism and neutral antagonism at cannabinoid CB1 receptors. *Life Sciences*, *76*(12), 1307–1324. http://dx.doi.org/10.1016/j.lfs.2004.10.025.

Pertwee, R. G., Howlett, A. C., Abood, M. E., Alexander, S. P., Di Marzo, V., Elphick, M. R., … Ross, R. A. (2010). International union of basic and clinical pharmacology. LXXIX. Cannabinoid receptors and their ligands: Beyond CB(1) and CB(2). *Pharmacological Reviews*, *62*(4), 588–631. http://dx.doi.org/10.1124/pr.110.003004. 62/4/588 [pii].

Phan, K. L., Angstadt, M., Golden, J., Onyewuenyi, I., Popovska, A., & de Wit, H. (2008). Cannabinoid modulation of amygdala reactivity to social signals of threat in humans. *Journal of Neuroscience*, *28*(10), 2313–2319. http://dx.doi.org/10.1523/JNEUROSCI.5603-07.2008. 28/10/2313 [pii].

Phillips, R. G., & LeDoux, J. E. (1992). Differential contribution of amygdala and hippocampus to cued and contextual fear conditioning. *Behavioral Neuroscience*, *106*(2), 274–285.

Pietrzak, R. H., Huang, Y., Corsi-Travali, S., Zheng, M. Q., Lin, S. F., Henry, S., & Neumeister, A. (2014). Cannabinoid type 1 receptor availability in the amygdala mediates threat processing in trauma survivors. *Neuropsychopharmacology*, *39*(11), 2519–2528. http://dx.doi.org/10.1038/npp.2014.110. npp2014110 [pii].

Plendl, W., & Wotjak, C. T. (2010). Dissociation of within- and between-session extinction of conditioned fear. *Journal of Neuroscience*, *30*(14), 4990–4998. http://dx.doi.org/10.1523/JNEUROSCI.6038-09.2010.

Puente, N., Elezgarai, I., Lafourcade, M., Reguero, L., Marsicano, G., Georges, F., & Grandes, P. (2010). Localization and function of the cannabinoid CB1 receptor in the anterolateral bed nucleus of the stria terminalis. *PloS One*, *5*(1), e8869. http://dx.doi.org/10.1371/journal.pone.0008869.

Qin, Z., Zhou, X., Pandey, N. R., Vecchiarelli, H. A., Stewart, C. A., Zhang, X., & Chen, H. H. (2015). Chronic stress induces anxiety via an amygdalar intracellular cascade that impairs endocannabinoid signaling. *Neuron*, *85*(6), 1319–1331. http://dx.doi.org/10.1016/j.neuron.2015.02.015.

Quirk, G. J., & Gehlert, D. R. (2003). Inhibition of the amygdala: Key to pathological states? *Annals of the New York Academy of Sciences*, *985*, 263–272.

Rabinak, C. A., Angstadt, M., Lyons, M., Mori, S., Milad, M. R., Liberzon, I., & Phan, K. L. (2014). Cannabinoid modulation of prefrontal-limbic activation during fear extinction learning and recall in humans. *Neurobiology of Learning and Memory*, *113*, 125–134. http://dx.doi.org/10.1016/j.nlm.2013.09.009. S1074-7427(13)00186-X [pii].

Rabinak, C. A., Angstadt, M., Sripada, C. S., Abelson, J. L., Liberzon, I., Milad, M. R., & Phan, K. L. (2013). Cannabinoid facilitation of fear extinction memory recall in humans. *Neuropharmacology*, *64*, 396–402. http://dx.doi.org/10.1016/j.neuropharm.2012.06.063. S0028-3908(12)00337-1 [pii].

Rademacher, D. J., Meier, S. E., Shi, L., Ho, W. S., Jarrahian, A., & Hillard, C. J. (2008). Effects of acute and repeated restraint stress on endocannabinoid content in the amygdala, ventral striatum, and medial prefrontal cortex in mice. *Neuropharmacology*, *54*(1), 108–116. http://dx.doi.org/10.1016/j.neuropharm.2007.06.012.

Ramikie, T. S., & Patel, S. (2012). Endocannabinoid signaling in the amygdala: Anatomy, synaptic signaling, behavior, and adaptations to stress. *Neuroscience*, *204*, 38–52. http://dx.doi.org/10.1016/j.neuroscience.2011.08.037.

Rea, K., Ford, G. K., Olango, W. M., Harhen, B., Roche, M., & Finn, D. P. (2014). Microinjection of 2-arachidonoyl glycerol into the rat ventral hippocampus differentially modulates contextually induced fear, depending on a persistent pain state. *European Journal of Neuroscience*, *39*(3), 435–443. http://dx.doi.org/10.1111/ejn.12452.

Reader, B. F., Jarrett, B. L., McKim, D. B., Wohleb, E. S., Godbout, J. P., & Sheridan, J. F. (2015). Peripheral and central effects of repeated social defeat stress: Monocyte trafficking, microglial activation, and anxiety. *Neuroscience*, *289*, 429–442. http://dx.doi.org/10.1016/j.neuroscience.2015.01.001.

Rei, D., Mason, X., Seo, J., Graff, J., Rudenko, A., Wang, J., & Tsai, L. H. (2015). Basolateral amygdala bidirectionally modulates stress-induced hippocampal learning and memory deficits through a p25/Cdk5-dependent pathway. *Proceedings of the National Academy of Sciences of the United States of America*, *112*(23), 7291–7296. http://dx.doi.org/10.1073/pnas.1415845112.

Reich, C. G., Mihalik, G. R., Iskander, A. N., Seckler, J. C., & Weiss, M. S. (2013). Adolescent chronic mild stress alters hippocampal CB1 receptor-mediated excitatory neurotransmission and plasticity. *Neuroscience*, *253*, 444–454. http://dx.doi.org/10.1016/j.neuroscience.2013.08.066. S0306-4522(13)00762-8 [pii].

Reich, C. G., Mohammadi, M. H., & Alger, B. E. (2008). Endocannabinoid modulation of fear responses: Learning and state-dependent performance effects. *Journal of Psychopharmacology*, *22*(7), 769–777. http://dx.doi.org/10.1177/0269881107083999.

Resstel, L. B., Lisboa, S. F., Aguiar, D. C., Correa, F. M., & Guimaraes, F. S. (2008). Activation of CB1 cannabinoid receptors in the dorsolateral periaqueductal gray reduces the expression of contextual fear conditioning in rats. *Psychopharmacology*, *198*(3), 405–411.

Resstel, L. B., Tavares, R. F., Lisboa, S. F., Joca, S. R., Correa, F. M., & Guimaraes, F. S. (2009). 5-HT1A receptors are involved in the cannabidiol-induced attenuation of behavioural and cardiovascular responses to acute restraint stress in rats. *British Journal of Pharmacology*, *156*(1), 181–188. http://dx.doi.org/10.1111/j.1476-5381.2008.00046.x. BPH046 [pii].

Rey, A. A., Purrio, M., Viveros, M. P., & Lutz, B. (2012). Biphasic effects of cannabinoids in anxiety responses: CB1 and GABA(B) receptors in the balance of GABAergic and glutamatergic neurotransmission. *Neuropsychopharmacology*, *37*(12), 2624–2634. http://dx.doi.org/10.1038/npp.2012.123.

Rinaldi-Carmona, M., Barth, F., Heaulme, M., Shire, D., Calandra, B., Congy, C., … Le Fur, G. (1994). SR141716A, a potent and selective antagonist of the brain cannabinoid receptor. *FEBS Letters*, *350*(2–3), 240–244.

Roberto, M., Cruz, M., Bajo, M., Siggins, G. R., Parsons, L. H., & Schweitzer, P. (2010). The endocannabinoid system tonically regulates inhibitory transmission and depresses the effect of ethanol in central amygdala. *Neuropsychopharmacology*, *35*(9), 1962–1972. http://dx.doi.org/10.1038/npp.2010.70.

Roche, M., O'Connor, E., Diskin, C., & Finn, D. P. (2007). The effect of CB(1) receptor antagonism in the right basolateral amygdala on conditioned fear and associated analgesia in rats. *European Journal of Neuroscience*, *26*(9), 2643–2653. http://dx.doi.org/10.1111/j.1460-9568.2007.05861.x.

Rodgers, R. J., Evans, P. M., & Murphy, A. (2005). Anxiogenic profile of AM-251, a selective cannabinoid CB1 receptor antagonist, in plus-maze-naive and plus-maze-experienced mice. *Behavioural Pharmacology*, *16*(5–6), 405–413.

Rodgers, R. J., Haller, J., Halasz, J., & Mikics, E. (2003). 'One-trial sensitization' to the anxiolytic-like effects of cannabinoid receptor antagonist SR141716A in the mouse elevated plus-maze. *European Journal of Neuroscience*, *17*(6), 1279–1286.

Roitman, P., Mechoulam, R., Cooper-Kazaz, R., & Shalev, A. (2014). Preliminary, open-label, pilot study of add-on oral Delta9-tetrahydrocannabinol in chronic post-traumatic

stress disorder. *Clinical Drug Investigation*, *34*(8), 587–591. http://dx.doi.org/10.1007/s40261-014-0212-3.
Romero-Sandoval, E. A., Horvath, R., Landry, R. P., & DeLeo, J. A. (2009). Cannabinoid receptor type 2 activation induces a microglial anti-inflammatory phenotype and reduces migration via MKP induction and ERK dephosphorylation. *Molecular Pain*, *5*, 25. http://dx.doi.org/10.1186/1744-8069-5-25.
Roohbakhsh, A., Keshavarz, S., Hasanein, P., Rezvani, M. E., & Moghaddam, A. H. (2009). Role of endocannabinoid system in the ventral hippocampus of rats in the modulation of anxiety-like behaviours. *Basic & Clinical Pharmacology & Toxicology*, *105*(5), 333–338. http://dx.doi.org/10.1111/j.1742-7843.2009.00449.x. PTO449 [pii].
Roohbakhsh, A., Moghaddam, A. H., Massoudi, R., & Zarrindast, M. R. (2007). Role of dorsal hippocampal cannabinoid receptors and nitric oxide in anxiety like behaviours in rats using the elevated plus-maze test. *Clinical and Experimental Pharmacology & Physiology*, *34*(3), 223–229. http://dx.doi.org/10.1111/j.1440-1681.2007.04576.x. CEP4576 [pii].
Roozendaal, B., Brunson, K. L., Holloway, B. L., McGaugh, J. L., & Baram, T. Z. (2002). Involvement of stress-released corticotropin-releasing hormone in the basolateral amygdala in regulating memory consolidation. *Proceedings of the National Academy of Sciences of the United States of America*, *99*(21), 13908–13913. http://dx.doi.org/10.1073/pnas.212504599.
Rose, J. E., & Woolsey, C. N. (1948). Structure and relations of limbic cortex and anterior thalamic nuclei in rabbit and cat. *Journal of Comparative Neurology*, *89*(3), 279–347.
Rossi, S., De Chiara, V., Musella, A., Kusayanagi, H., Mataluni, G., Bernardi, G., & Centonze, D. (2008). Chronic psychoemotional stress impairs cannabinoid-receptor-mediated control of GABA transmission in the striatum. *Journal of Neuroscience*, *28*(29), 7284–7292. http://dx.doi.org/10.1523/JNEUROSCI.5346-07.2008.
Rossi, S., Furlan, R., De Chiara, V., Motta, C., Studer, V., Mori, F., & Centonze, D. (2012). Interleukin-1beta causes synaptic hyperexcitability in multiple sclerosis. *Annals of Neurology*, *71*(1), 76–83. http://dx.doi.org/10.1002/ana.22512.
Rossi, S., Sacchetti, L., Napolitano, F., De Chiara, V., Motta, C., Studer, V., & Centonze, D. (2012). Interleukin-1beta causes anxiety by interacting with the endocannabinoid system. *Journal of Neuroscience*, *32*(40), 13896–13905. http://dx.doi.org/10.1523/JNEUROSCI.1515-12.2012.
Rubino, T., Guidali, C., Vigano, D., Realini, N., Valenti, M., Massi, P., & Parolaro, D. (2008). CB1 receptor stimulation in specific brain areas differently modulate anxiety-related behaviour. *Neuropharmacology*, *54*(1), 151–160. http://dx.doi.org/10.1016/j.neuropharm.2007.06.024.
Rubino, T., Realini, N., Castiglioni, C., Guidali, C., Vigano, D., Marras, E., & Parolaro, D. (2008). Role in anxiety behavior of the endocannabinoid system in the prefrontal cortex. *Cerebral Cortex*, *18*(6), 1292–1301.
Rubino, T., Sala, M., Vigano, D., Braida, D., Castiglioni, C., Limonta, V., & Parolaro, D. (2007). Cellular mechanisms underlying the anxiolytic effect of low doses of peripheral delta9-tetrahydrocannabinol in rats. *Neuropsychopharmacology*, *32*(9), 2036–2045. http://dx.doi.org/10.1038/sj.npp.1301330.
Ruehle, S., Remmers, F., Romo-Parra, H., Massa, F., Wickert, M., Wortge, S., & Lutz, B. (2013). Cannabinoid CB1 receptor in dorsal telencephalic glutamatergic neurons: Distinctive sufficiency for hippocampus-dependent and amygdala-dependent synaptic and behavioral functions. *Journal of Neuroscience*, *33*(25), 10264–10277. http://dx.doi.org/10.1523/JNEUROSCI.4171-12.2013. 33/25/10264 [pii].
Ruehle, S., Rey, A. A., Remmers, F., & Lutz, B. (2012). The endocannabinoid system in anxiety, fear memory and habituation. *Journal of Psychopharmacology*, *26*(1), 23–39. http://dx.doi.org/10.1177/0269881111408958.

Rutkowska, M., Jamontt, J., & Gliniak, H. (2006). Effects of cannabinoids on the anxiety-like response in mice. *Pharmacological Reports*, *58*(2), 200–206.

Sah, P., Faber, E. S., Lopez De Armentia, M., & Power, J. (2003). The amygdaloid complex: Anatomy and physiology. *Physiological Reviews*, *83*(3), 803–834. http://dx.doi.org/10.1152/physrev.00002.2003. 83/3/803 [pii].

Santos, C. J., Stern, C. A., & Bertoglio, L. J. (2008). Attenuation of anxiety-related behaviour after the antagonism of transient receptor potential vanilloid type 1 channels in the rat ventral hippocampus. *Behavioural Pharmacology*, *19*(4), 357–360.

Schacht, J. P., Hutchison, K. E., & Filbey, F. M. (2012). Associations between cannabinoid receptor-1 (CNR1) variation and hippocampus and amygdala volumes in heavy cannabis users. *Neuropsychopharmacology*, *37*(11), 2368–2376. http://dx.doi.org/10.1038/npp.2012.92.

Schafe, G. E., Bauer, E. P., Rosis, S., Farb, C. R., Rodrigues, S. M., & LeDoux, J. E. (2005). Memory consolidation of Pavlovian fear conditioning requires nitric oxide signaling in the lateral amygdala. *European Journal of Neuroscience*, *22*(1), 201–211.

Schneck, N., Miller, J. M., Delorenzo, C., Kikuchi, T., Sublette, M. E., Oquendo, M. A., & Parsey, R. V. (2016). Relationship of the serotonin transporter gene promoter polymorphism (5-HTTLPR) genotype and serotonin transporter binding to neural processing of negative emotional stimuli. *Journal of Affective Disorders*, *190*, 494–498. http://dx.doi.org/10.1016/j.jad.2015.10.047.

Schramm-Sapyta, N. L., Cha, Y. M., Chaudhry, S., Wilson, W. A., Swartzwelder, H. S., & Kuhn, C. M. (2007). Differential anxiogenic, aversive, and locomotor effects of THC in adolescent and adult rats. *Psychopharmacology*, *191*(4), 867–877. http://dx.doi.org/10.1007/s00213-006-0676-9.

Sciolino, N. R., Zhou, W., & Hohmann, A. G. (2011). Enhancement of endocannabinoid signaling with JZL184, an inhibitor of the 2-arachidonoylglycerol hydrolyzing enzyme monoacylglycerol lipase, produces anxiolytic effects under conditions of high environmental aversiveness in rats. *Pharmacological Research*, *64*(3), 226–234. http://dx.doi.org/10.1016/j.phrs.2011.04.010.

Scott, S. K., Young, A. W., Calder, A. J., Hellawell, D. J., Aggleton, J. P., & Johnson, M. (1997). Impaired auditory recognition of fear and anger following bilateral amygdala lesions. *Nature*, *385*(6613), 254–257. http://dx.doi.org/10.1038/385254a0.

Seidenbecher, T., Laxmi, T. R., Stork, O., & Pape, H. C. (2003). Amygdalar and hippocampal theta rhythm synchronization during fear memory retrieval. *Science*, *301*(5634), 846–850. http://dx.doi.org/10.1126/science.1085818.

Shammah-Lagnado, S. J., Beltramino, C. A., McDonald, A. J., Miselis, R. R., Yang, M., de Olmos, J., … Alheid, G. F. (2000). Supracapsular bed nucleus of the stria terminalis contains central and medial extended amygdala elements: Evidence from anterograde and retrograde tracing experiments in the rat. *Journal of Comparative Neurology*, *422*(4), 533–555. http://dx.doi.org/10.1002/1096-9861(20000710)422:4<533::AID-CNE5>3.0.CO;2-7. [pii].

Shonesy, B. C., Bluett, R. J., Ramikie, T. S., Baldi, R., Hermanson, D. J., Kingsley, P. J., & Patel, S. (2014). Genetic disruption of 2-arachidonoylglycerol synthesis reveals a key role for endocannabinoid signaling in anxiety modulation. *Cell Reports*, *9*(5), 1644–1653. http://dx.doi.org/10.1016/j.celrep.2014.11.001.

Silva, B. A., Mattucci, C., Krzywkowski, P., Murana, E., Illarionova, A., Grinevich, V., & Gross, C. T. (2013). Independent hypothalamic circuits for social and predator fear. *Nature Neuroscience*, *16*(12), 1731–1733. http://dx.doi.org/10.1038/nn.3573. nn.3573 [pii].

Sink, K. S., Segovia, K. N., Sink, J., Randall, P. A., Collins, L. E., Correa, M., & Salamone, J. D. (2010). Potential anxiogenic effects of cannabinoid CB1 receptor antagonists/inverse agonists in rats: Comparisons between AM4113, AM251, and the benzodiazepine inverse agonist FG-7142. *European Neuropsychopharmacology*, *20*(2), 112–122. http://dx.doi.org/10.1016/j.euroneuro.2009.11.002.

Sipe, J. C., Scott, T. M., Murray, S., Harismendy, O., Simon, G. M., Cravatt, B. F., & Waalen, J. (2010). Biomarkers of endocannabinoid system activation in severe obesity. *PloS One*, *5*(1), e8792. http://dx.doi.org/10.1371/journal.pone.0008792.

Slanina, K. A., & Schweitzer, P. (2005). Inhibition of cyclooxygenase-2 elicits a CB1-mediated decrease of excitatory transmission in rat CA1 hippocampus. *Neuropharmacology*, *49*(5), 653–659. http://dx.doi.org/10.1016/j.neuropharm.2005.04.019.

Soria-Gomez, E., Busquets-Garcia, A., Hu, F., Mehidi, A., Cannich, A., Roux, L., & Marsicano, G. (2015). Habenular CB1 receptors control the expression of aversive memories. *Neuron*, *88*(2), 306–313. http://dx.doi.org/10.1016/j.neuron.2015.08.035.

Sotres-Bayon, F., & Quirk, G. J. (2010). Prefrontal control of fear: More than just extinction. *Current Opinion in Neurobiology*, *20*(2), 231–235. http://dx.doi.org/10.1016/j.conb.2010.02.005.

Spiacci, G. B. L., Antero, L. A., Reis, D. G., Lisboa, S. F., & Resstel, L. B. (2016). Dorsal hippocampus cannabinoid type 1 receptors modulate the expression of contextual fear conditioning in rats: Involvement of local glutamatergic/nitrergic and GABAergic neurotransmissions. *European Neuropsychopharmacology*, *26*(10), 1579–1589. http://dx.doi.org/10.1016/j.euroneuro.2016.08.010.

Steindel, F., Lerner, R., Haring, M., Ruehle, S., Marsicano, G., Lutz, B., & Monory, K. (2013). Neuron-type specific cannabinoid-mediated G protein signalling in mouse hippocampus. *Journal of Neurochemistry*, *124*(6), 795–807. http://dx.doi.org/10.1111/jnc.12137.

Steiner, M. A., Wanisch, K., Monory, K., Marsicano, G., Borroni, E., Bachli, H., & Wotjak, C. T. (2008). Impaired cannabinoid receptor type 1 signaling interferes with stress-coping behavior in mice. *The Pharmacogenomics Journal*, *8*(3), 196–208. http://dx.doi.org/10.1038/sj.tpj.6500466. 6500466 [pii].

Stella, N. (2009). Endocannabinoid signaling in microglial cells. *Neuropharmacology*, *56*(Suppl. 1), 244–253. http://dx.doi.org/10.1016/j.neuropharm.2008.07.037.

Sullivan, G. M., Apergis, J., Bush, D. E., Johnson, L. R., Hou, M., & Ledoux, J. E. (2004). Lesions in the bed nucleus of the stria terminalis disrupt corticosterone and freezing responses elicited by a contextual but not by a specific cue-conditioned fear stimulus. *Neuroscience*, *128*(1), 7–14.

Suzuki, A., Josselyn, S. A., Frankland, P. W., Masushige, S., Silva, A. J., & Kida, S. (2004). Memory reconsolidation and extinction have distinct temporal and biochemical signatures. *Journal of Neuroscience*, *24*(20), 4787–4795. http://dx.doi.org/10.1523/JNEUROSCI.5491-03.2004.

Sztainberg, Y., Kuperman, Y., Tsoory, M., Lebow, M., & Chen, A. (2010). The anxiolytic effect of environmental enrichment is mediated via amygdalar CRF receptor type 1. *Molecular Psychiatry*, *15*(9), 905–917. http://dx.doi.org/10.1038/mp.2009.151.

Tan, H., Lauzon, N. M., Bishop, S. F., Chi, N., Bechard, M., & Laviolette, S. R. (2011). Cannabinoid transmission in the basolateral amygdala modulates fear memory formation via functional inputs to the prelimbic cortex. *Journal of Neuroscience*, *31*(14), 5300–5312. http://dx.doi.org/10.1523/JNEUROSCI.4718-10.2011.

Terzian, A. L., Aguiar, D. C., Guimaraes, F. S., & Moreira, F. A. (2009). Modulation of anxiety-like behaviour by transient receptor potential vanilloid type 1 (TRPV1) channels located in the dorsolateral periaqueductal gray. *European Neuropsychopharmacology*, *19*(3), 188–195. http://dx.doi.org/10.1016/j.euroneuro.2008.11.004.

Terzian, A. L., dos Reis, D. G., Guimaraes, F. S., Correa, F. M., & Resstel, L. B. (2014). Medial prefrontal cortex transient receptor potential vanilloid type 1 (TRPV1) in the expression of contextual fear conditioning in wistar rats. *Psychopharmacology*, *231*(1), 149–157. http://dx.doi.org/10.1007/s00213-013-3211-9.

Terzian, A. L., Drago, F., Wotjak, C. T., & Micale, V. (2011). The dopamine and cannabinoid interaction in the modulation of emotions and cognition: Assessing the role of cannabinoid CB1 receptor in neurons expressing dopamine D1 receptors. *Frontiers in Behavioral Neuroscience*, *5*, 49. http://dx.doi.org/10.3389/fnbeh.2011.00049.

Terzian, A. L., Micale, V., & Wotjak, C. T. (2014). Cannabinoid receptor type 1 receptors on GABAergic vs. glutamatergic neurons differentially gate sex-dependent social interest in mice. *European Journal of Neuroscience*, *40*(1), 2293–2298. http://dx.doi.org/10.1111/ejn.12561.

Tham, C. S., Whitaker, J., Luo, L., & Webb, M. (2007). Inhibition of microglial fatty acid amide hydrolase modulates LPS stimulated release of inflammatory mediators. *FEBS Letters*, *581*(16), 2899–2904. http://dx.doi.org/10.1016/j.febslet.2007.05.037.

Thiemann, G., Watt, C. A., Ledent, C., Molleman, A., & Hasenohrl, R. U. (2009). Modulation of anxiety by acute blockade and genetic deletion of the CB(1) cannabinoid receptor in mice together with biogenic amine changes in the forebrain. *Behavioural Brain Research*, *200*(1), 60–67. http://dx.doi.org/10.1016/j.bbr.2008.12.035.

Todd, S. M., & Arnold, J. C. (2016). Neural correlates of interactions between cannabidiol and delta(9)-tetrahydrocannabinol in mice: Implications for medical cannabis. *British Journal of Pharmacology*, *173*(1), 53–65. http://dx.doi.org/10.1111/bph.13333.

Toth, A., Boczan, J., Kedei, N., Lizanecz, E., Bagi, Z., Papp, Z., & Blumberg, P. M. (2005). Expression and distribution of vanilloid receptor 1 (TRPV1) in the adult rat brain. *Brain Research Molecular Brain Research*, *135*(1–2), 162–168. http://dx.doi.org/10.1016/j.molbrainres.2004.12.003. S0169-328X(04)00624-2 [pii].

Tovote, P., Fadok, J. P., & Luthi, A. (2015). Neuronal circuits for fear and anxiety. *Nature Reviews. Neuroscience*, *16*(6), 317–331. http://dx.doi.org/10.1038/nrn3945. nrn3945 [pii].

Tsou, K., Brown, S., Sanudo-Pena, M. C., Mackie, K., & Walker, J. M. (1998). Immunohistochemical distribution of cannabinoid CB1 receptors in the rat central nervous system. *Neuroscience*, *83*(2), 393–411. S0306452297004363 [pii].

Tsou, K., Nogueron, M. I., Muthian, S., Sanudo-Pena, M. C., Hillard, C. J., Deutsch, D. G., & Walker, J. M. (1998). Fatty acid amide hydrolase is located preferentially in large neurons in the rat central nervous system as revealed by immunohistochemistry. *Neuroscience Letters*, *254*(3), 137–140.

Tzavara, E. T., Davis, R. J., Perry, K. W., Li, X., Salhoff, C., Bymaster, F. P., & Nomikos, G. G. (2003). The CB1 receptor antagonist SR141716A selectively increases monoaminergic neurotransmission in the medial prefrontal cortex: Implications for therapeutic actions. *British Journal of Pharmacology*, *138*(4), 544–553. http://dx.doi.org/10.1038/sj.bjp.0705100.

Uliana, D. L., Hott, S. C., Lisboa, S. F., & Resstel, L. B. (2016). Dorsolateral periaqueductal gray matter CB1 and TRPV1 receptors exert opposite modulation on expression of contextual fear conditioning. *Neuropharmacology*, *103*, 257–269. http://dx.doi.org/10.1016/j.neuropharm.2015.12.020.

Ulrich-Lai, Y. M., & Herman, J. P. (2009). Neural regulation of endocrine and autonomic stress responses. *Nature Reviews. Neuroscience*, *10*(6), 397–409. http://dx.doi.org/10.1038/nrn2647.

Uriguen, L., Perez-Rial, S., Ledent, C., Palomo, T., & Manzanares, J. (2004). Impaired action of anxiolytic drugs in mice deficient in cannabinoid CB1 receptors. *Neuropharmacology*, *46*(7), 966–973. http://dx.doi.org/10.1016/j.neuropharm.2004.01.003.

Uylings, H. B., Groenewegen, H. J., & Kolb, B. (2003). Do rats have a prefrontal cortex? *Behavioural Brain Research*, *146*(1–2), 3–17.

Uylings, H. B., & van Eden, C. G. (1990). Qualitative and quantitative comparison of the prefrontal cortex in rat and in primates, including humans. *Progress in Brain Research*, *85*, 31–62.

Valjent, E., Mitchell, J. M., Besson, M. J., Caboche, J., & Maldonado, R. (2002). Behavioural and biochemical evidence for interactions between delta 9-tetrahydrocannabinol and nicotine. *British Journal of Pharmacology, 135*(2), 564–578. http://dx.doi.org/10.1038/sj.bjp.0704479.

Varga, B., Kassai, F., & Gyertyan, I. (2012). Interactions of CB1 and mGlu5 receptor antagonists in food intake, anxiety and memory models in rats. *Pharmacology, Biochemistry, and Behavior, 103*(2), 425–430. http://dx.doi.org/10.1016/j.pbb.2012.09.016.

Vazquez-Borsetti, P., Celada, P., Cortes, R., & Artigas, F. (2011). Simultaneous projections from prefrontal cortex to dopaminergic and serotonergic nuclei. *International Journal of Neuropsychopharmacology, 14*(3), 289–302. http://dx.doi.org/10.1017/S1461145710000349.

Vertes, R. P. (2006). Interactions among the medial prefrontal cortex, hippocampus and midline thalamus in emotional and cognitive processing in the rat. *Neuroscience, 142*(1), 1–20. http://dx.doi.org/10.1016/j.neuroscience.2006.06.027. S0306-4522(06)00858-X [pii].

Viana, T. G., Hott, S. C., Resstel, L. B., Aguiar, D. C., & Moreira, F. A. (2015). Anti-aversive role of the endocannabinoid system in the periaqueductal gray stimulation model of panic attacks in rats. *Psychopharmacology, 232*(9), 1545–1553. http://dx.doi.org/10.1007/s00213-014-3793-x.

Vincent, S. R., & Kimura, H. (1992). Histochemical mapping of nitric oxide synthase in the rat brain. *Neuroscience, 46*(4), 755–784.

Viveros, M. P., Marco, E. M., & File, S. E. (2005). Endocannabinoid system and stress and anxiety responses. *Pharmacology, Biochemistry, and Behavior, 81*(2), 331–342. http://dx.doi.org/10.1016/j.pbb.2005.01.029. S0091-3057(05)00134-6 [pii].

Vouimba, R. M., & Richter-Levin, G. (2005). Physiological dissociation in hippocampal subregions in response to amygdala stimulation. *Cerebral Cortex, 15*(11), 1815–1821. http://dx.doi.org/10.1093/cercor/bhi058.

Walker, J. M., Huang, S. M., Strangman, N. M., Tsou, K., & Sanudo-Pena, M. C. (1999). Pain modulation by release of the endogenous cannabinoid anandamide. *Proceedings of the National Academy of Sciences of the United States of America, 96*(21), 12198–12203.

Walker, D. L., Toufexis, D. J., & Davis, M. (2003). Role of the bed nucleus of the stria terminalis versus the amygdala in fear, stress, and anxiety. *European Journal of Pharmacology, 463*(1–3), 199–216. S0014299903012822 [pii].

Walter, L., Franklin, A., Witting, A., Wade, C., Xie, Y., Kunos, G., & Stella, N. (2003). Nonpsychotropic cannabinoid receptors regulate microglial cell migration. *Journal of Neuroscience, 23*(4), 1398–1405.

Wilson, R. I., Kunos, G., & Nicoll, R. A. (2001). Presynaptic specificity of endocannabinoid signaling in the hippocampus. *Neuron, 31*(3), 453–462. S0896-6273(01)00372-5 [pii].

Wolf, S. A., Bick-Sander, A., Fabel, K., Leal-Galicia, P., Tauber, S., Ramirez-Rodriguez, G., & Kempermann, G. (2010). Cannabinoid receptor CB1 mediates baseline and activity-induced survival of new neurons in adult hippocampal neurogenesis. *Cell Communication and Signaling, 8*, 12. http://dx.doi.org/10.1186/1478-811X-8-12. 1478-811X-8-12 [pii].

Wultsch, T., Chourbaji, S., Fritzen, S., Kittel, S., Grunblatt, E., Gerlach, M., & Reif, A. (2007). Behavioural and expressional phenotyping of nitric oxide synthase-I knockdown animals. *Journal of Neural Transmission, Supplementum*, (72), 69–85.

Yirmiya, R., Rimmerman, N., & Reshef, R. (2015). Depression as a microglial disease. *Trends in Neurosciences, 38*(10), 637–658. http://dx.doi.org/10.1016/j.tins.2015.08.001.

Zarrindast, M. R., Nasehi, M., Piri, M., & Bina, P. (2010). Anxiety-like behavior induced by histaminergic agents can be prevented by cannabinoidergic WIN55,212-2 injected into the dorsal hippocampus in mice. *Pharmacology, Biochemistry, and Behavior, 94*(3), 387–396. http://dx.doi.org/10.1016/j.pbb.2009.09.021. S0091-3057(09)00290-1 [pii].

Zarrindast, M. R., Sarahroodi, S., Arzi, A., Khodayar, M. J., Taheri-Shalmani, S., & Rezayof, A. (2008). Cannabinoid CB1 receptors of the rat central amygdala mediate

anxiety-like behavior: Interaction with the opioid system. *Behavioural Pharmacology, 19*(7), 716–723. http://dx.doi.org/10.1097/FBP.0b013e3283123c83.
Zhu, P. J., & Lovinger, D. M. (2005). Retrograde endocannabinoid signaling in a postsynaptic neuron/synaptic bouton preparation from basolateral amygdala. *Journal of Neuroscience, 25*(26), 6199–6207. http://dx.doi.org/10.1523/JNEUROSCI.1148-05.2005.
Ziv, Y., Ron, N., Butovsky, O., Landa, G., Sudai, E., Greenberg, N., & Schwartz, M. (2006). Immune cells contribute to the maintenance of neurogenesis and spatial learning abilities in adulthood. *Nature Neuroscience, 9*(2), 268–275. http://dx.doi.org/10.1038/nn1629.
Zoppi, S., Madrigal, J. L., Caso, J. R., Garcia-Gutierrez, M. S., Manzanares, J., Leza, J. C., & Garcia-Bueno, B. (2014). Regulatory role of the cannabinoid CB2 receptor in stress-induced neuroinflammation in mice. *British Journal of Pharmacology, 171*(11), 2814–2826. http://dx.doi.org/10.1111/bph.12607.
Zoppi, S., Perez Nievas, B. G., Madrigal, J. L., Manzanares, J., Leza, J. C., & Garcia-Bueno, B. (2011). Regulatory role of cannabinoid receptor 1 in stress-induced excitotoxicity and neuroinflammation. *Neuropsychopharmacology, 36*(4), 805–818. http://dx.doi.org/10.1038/npp.2010.214. npp2010214 [pii].
Zuardi, A. W., Cosme, R. A., Graeff, F. G., & Guimaraes, F. S. (1993). Effects of ipsapirone and cannabidiol on human experimental anxiety. *Journal of Psychopharmacology*, 7(1 Suppl.), 82–88. http://dx.doi.org/10.1177/026988119300700112. 7/1_suppl/82 [pii].
Zuardi, A. W., Shirakawa, I., Finkelfarb, E., & Karniol, I. G. (1982). Action of cannabidiol on the anxiety and other effects produced by delta 9-THC in normal subjects. *Psychopharmacology, 76*(3), 245–250.
Zygmunt, P. M., Petersson, J., Andersson, D. A., Chuang, H., Sorgard, M., Di Marzo, V., & Hogestatt, E. D. (1999). Vanilloid receptors on sensory nerves mediate the vasodilator action of anandamide. *Nature, 400*(6743), 452–457. http://dx.doi.org/10.1038/22761.

CHAPTER NINE

The Melatonergic System in Anxiety Disorders and the Role of Melatonin in Conditional Fear

F. Huang*[,1], Z. Yang*, C.-Q. Li[†]
*Yiyang Medical College, Yiyang, Hunan, PR China
[†]School of Basic Medical Sciences, Central South University, Changsha, Hunan, PR China
[1]Corresponding author: e-mail address: 874558215@qq.com

Contents

Abstract

Background: Resistance to extinction of certain conditioned responses forms the basis of anxieties, phobias, and compulsions. There has been an available effective means of extinction-based exposure psychotherapy for the treatment of anxiety disorders, such as posttraumatic stress disorder (PTSD) that has been hypothesized to result from impaired extinction of fear memory. PTSD is considered as a memory disorder within a Pavlovian fear conditioning and extinction framework. Therefore, the aim of this review was to report the preclinical profile of melatonin, a pineal gland hormone, as a potential pharmacological option in the treatment of anxiety disorders such as PTSD, tested with the Pavlovian fear conditioning paradigm.

Vitamins and Hormones, Volume 103
ISSN 0083-6729
http://dx.doi.org/10.1016/bs.vh.2016.09.003

Methods: We performed a literature review regarding studies that evaluated the effects of melatonin on fear conditioning and fear extinction.
Results: Results showed that a single administration 30 min before conditioning has no effect on the acquisition of cued fear, but impaired contextual fear conditioning. Compared to rats injected with vehicle, rats injected with melatonin 30 min before extinction training presented a significant lower freezing during both extinction training and extinction test phases. However, melatonin injected immediately after extinction training was ineffective on extinction learning.
Conclusion: Melatonin impaired contextual fear conditioning, a hippocampus-dependent task. On the contrary, melatonin facilitates the extinction of conditional cued fear without affecting its acquisition or expression, and melatonin facilitates cued fear extinction only when it is present during extinction training. Although further studies are necessary, the research undertaken until now shows that melatonin modulates fear conditioning and fear extinction and consequently melatonin may serve as an agent for the treatment of PTSD.

1. INTRODUCTION

Anxiety disorder (or generalized anxiety, as per DSM-IV) (American Psychiatric Association, 1994) is a chronic and disabling disorder with excessive, uncontrollable, and often irrational worry associated with physical symptoms, including fatigue, fidgeting, headaches, nausea, numbness, and sweating.

Pathological anxiety may reflect a failure to extinguish conditioned fear. In anxiety disorders such as posttraumatic stress disorder (PTSD), this type of conditioned fear fails to extinguish, and individuals with PTSD reexperience the event for decades after danger has passed (Hayes, VanElzakker, & Shin, 2012). PTSD is a severe anxiety disorder suffered by approximately 10–30% of individuals who have experienced or witnessed a psychologically traumatic event (de Vries & Olff, 2009; Dohrenwend et al., 2006) such as sexual assault, combat, natural disaster, motor vehicle accident, or witnessing the death or serious injury of another individual. Treatments for PTSD and other anxiety disorders have focused on exposure therapy that involves repeated exposure to trauma-related cues (Hofmann, Sawyer, Korte, & Smits, 2009; Hofmann & Smits, 2008).

2. ANXIETY AND MELATONERGIC SYSTEM

Studies involving the pharmacologic manipulation have been shown to be effective in the treatment of anxiety disorders, including benzodiazepines, selective serotonin reuptake inhibitors, serotonin noradrenaline reuptake inhibitors, tricyclic antidepressants, and pregabalin (Baldwin et al., 2014). However, use of these drugs can lead to side effects such as nausea, sexual dysfunction, dependence, and cognitive problems (Buoli, Caldiroli, Caletti, Paoli, & Altamura, 2013). For this reason, there is a growing interest in exploring new pharmacological approaches for the treatment of anxiety. Among these approaches, the melatonergic system has gained considerable attention.

The neurohormone melatonin (*N*-acetyl-5-methoxytryptamine) is prominently synthesized in the pineal gland. Melatonin secretion into the bloodstream occurs primarily at night, with negligible levels secreted during the day, attaining peak concentrations of plasma melatonin between 02:00 and 04:00 h (Pandi-Perumal et al., 2008). This hormone plays an important role in the regulation of circadian and sleep–wake system, melatonin also has several pharmacological effects including sedative, antioxidant, anxiolytic, antidepressant, anticonvulsant, and analgesic activities (Mantovani et al., 2006; Reiter & Korkmaz, 2008).

Effects of melatonin are mediated by the activation of two G protein coupled receptors, MT1 (Mel1a) and MT2 (Mel1b) (Reppert, Weaver, & Godson, 1996). The "MT3 receptor" is an enzyme, quinone reductase, and belongs to a group of reductases that participate in the protection against oxidative stress by preventing electron transfer reactions of quinones. Within the CNS, MT1 and MT2 receptors have been localized in the suprachiasmatic nucleus, the cortex, thalamus and cerebellar cortex, the hippocampus. MT1 receptors are also found in several other hypothalamic nuclei and brain areas, such as the caudate putamen, the nucleus accumbens, the *substantia nigra*, and the ventrotegmental area (Brunner et al., 2006; Dubocovich & Markowska, 2005; Wu et al., 2006).

It has been demonstrated that melatonin and its analogs have anxiolytic properties in validated paradigms of anxiety behavior in animals: the tested animal models were the Vogel conflict procedure, the elevated plus-maze procedure (EPMT), the conditioned footshock-induced ultrasonic vocalizations test, the social interaction test, the plus-maze test, the cumulative burying behavior paradigm, hole board test, novelty suppressed feeding test (NSFT), and open field test (OFT) (Boulle et al., 2014; Bustamante-García, Lira-Rocha, Espejo-González, Gómez-Martínez, & Picazo, 2014; Golombek, Martini, & Cardinali, 1993; Golus & King, 1981; Millan, Brocco, Gobert, & Dekeyne, 2005; Ochoa-Sanchez et al., 2012; Papp, Litwa, Gruca, & Mocaër, 2006; Pierrefiche, Zerbib, & Laborit, 1993; Tian et al., 2010) (Table 1).

Table 1 Summary of Anxiolytic Effects of Melatonin and Its Analogs in Animals

Authors	Animals	Drugs	Dosages (mg/kg)	Animal Models	Results
Golus and King (1981)	Rats	Melatonin	1	Open field test (OFT)	Increased the activity within the central area of the OFT
Golombek et al. (1993)	Rats	Melatonin	1	Elevated plus-maze test (EPMT)	Increased the time spent in the open arms of the EPMT
Millan et al. (2005)	Rats	Agomelatine	2.5–80	Social interaction test Vogel conflict test Plus-maze test	Enhanced the time devoted to active social interaction Increased the punished responses Enhanced percentage entries into open arms
Papp et al. (2006)	Rats	Agomelatine	10–75	Vogel conflict test Conditioned footshock-induced ultrasonic vocalization test	Increased the number of punished responses Decreased vocalization time
		Agomelatine Melatonin	10–75 10–75	EPMT	Enhanced open arms Exploration
Ochoa-Sanchez et al. (2012)	Rats	UCM765 Melatonin	10 20	EPMT Novelty suppressed feeding test (NSFT) OFT	Increased the time spent in the open arm of the EPMT Reduced the latency to eat in a novel environment Increased the activity within the central area of the OFT

Table 1 Summary of Anxiolytic Effects of Melatonin and Its Analogs in Animals—cont'd

Authors	Animals	Drugs	Dosages (mg/kg)	Animal Models	Results
Bustamante-García et al. (2014)	Rats	M3C Melatonin	0.5, 1.0, 2.0, 2.5 2.0	The cumulative burying Behavior paradigm	Decreased in burying Behavior
		M3C Melatonin	2.0 and 2.5 2.0	Elevated plus-maze	Increased the percentage of time spent on the open arms of the EPM
Pierrefiche et al. (1993)	Mice	Melatonin	0.5–5.0	Hole board test Unconditioned conflict test	Decreased head-dip performance Increased the time spent in the lit box as well as the number of transitions between the two compartments
Tian et al. (2010)	Mice	Neu-P11 Melatonin	50 50	EPMT	Increased the percentage of time spent in the open arms

Clinically, MLT has elicited anxiolytic effects (Caumo, Levandovski, & Hidalgo, 2009; Srinivasan et al., 2006). It has also been successfully used to facilitate the discontinuation of benzodiazepine therapy in elderly patients (Garfinkel, Zisapel, Wainstein, & Laudon, 1999) and to elicit anxiolytic effects in cataract surgery patients (Khezri & Merate, 2013). In addition, ramelteon and agomelatine, two nonselective MT1/MT2 receptor agonists have produced anxiolytic effects in humans (den Boer, Bosker, & Meesters, 2006; Gross, Nourse, & Wasser, 2009; Stein, Ahokas, & de Bodinat, 2008).

3. PTSD AND PAVLOVIAN FEAR CONDITIONING AND EXTINCTION

There has been an available effective means of extinction-based exposure psychotherapy for the treatment of anxiety disorders, such as PTSD (Bentz, Michael, de Quervain, & Wilhelm, 2010; Rothbaum & Davis,

2003) that has been hypothesized to result from impaired extinction of fear memory (Bremner et al., 2005; Milad, Rauch, Pitman, & Quirk, 2006). PTSD is considered as a memory disorder, within a Pavlovian fear conditioning and extinction framework (Jovanovic & Ressler, 2010; Shin & Handwerger, 2009).

Pavlovian fear conditioning involves pairing an initially neutral conditioned stimulus (CS) such as a tone or context with an aversive unconditioned stimulus (US) like a footshock. After several pairings of these stimuli, the CS comes to elicit conditional fear responses such as defensive behavioral responses (e.g., freezing). After conditioning, repeated presentations of the CS in the absence of the US results in a progressive reduction of the conditioned response. This process is called extinction. Numerous studies have suggested that extinction is a form of new, context-dependent learning (Bouton, 1993; Bouton, Westbrook, Corcoran, & Maren, 2006). These studies suggest that the memory formed during conditioning is not eliminated during extinction but rather is suppressed by extinction learning, which likely attribute to plasticity at distinct synapses from those mediating acquisition.

PTSD is associated with increased acquisition of fear conditioning and a failure to extinguish or maintain extinction of conditioned fear (VanElzakker, Dahlgren, Davis, Dubois, & Shin, 2014). Therefore, the aim of this literature review was to expose the studies conducted regarding fear learning and extinction with melatonin using the Pavlovian fear conditioning paradigm (Huang, Yang, Liu, & Li, 2014; Yang, Li, & Huang, 2013).

4. METHODS

First, we examined the effects of systemic injections of melatonin with good penetration through the blood–brain barrier on contextual fear conditioning and fear expression in rats. Second, we examined the effects of melatonin administration on the acquisition, expression, and extinction of conditional cued fear in rats.

4.1 Contextual Fear Conditioning and Test

The behavioral procedure involved three phases: habituation, fear conditioning, and testing. During habituation phase, rats were habituated to the conditioning chamber (CS) for 20 min with no stimuli presented. Twenty-four hours later (fear conditioning phase), rats received a 1 s 0.4 mA footshock (US) beginning 3 min after being placed in the chamber. The training session consisted of five of these conditioned and unconditioned stimulus (CS–US) pairings with an intertrial interval of 60 s. The rats remained in the training box for 60 s following the last CS–US pairing, after which they were returned to the home cages. Freezing was used as the measure of fear and is characterized by cessation of movement except that required for respiration (Blanchard & Blanchard, 1969). Freezing activity during intertrial interval was scored with a digital stopwatch from videotapes. Observers scoring freezing were blind to the treatments. Freezing is presented as the percent time spent freezing (time spent freezing/total time × 100). Twenty-four hours after fear conditioning (testing phase), rats were placed in the original training chamber for 5 min without any shock, and freezing behavior was scored during the entire duration of context exposure.

4.2 Cued Fear Acquisition

The behavioral procedure involved three phases: habituation (context A), fear conditioning (context A), and testing (context B), each separated by 24 h to allow for memory consolidation. During habituation phase (Day 0), rats were habituated to the conditioning chamber (context A) for 20 min with no stimuli presented. On day 1 (fear conditioning phase), rats were allowed to explore the chamber for 3 min. At the end of 3 min rats were subjected to five trials of audio tone (CS) and foot shock (US) with an intertrial interval of 60 s. Audio tone (4 kHz, 80 dB, 20 s duration) was followed immediately by a footshock (0.5 mA, 0.5 s duration) from the metal grid floor. The rats remained in the training box for 60 s following the last CS–US pairing, after which they were returned to the home cages. On day 2 (testing phase), after a 3 min acclimation, rats received five tone alone presentations with an intertrial interval of 60 s.

4.3 Cued Fear Extinction

The behavioral procedure involved four phases: habituation (context A), fear conditioning (context A), extinction training (context B), and extinction test (context B), each separated by 24 h to allow for memory consolidation. In both experiments, cued fear was induced in nondrugged, naïve rats described earlier. Rats were matched into two groups that received either melatonin or vehicle based on freezing scores during fear conditioning. Twenty-four hours after fear conditioning (day 2, extinction training phase), rats were placed in context B and were allowed to acclimate for

3 min. Following this, rats received 14 tone (4 kHz, 80 dB, 20 s duration) alone presentations with an intertrial interval of 60 s. The rats were immediately returned to their home cages 60 s after the last tone presentation. On day 3 (extinction test phase), rats received 14 tone alone presentations in context B as described on day 2.

5. RESULTS

5.1 Melatonin Impaired Contextual Fear Conditioning and Did Not Affect Fear Expression in Rats

Melatonin or vehicle was injected intraperitoneally (i.p.) at dose of 2.5 mg/kg 30 min before contextual fear conditioning. During the contextual fear conditioning phase, there was a significant difference between groups (group, $F\ (1,\ 16) = 123.282$, $p < 0.001$; group*time block, $F\ (5,\ 80) = 39.175$, $p < 0.001$), and the melatonin group presented a significant lower freezing in comparison to the vehicle group. During the test phase, a significantly lower freezing was also observed in the melatonin group ($p < 0.01$). These data suggested that melatonin impaired contextual fear conditioning.

However, when melatonin or vehicle was injected 30 min before testing, there was no significant difference between groups, both during the conditioning phase and the test phase, indicating that melatonin did not affect fear expression.

5.2 Melatonin Had no Effect on the Acquisition and Retention of Conditional Cued Fear Response

Melatonin or vehicle was injected i.p. at dose of 2.5 mg/kg 30 min before cued fear conditioning. During the conditioning phase, the melatonin and vehicle groups showed equivalent fear learning. At 24 h postconditioning, the rats were presented with the CS in a novel context. There was no a significant group effect of percent freezing. These data suggest that melatonin does not affect the acquisition and retention of cued fear.

5.3 Preextinction Injection of Melatonin Facilitated the Acquisition and Retention of Cued Fear Extinction

Next, we assessed the effects of melatonin on the expression and extinction of cued fear memory. Two groups of naïve rats underwent fear conditioning. One day after training, rats were injected with melatonin or vehicle 30 min before extinction training. The next day, they were subjected to a drug-free extinction test. During the extinction training phase, a significantly lower freezing was observed in the melatonin group in comparison to the vehicle group (group, F (1, 18) = 749.594, $p < 0.001$; group × trail, F (13, 234) = 28.065, $p < 0.001$). During the extinction test phase, the melatonin group presented a significant lower freezing ($p < 0.001$). These results suggest that injection of melatonin 30 min before the extinction training facilitates the acquisition and retention of extinction learning.

We also administered melatonin immediately after extinction training to determine the effects of melatonin on extinction when it is administrated at different time points. During the extinction training phase and test phase, there was no a significant effect of group (F (1, 10) = 0.049, $p > 0.05$) and no significant trial-by-group interaction (F (13, 130) = 1.566, $p > 0.05$). These results suggest that melatonin administration immediately after extinction training has no effect on the extinction of conditional cued fear.

6. DISCUSSION

In our literature review, melatonin exerts opposing influences on the acquisition and extinction of conditional fear is in accordance with many other studies showing that acquisition and extinction are distinct learning processes. There exist behavioral, systems, and molecular differences between acquisition and extinction (Bouton et al., 2006; Lattal, Radulovic, & Lukowiak, 2006; Lai, Franke, & Gan, 2012; Marsicano et al., 2002). Furthermore, in line with this finding, it has been found that melatonin facilitates the extinction of active avoidance reflex, whereas memory acquisition is not influenced (Kovács, Gajari, Telegdy, & Lissak, 1974).

In this review, on one hand, melatonin impaired contextual fear conditioning in rats. Considerable evidence suggests that the amygdala is an important site of the neural circuits related to both cued and contextual fear conditioning. However, the hippocampus is usually required only for contextual task (LeDoux, 2000; Maren, 2001). In the brain, melatonin receptors [MT(1)/MT(2)] have been found in regions implicated in cognition and memory, such as the hippocampus (Pandi-Perumal et al., 2008). In this review, on the other hand, melatonin facilitated the acquisition and retention of cued fear extinction in rats. Converging evidence has identified a network of brain structures including the amygdala, prefrontal cortex, and hippocampus (Cho, Deisseroth, & Bolshakov, 2013; Maren & Quirk, 2004; Sotres-Bayon, Bush, & LeDoux, 2004, 2007) that supports the acquisition, storage, retrieval, and contextual modulation of fear extinction. In the brain, binding sites of melatonin have been found in the amygdala, hippocampus, and prefrontal cortex (Ekmekcioglu, 2006; Savaskan et al., 2005; Uz et al., 2005), three regions that are involved in fear extinction (Bouton et al., 2006, Sotres-Bayon et al., 2004).

Although the mechanisms through which melatonin acts on extinction remains to be determined, the effects of melatonin may be through the direct modulation of memory formation circuits. Melatonin has shown to play an important role in structural remodeling of synaptic connections during memory and learning processes (Baydas et al., 2002). Other researches also demonstrated the ability of melatonin to modulate neuronal firing in the hippocampus and other brain regions (Baydas, Ozveren, Akdemir, Tuzcu, & Yasar, 2005; Gorfine & Zisapel, 2007; Wang, Suthana, Chaudhury, Weaver, & Colwell, 2005). Further, a previous research showed that MT(2) receptor knockout mice demonstrate a significantly reduced long-term potentiation as well as impaired memory performance tested in an elevated plus-maze paradigm (Larson et al., 2006). Thus, melatonin may regulate learning and memory through its influence on synaptic connections in central nervous system neurons. Alternatively, melatonin may have an indirect effect on memory formation via some neurotrophin such as brain-derived neurotrophic factor (BDNF). Melatonin has been shown to increase the production of BDNF (Kong et al., 2008) that may play an important role in cued fear extinction (Heldt, Stanek, Chhatwal, & Ressler, 2007). Further, melatonin modulates neurotransmitters such as gamma-amino butyric acid (Rosenstein & Cardinali, 1986) and glutamate (Vimala, Bhutada, & Patel, 2014) which are involved in extinction learning (Davis & Myers, 2002).

There has been an available effective means of extinction-based exposure psychotherapy for the treatment of anxiety disorders, such as PTSD (Bentz et al., 2010; Rothbaum & Davis, 2003) that has been hypothesized to result from impaired extinction of fear memory (Bremner et al., 2005; Milad et al., 2006). Furthermore, decreased melatonin levels in patients with PTSD were reported in clinical studies (McFarlane, Barton, Briggs, & Kennaway, 2010). Therefore, facilitating effects of melatonin on fear extinction suggest that melatonin may serve as an agent for the treatment of anxiety disorders such as PTSD. Further studies will need to investigate the neurobiological mechanism through which melatonin modulates anxiety disorders.

ACKNOWLEDGMENTS

This work was supported by National Natural Science Foundation of China (31171151 and 31371212) and Scientific Research Fund of Hunan Provincial Education Department (14B184).

REFERENCES

American Psychiatric Association (1994). *Diagnostic and statistical manual of mental disorders* (4th ed.). Washington (DC): American Psychiatric Press.

Baldwin, D. S., Anderson, I. M., Nutt, D. J., Allgulander, C., Bandelow, B., den Boer, J. A., et al. (2014). Evidence-based pharmacological treatment of anxiety disorders, post-traumatic stress disorder and obsessive-compulsive disorder: A revision of the 2005 guidelines from the British association for psychopharmacology. *Journal of Psychopharmacology, 28*(5), 403–439.

Baydas, G., Nedzvetsky, V. S., Nerush, P. A., Kırıchenko, S. V., Demchenko, H. M., & Reiter, R. J. (2002). A novel role for melatonin: Regulation of the expression of cell adhesion molecules in the hippocampus, cortex and cerebellum. *Neuroscience Letters, 326*, 109–112.

Baydas, G., Ozveren, F., Akdemir, I., Tuzcu, M., & Yasar, A. (2005). Learning and memory deficits in rats induced by chronic thinner exposure are reversed by melatonin. *Journal of Pineal Research, 39*, 50–56.

Bentz, D., Michael, T., de Quervain, D. J., & Wilhelm, F. H. (2010). Enhancing exposure therapy for anxiety disorders with glucocorticoids: From basic mechanisms of emotional learning to clinical applications. *Journal of Anxiety Disorders, 24*, 223–230.

Blanchard, D. C., & Blanchard, R. J. (1969). Crouching as an index of fear. *Journal of Comparative Physiology and Psychology, 67*, 370–375.

Boulle, F., Massart, R., Stragier, E., Païzanis, E., Zaidan, L., Marday, S., et al. (2014). Hippocampal and behavioral dysfunctions in a mouse model of environmental stress: Normalization by agomelatine. *Translational Psychiatry, 4*, e485.

Bouton, M. E. (1993). Context, time, and memory retrieval in the interference paradigms of Pavlovian learning. *Psychological Bulletin, 114*, 80–99.

Bouton, M. E., Westbrook, R. F., Corcoran, K. A., & Maren, S. (2006). Contextual and temporal modulation of extinction: Behavioral and biological mechanisms. *Biological Psychiatry, 60*, 352–360.

Bremner, J. D., Vermetten, E., Schmahl, C., Vaccarino, V., Vythilingam, M., Afzal, N., et al. (2005). Positron emission tomographic imaging of neural correlates of a fear acquisition and extinction paradigm in women with childhood sexual-abuse-related post-traumatic stress disorder. *Psychological Medicine, 35*, 791–806.

Brunner, P., Sozer-Topcular, N., Jockers, R., Ravid, R., Angeloni, D., Fraschini, F., et al. (2006). Pineal and cortical melatonin receptors MT1 and MT2 are decreased in Alzheimer's disease. *European Journal of Histochemistry, 50*, 311–316.

Buoli, M., Caldiroli, A., Caletti, E., Paoli, R. A., & Altamura, A. C. (2013). New approaches to the pharmacological management of generalized anxiety disorder. *Expert Opinion on Pharmacotherapy, 14*(2), 175–184.

Bustamante-García, R., Lira-Rocha, A. S., Espejo-González, O., Gómez-Martínez, A. E., & Picazo, O. (2014). Anxiolytic-like effects of a new 1-N substituted analog of melatonin in pinealectomized rats. *Progress in Neuropsychopharmacology and Biological Psychiatry, 51*, 133–139.

Caumo, W., Levandovski, R., & Hidalgo, M. P. L. (2009). Preoperative anxiolytic effect of melatonin and clonidine on postoperative pain and morphine consumption in patients undergoing abdominal hysterectomy: A double-blind, randomized, placebo-controlled study. *The Journal of Pain, 10*, 100–108.

Cho, J. H., Deisseroth, K., & Bolshakov, V. Y. (2013). Synaptic encoding of fear extinction in mPFC-amygdala circuits. *Neuron, 80*(6), 1491–1507.

Davis, M., & Myers, K. M. (2002). The role of glutamate and gamma-aminobutyric acid in fear extinction: Clinical implications for exposure therapy. *Biological Psychiatry, 52*(10), 998–1007.

de Vries, G.-J., & Olff, M. (2009). The lifetime prevalence of traumatic events and posttraumatic stress disorder in the Netherlands. *Journal of Traumatic Stress*, *22*(4), 259–267.

den Boer, J., Bosker, F., & Meesters, Y. (2006). Clinical efficacy of agomelatine in depression: The evidence. *International Clinical Psychopharmacology*, *21*(Suppl. 1), S21–S24.

Dohrenwend, B. P., Turner, J. B., Turse, N. A., Adams, B. G., Koenen, K. C., & Marshall, R. (2006). The psychological risks of Vietnam for U.S. veterans: A revisit with new data and methods. *Science*, *313*(5789), 979–982.

Dubocovich, M. L., & Markowska, M. (2005). Functional MT1 and MT2 melatonin receptors in mammals. *Endocrine*, *27*, 101–110.

Ekmekcioglu, C. (2006). Melatonin receptors in humans: Biological role and clinical relevance. *Biomedicine and Pharmacotherapy*, *60*, 97–108.

Garfinkel, D., Zisapel, N., Wainstein, J., & Laudon, M. (1999). Facilitation of benzodiazepine discontinuation by melatonin: A new clinical approach. *Archives of Internal Medicine*, *159*, 2456–2460.

Golombek, D. A., Martini, M., & Cardinali, D. P. (1993). Melatonin as an anxiolytic in rats: Time dependence and interaction with the central GABAergic system. *European Journal of Pharmacology*, *237*, 231–236.

Golus, P., & King, M. G. (1981). The effects of melatonin on open field behavior. *Pharmacology Biochemistry and Behavior*, *15*, 883–885.

Gorfine, T., & Zisapel, N. (2007). Melatonin and the human hippocampus, a time dependent interplay. *Journal of Pineal Research*, *43*, 80–86.

Gross, P., Nourse, R., & Wasser, T. (2009). Ramelteon for insomnia symptoms in a community sample of adults with generalized anxiety disorder: An open label study. *Journal of Clinical Sleep Medicine*, *5*, 28–33.

Hayes, J. P., VanElzakker, M. B., & Shin, L. M. (2012). Emotion and cognition interactions in PTSD: A review of neurocognitive and neuroimaging studies. *Frontiers in Integrative Neuroscience*, *6*, 89.

Heldt, S. A., Stanek, L., Chhatwal, J. P., & Ressler, K. J. (2007). Hippocampus-specific deletion of BDNF in adult mice impairs spatial memory and extinction of aversive memories. *Molecular Psychiatry*, *12*, 656–670.

Hofmann, S. G., Sawyer, A. T., Korte, K. J., & Smits, J. A. J. (2009). Is it beneficial to add pharmacotherapy to cognitive-behavioral therapy when treating anxiety disorders? *International Journal of Cognitive Therapy*, *2*(2), 160–175.

Hofmann, S. G., & Smits, J. A. J. (2008). Cognitive-behavioral therapy for adult anxiety disorders: A meta-analysis of randomized placebo-controlled trials. *The Journal of Clinical Psychiatry*, *69*(4), 621–632.

Huang, F., Yang, Z., Liu, X., & Li, C. Q. (2014). Melatonin facilitates extinction, but not acquisition or expression, of conditional cued fear in rats. *BMC Neuroscience*, *15*, 86.

Jovanovic, T., & Ressler, K. J. (2010). How the neurocircuitry and genetics of fear inhibition may inform our understanding of PTSD. *The American Journal of Psychiatry*, *167*(6), 648–662.

Khezri, M. B., & Merate, H. (2013). The effects of melatonin on anxiety and pain scores of patients, intraocular pressure, and operating conditions during cataract surgery under topical anesthesia. *Indian Journal of Ophthalmology*, *61*, 319–324.

Kong, X., Li, X., Cai, Z., Yang, N., Liu, Y., Shu, J., et al. (2008). Melatonin regulates the viability and differentiation of rat midbrain neural stem cells. *Cellular and Molecular Neurobiology*, *28*, 569–579.

Kovács, G. L., Gajari, I., Telegdy, G., & Lissak, K. (1974). Effect of melatonin and pinealectomy on avoidance and exploratory activity in the rat. *Physiology & Behavior*, *13*, 349–355.

Lai, C. S., Franke, T. F., & Gan, W. B. (2012). Opposite effects of fear conditioning and extinction on dendritic spine remodelling. *Nature, 483*(7387), 87–91.

Larson, J., Jessen, R. E., Uz, T., Arslan, A. D., Kurtuncu, M., Imbesi, M., et al. (2006). Impaired hippocampal long-term potentiation in melatonin MT2 receptor-deficient mice. *Neuroscience Letters, 393*, 23–26.

Lattal, K. M., Radulovic, J., & Lukowiak, K. (2006). Extinction: Does it or doesn't it? The requirement of altered gene activity and new protein synthesis. *Biological Psychiatry, 60*, 344–351.

LeDoux, J. E. (2000). Emotion circuits in the brain. *Annual Review of Neuroscience, 23*, 155–184.

Mantovani, M., Kaster, M. P., Pertile, R., Calixto, J. B., Rodrigues, A. L., & Santos, A. R. (2006). Mechanisms involved in the antinociception caused by melatonin in mice. *Journal of Pineal Research, 41*(4), 382–389.

Maren, S. (2001). Neurobiology of Pavlovian fear conditioning. *Annual Review of Neuroscience, 24*, 897–931.

Maren, S., & Quirk, G. J. (2004). Neuronal signalling of fear memory. *Nature Reviews. Neuroscience, 5*, 844–852.

Marsicano, G., Wotjak, C. T., Azad, S. C., Bisogno, T., Rammes, G., Cascio, M. G., et al. (2002). The endogenous cannabinoid system controls extinction of aversive memories. *Nature, 418*(6897), 530–534.

McFarlane, A. C., Barton, C. A., Briggs, N., & Kennaway, D. J. (2010). The relationship between urinary melatonin metabolite excretion and posttraumatic symptoms following traumatic injury. *Journal of Affective Disorders, 127*(1–3), 365–369.

Milad, M. R., Rauch, S. L., Pitman, R. K., & Quirk, G. J. (2006). Fear extinction in rats: Implications for human brain imaging and anxiety disorders. *Biological Psychology, 73*, 61–71.

Millan, M. J., Brocco, M., Gobert, A. M., & Dekeyne, A. (2005). Anxiolytic properties of agomelatine, an antidepressant with melatoninergic and serotonergic properties: Role of 5-HT2C receptor blockade. *Psychopharmacology, 177*, 448–458.

Ochoa-Sanchez, R., Rainer, Q., Comai, S., Spadoni, G., Bedini, A., Rivara, S., et al. (2012). Anxiolytic effects of the melatonin MT(2) receptor partial agonist UCM765: Comparison with melatonin and diazepam. *Progress in Neuropsychopharmacology and Biological Psychiatry, 39*(2), 318–325.

Pandi-Perumal, S. R., Trakht, I., Srinivasan, V., Spence, D. W., Maestroni, G. J., Zisapel, N., et al. (2008). Physiological effects of melatonin: Role of melatonin receptors and signal transduction pathways. *Progress in Neurobiology, 85*(3), 335–353.

Papp, M., Litwa, E., Gruca, P., & Mocaër, E. (2006). Anxiolytic-like activity of agomelatine and melatonin in three animal models of anxiety. *Behavioural Pharmacology, 17*, 9–18.

Pierrefiche, G., Zerbib, R., & Laborit, H. (1993). Anxiolytic activity of melatonin in mice: Involvement of benzodiazepine receptors. *Research Communications in Chemical Pathology and Pharmacology, 82*(2), 131–142.

Reiter, R. J., & Korkmaz, A. (2008). Clinical aspects of melatonin. *Saudi Medical Journal, 29*(11), 1537–1547.

Reppert, S. M., Weaver, D. R., & Godson, C. (1996). Melatonin receptors step into the light: Cloning and classification of subtypes. *Trends in Pharmacological Sciences, 17*(3), 100–102.

Rosenstein, R. E., & Cardinali, D. P. (1986). Melatonin increases in vivo GABA accumulation in rat hypothalamus, cerebellum, cerebral cortex and pineal gland. *Brain Research, 398*(2), 403–406.

Rothbaum, B. O., & Davis, M. (2003). Applying learning principles to the treatment of post-trauma reactions. *The Annals of the New York Academy of Sciences, 1008*, 112–121.

Savaskan, E., Ayoub, M. A., Ravid, R., Angeloni, D., Franchini, F., Meier, F., et al. (2005). Reduced hippocampal MT2 melatonin receptor expression in Alzheimer's disease. *Journal of Pineal Research*, *38*, 10–16.

Shin, L. M., & Handwerger, K. (2009). Is posttraumatic stress disorder a stress-induced fear circuitry disorder? *Journal of Traumatic Stress*, *22*(5), 409–415.

Sotres-Bayon, F., Bush, D. E., & LeDoux, J. E. (2004). Emotional perseveration: An update on prefrontal-amygdala interactions in fear extinction. *Learning and Memory*, *11*, 525–535.

Sotres-Bayon, F., Bush, D. E., & LeDoux, J. E. (2007). Acquisition of fear extinction requires activation of NR2B-containing NMDA receptors in the lateral amygdala. *Neuropsychopharmacology*, *32*, 1929–1940.

Srinivasan, V., Smits, M., Spence, W., Lowe, A. D., Kayumov, L., Pandi-Perumal, S. R., et al. (2006). Melatonin in mood disorders. *The World Journal of Biological Psychiatry*, *7*, 138–151.

Stein, D., Ahokas, A., & de Bodinat, C. (2008). Efficacy of agomelatine in generalized anxiety disorder: A randomized, double-blind, placebo-controlled study. *Journal of Clinical Psychopharmacology*, *28*, 561–566.

Tian, S. W., Laudon, M., Han, L., Gao, J., Huang, F. L., Yang, Y. F., et al. (2010). Antidepressant- and anxiolytic effects of the novel melatonin agonist Neu-P11 in rodent models. *Acta Pharmacologica Sinica*, *31*(7), 775–783.

Uz, T., Arslan, A. D., Kurtuncu, M., Imbesi, M., Akhisaroglu, M., Dwivedi, Y., et al. (2005). The regional and cellular expression profile of the melatonin receptor MT1 in the central dopaminergic system. *Brain Research. Molecular Brain Research*, *136*, 45–53.

VanElzakker, M. B., Dahlgren, M. K., Davis, F. C., Dubois, S., & Shin, L. M. (2014). From Pavlov to PTSD: The extinction of conditioned fear in rodents, humans, and anxiety disorders. *Neurobiology of Learning and Memory*, *113*, 3–18.

Vimala, P. V., Bhutada, P. S., & Patel, F. R. (2014). Therapeutic potential of agomelatine in epilepsy and epileptic complications. *Medical Hypotheses*, *82*(1), 105–110.

Wang, L. M., Suthana, N. A., Chaudhury, D., Weaver, D. R., & Colwell, C. S. (2005). Melatonin inhibits hippocampal long-term potentiation. *The European Journal of Neuroscience*, *22*, 2231–2237.

Wu, Y. H., Zhou, J. N., Balesar, R., Unmehopa, U., Bao, A., Jockers, R., et al. (2006). Distribution of MT1 melatonin receptor immunoreactivity in the human hypothalamus and pituitary gland: Colocalization of MT1 with vasopressin, oxytocin, and corticotropin-releasing hormone. *The Journal of Comparative Neurology*, *499*, 897–910.

Yang, Z., Li, C. Q., & Huang, F. (2013). Melatonin impaired acquisition but not expression of contextual fear in rats. *Neuroscience Letters*, *552*, 10–14.

CHAPTER TEN

The Role of Elements in Anxiety

K. Młyniec[1], M. Gaweł, U. Doboszewska, G. Starowicz, G. Nowak
Faculty of Pharmacy, Jagiellonian University Medical College, Krakow, Poland
[1]Coresponding author: e-mail address: katarzyna.mlyniec@uj.edu.pl

Contents

Abstract

Elements (bioelements) are necessary factors required for the physiological function of organisms. They are critically involved in fundamental processes of life. Extra- and intracellular message and metabolic pathway factors as well as structural components include one or many elements in their functional structure. Recent years have seen an intensification in terms of knowledge gained about the roles of elements in anxiety disorders. In this chapter we present a review of the most important current data concerning the involvement of zinc, magnesium, copper, lithium, iron, and manganese, and their deficiency, in the pathophysiology and treatment of anxiety.

1. INTRODUCTION

Appropriate levels of essential elements are required for living organisms. These elements are crucial in numerous physiological processes. Their excess or deficiency leads to various pathological conditions, including

Vitamins and Hormones, Volume 103
ISSN 0083-6729
http://dx.doi.org/10.1016/bs.vh.2016.09.002

psychiatric disorders. As shown by Islam et al. (2013), trace elements play an important role in anxiety and may be a prognostic marker of that illness as well as a tool for successful treatment. There is evidence that dietary deficiency in some elements leads to the development of anxiety-like behavior. Preclinical as well as clinical studies have shown that supplementation of essential elements enhanced anxiolytic effects. Moreover, some of them administered alone reduced symptoms of anxiety. Possible anxiolytic-like properties may result from the regulatory function of trace elements in the body, mainly in the central nervous system, including regions that are involved in anxiety. It has been found that trace elements can modulate neurotransmission. This includes serotonergic, noradrenergic, dopaminergic, GABAergic, and glutamatergic transmission. Deficiency of some elements manifests in excessive glutamate release and hyperactivation of the hypothalamic–pituitary–adrenal (HPA) axis, both observed in the pathophysiology of anxiety.

2. ANIMAL MODELS OF ANXIETY

Translating human psychiatric symptoms into animal tests remains a challenge for science in the present day, especially in view of the constantly evolving theories of anxiety and depression. Often failing to reproduce the complexity of human disorders, lacking in specificity or showing unwanted selectivity to particular neuromediatory systems, animal tests mimic, rather than model, real psychiatric states. Nevertheless, more than 30 models of anxiety, including trials based on physiological or endocrine responses to stress and behavioral models (conditioned and unconditioned), have proven to be crucial for preclinical research on the neurobiology of psychiatric disorders (Kalueff & Nutt, 2007; Rodgers, Cao, Dalvi, & Holmes, 1997). Some of them are presented in Table 1.

Most of the unconditioned responses induced by exposure to a new environment involve typical approach–avoidance conflict, simultaneously evoking both fear and curiosity in animals. The elevated plus maze (EPM) is undoubtedly one of the most popular animal models of anxiety. The test is based on the natural tendency of rodents to explore novel environments and their aversion to open spaces. The plus-shaped apparatus is raised 40–70 cm above the floor and is composed of two open and two enclosed arms. The enclosed-arm entries reflect the animal's locomotor activity, while the percentage of entries and time spent in the open

Table 1 The Table Summarizes the Most Frequently Used Behavioral Rodent Tests

Unconditioned Responses	**Conditioned Responses**
Elevated plus maze (zero/T maze)	Four-plate test
Light–dark exploration	Geller–Seifter conflict
Open-field	Vogel conflict
Hole-board	Two-way avoidance conditioning
Social interaction	Acoustic startle
Hyponeophagia	Fear-potentiated startle
Predator-based models	

Unconditioned responses include ethologically based paradigms and involve spontaneous or natural reactions (innate fear/avoidance behavior) to stress stimuli, whereas conditioned ones entail animals' responses to stimuli associated with pain or discomfort (learned/punished responses).

arms represent measures of anxiety. Animals exhibiting anxiolytic activity spend increased time in the open arms and have an increased number of entries into the open arms (Flint, 2003; Rodgers et al., 1997).

The light/dark model (L/D) is conceptually similar to the EPM. The test is based on the conflict between a rodent's spontaneous exploratory behavior and its innate aversion to brightly illuminated areas. The model allows mice to freely explore two interconnected chambers, different in color (white/black) and in lighting (light/dark). The most commonly used parameters to assess the anxiolytic activity are the number of transitions between the two compartments and the time spent in the light chamber (this parameter provides the most consistent dose–effect responses with different compounds). Other variables, such as latency before entering/exiting the dark chamber and movement in each compartment, are also measured in the test. A rodent exhibiting a decrease in the time spent in the lit compartment is believed to possess an increased level of anxiety. Compounds with anxiolytic properties increase the number of transitions between the two chambers and decrease the number of attempts at entry in the bright area (mice pass into the lit chamber without hesitation) (Bourin, 2015; Finn, Rutledge-Gorman, & Crabbe, 2003).

Although EPM and L/D models seem to be better analogs of human anxiety and permit a rapid screening of anxiety-modulating drugs without former training, food/water deprivation or shock administration, they cannot directly discriminate between decreased anxiety-related avoidance and an increased novelty-seeking approach. Moreover, ethological models present individual differences and variable behavioral baseline levels. In contrast to unconditioned models, most of the conditioned ones require former training and pairing of a primary neutral stimulus with an aversive one. During the test session, animals respond to conditioned stimuli. According to Handley, the neuronal control of anxiety can differ depending on whether the interpretation of an aversive signal is innate or learned (Handley, 1995).

The apparatus in one of the conditioned models (i.e., the four-plate [FP] test) consists of a test box with the floor made of four identical rectangular metal plates separated from one another by a gap. The test is based on the suppression of the rodent's exploratory behavior by the delivery of a mild electric foot shock every time the animal crosses from one metal plate to another. Anxiolytics such as benzodiazepines (BZDs) increase the number of punished plate crossings performed by animals (Bourin, 2015).

In the Vogel conflict test (VCT), thirsty animals are punished with mild electric shocks while drinking water. This leads to a significant reduction of water consumption in water-deprived control animals. Whatever the differences in water deprivation and punishment schedule, anxiolytic drugs increase the number of punished licks and reestablish drinking responses (Bourin, Petit-Demoulière, Dhonnchadha, & Hascöet, 2007; Campos, Fogaça, Aguiar, & Guimarães, 2013; Razafsha et al., 2013).

Conditioned fear can also be evaluated in the fear-potentiated startle (FPS) test. The paradigm requires an initial learning session in which a neutral stimulus (conditioned stimulus, CS), such as light, is paired with an aversive stimulus, such as an electric shock to the foot. During the test session, reexposure to the CS evokes a conditioned fear response in animals, which can be characterized by freezing (complete immobility), a reflex expression (characterized by FPS), autonomic (increase in heart rate and arterial pressure), and endocrine (stress-related hormone release) responses. Many studies have demonstrated that anxiolytics such as BZDs reduce the freezing behavior and the startle amplitude with no change in the baseline level of the startle (Bourin et al., 2007; Campos et al., 2013; Razafsha et al., 2013).

Despite the difficulties in developing functional and validated models of anxiety, when they are developed, these are substantial models for preclinical research on the neurobiology of psychiatric disorders and on the targets for therapy. Some of them may be more relevant for one type of disorder than for another, as none of the models can mirror the multiple and diverse pathomechanisms of anxiety.

3. ZINC

Zinc is one of the most important trace elements in living organisms. Zinc plays an important regulatory role in the function of transcriptional and structural as well as enzymatic proteins. Its daily requirement depends on age and gender. Women require 8–9 mg, and men 11 mg. The main sources of this element include oysters, as well as red meat and whole grains (Fig. 1). Zinc is essential for the development and function of many tissues, including the brain. Deficiency of this element is linked to many diseases, including psychiatric illnesses such as depression and anxiety (Maes et al., 1997; Młyniec et al., 2012; Swardfager et al., 2013). Zinc has been found to produce both antidepressant- and anxiolytic-like effects. Acute administration of zinc hydroaspartate at doses of 32.5 and 16.25 mg/kg increased the percentage of entries into the open arms of the EPM in rats (Partyka et al., 2011). Moreover, the authors have investigated potential anxiolytic properties of zinc in other commonly used tests, such as the FP test and stress-induced hyperthermia in mice. They found a significantly increased number of punished crossings and attenuated stress-induced hyperthermia in mice following acute zinc hydroaspartate treatment. In other studies, 7-day zinc administration at doses of 15 and 20 mg/kg significantly increased the

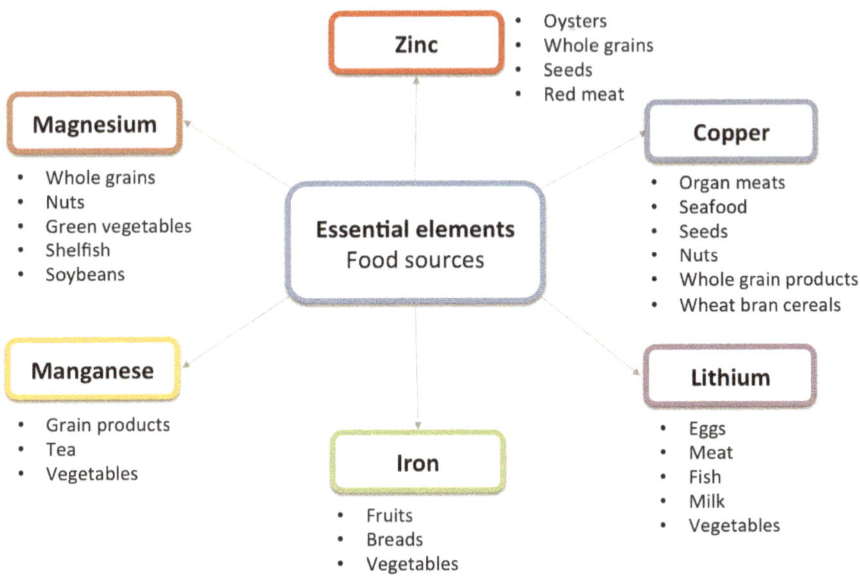

Fig. 1 Food sources of selected essential elements.

number of open-arm entries as well as the amount of time spent in the open arms in the EPM test (Joshi, Akhtar, Najmi, Khuroo, & Goswami, 2012). Another salt of zinc—namely, zinc chloride—at doses of 20 and 30 mg/kg showed an anxiolytic-like effect in the VCT (Abdel-Maksoud, Abdel-hamid, & Moustafa, 2012).

It is postulated that the possible anxiolytic-like effect of zinc may result from its influence on glutamatergic and GABAergic transmission. Zinc modulates several important receptors, such as GABA, NMDA, and GPR39. Zinc attenuates glutamatergic transmission via the binding site located at the NMDA receptor and plays a part in maintaining homeostasis between excitatory and inhibitory neurotransmission. A study by Abdel-Maksoud et al. (2012) has shown a synergistic effect of zinc and NMDA antagonists. CGP 37849 administered with zinc, both in noneffective doses, produced anxiolytic-like activity as measured by the VCT. The authors found similar results following MK-801 and zinc treatment at doses that were inactive in the VCT. Moreover, they found that anxiolytic-like effects evoked by zinc per se were antagonized by NMDA administration at a dose of 75 mg/kg. These results show that the anxiolytic properties of zinc may be due to glutamatergic inhibition.

3.1 Zinc Deficiency

Zinc deficiency can lead to various conditions, including psychiatric disturbances such as depression and anxiety (Grønli, Kvamme, Friborg, & Wynn, 2013). Preclinical studies using a zinc-deficient diet showed that lack of zinc has an important influence on anxiogenic-like behavior (Młyniec et al., 2014). Two weeks of zinc deprivation (zinc-deficient diet: 2.7 mg Zn/kg; zinc-adequate diet: 44 mg Zn/kg) in rats caused anxiety-like behavior as measured by the open-field test. Decreased frequency of line crossing as well as decreased time spent grooming were observed. Moreover, zinc-deficient rats showed decreased time spent in the open arms of the plus maze test (Takeda, Tamano, Kan, Itoh, & Oku, 2007). A possible explanation of the observed behavior following a 2-week diet low in zinc, according to the authors, is based on hyperactivity of the HPA axis (Takeda et al., 2007). In their study, significantly higher corticosterone levels were found in zinc-deprived rats. Similar conclusions were drawn from a study by Chu, Mouat, Harris, Coffield, and Grider (2003), in which thigmotaxia related to anxiety-like behavior and high corticosterone levels were observed in the zinc-deficient group. From a neurobiological point of view, this may be correlated to glutamatergic neurotransmission, since elevated glucocorticoids inhibit activity of glutamatergic transporters EAAT1, leading to glutamate accumulation in the synaptic cleft of the hippocampus. Abnormal glutamate release in the absence of zinc has been found in young rats. Zinc restriction causes excessive glutamate release in the hippocampus of these animals, probably due to calcium dyshomeostasis (Takeda, Tamano, Itoh, & Oku, 2008).

In the study by Whittle, Lubec, and Singewald (2009), mice received a diet containing 40% of the recommended daily intake requirement for zinc. After 3 weeks of administration of this diet, mice displayed enhanced latencies to eat in the novelty suppressed feeding test, an effect that was reversed by chronic desipramine and *Hypericum perforatum* treatments. For the zinc-deficient group vs the control group, no changes in the light/dark test were observed, and there was only a tendency for an anxiogenic effect in the EPM. However, the authors found significant molecular changes in the amygdala of zinc-deficient animals, which is strongly correlated to psychiatric disorders such as depression and anxiety. Zinc deficiency caused the hyperactivation of neurons in the basolateral amygdala, a phenomenon that was normalized following desipramine treatment (Whittle et al., 2009).

3.2 GPR39 Zinc Receptor

Zinc was found to modulate not only the GABA- and glutamatergic receptors, but also metabotropic GPR39, which belongs to the ghrelin family. GPR39, known as the zinc receptor, is widely expressed in the central nervous system, including the hippocampus and the amygdala, both of which are directly linked to anxiety behavior (Jackson, Nothacker, & Civelli, 2006; Młyniec, Singewald, Holst, & Nowak, 2015). In the absence of GPR39, anxiety-like behavior is observed. GPR39 knockout mice showed a significantly decreased number of entries into the lit compartment, as well as a significantly decreased number of line crossings, and significantly increased immobility time, all as measured by the light/dark test (Młyniec, Budziszewska, Holst, Ostachowicz, & Nowak, 2015). In this study some tendencies were observed for decreased time spent in the lit compartment and increased freezing time. GPR39 may have an influence on GABAergic transmission, and this is a possible explanation of the link between GPR39 and anxiety. Activation of GPR39 significantly increases the expression and activity of KCC2 (K^+/Cl^- cotransporter 2) in the CA3 region of the hippocampus (Chorin et al., 2011). This strong inhibitory effect on transmission occurs by induction of the hyperpolarizing shift in the $GABA_A$ reversal potential. An influence of GPR39 activation on glutamatergic transmission has also been found. GPR39 is required for triggering the synthesis of 2-arachidonoyglycerol (2-AG). Interestingly, no synthesis of 2-AG has been found in mice lacking GPR39 (Perez-Rosello et al., 2013). This means that excesseive glutamate release may be observed under GPR39-deficient conditions. Administration of NMDA for 10 days caused a significant reduction of hippocampal GPR39 expression (Młyniec, Gaweł, Librowski, et al., 2015). On the other hand, no changes in the hippocampal glutamate levels were observed in GPR39 knockout mice (Młyniec, Gaweł, Librowski, et al., 2015), so additional studies are required here.

4. MAGNESIUM

Magnesium (Mg) is the fourth most abundant essential element in the body and the second most abundant intracellular cation. An adult body contains about 24 g of Mg, with approximately 50% present in the bones. The rest is located in the intracellular space of soft tissues. Only 1% of total body Mg is found in the blood (Jahnen-Dechent & Ketteler, 2012). Intracellular Mg is needed for more than 300 biochemical reactions, many of which require energy production. It plays an important role in protein and nucleic acid synthesis and is essential in maintaining Ca, K, and Na homeostasis. It is necessary for proper muscle and nerve function, normal heart rhythm, and bone formation (Jahnen-Dechent & Ketteler, 2012). Dietary Mg is absorbed mainly in the small intestine and is excreted through the kidneys (Seo & Park, 2008). The recommended daily allowance (RDA) is 300 mg (European Commission, 1990). In healthy adults, plasma Mg level ranges from 0.7 to 1 mmol/L (1.5–2 mEq/L; 17–24 mg/L) (Williamson, Snyder, & Wallach, 2011). Therapeutically, Mg can be administered p.o., i.v., or i.m. Commonly used Mg compounds include Mg-aspartate, -chloride, -gluconate, -lactate, -sulfate, and -oxide (Grober, Schmidt, & Kisters, 2015). Therapy with Mg was found to be beneficial in severe hypomagnesemia, pregnancy toxemia (to prevent and control seizures associated with preeclampsia and eclampsia), asthma, arrhythmias (torsades de pointes, atrial fibrillation), migraine, etc. (Kaye & O'Sullivan, 2002; Tong & Rude, 2005). There is also a large amount of evidence for the role of Mg in the pathophysiology and treatment of neurological diseases (migraine, pain, traumatic brain injury, cerebral ischemia, stroke) and psychiatric diseases (depression, psychoses, hyperactivity, and autism in children, addiction, as well as anxiety) (Vink & Nechifor, 2011).

4.1 Anxiolytic Activity of Magnesium

Magnesium antagonizes the NMDA receptor (Zhu & Auerbach, 2001) and exhibits $GABA_A$ receptor agonist activity (Schwartz, Wagner, Yu, & Martin, 1994). These mechanisms of action may suggest its anxiolytic activity, which has been confirmed in preclinical paradigms. Mg when administered in an acute intraperitoneal (i.p.) manner at doses of 20 and 30 mg/kg, but not at 10 mg/kg, was shown to be active in the EPM test in mice (Poleszak et al., 2004). In addition, both chronic and acute i.p. administration of Mg-hydroaspartate (30 mg Mg/kg) increased time spent in the open arms in this test, which was associated with an increase in serum magnesium concentration (Poleszak et al., 2004). The anxiolytic-like effect of Mg (20 mg/kg) in this test was antagonized by a $GABA_A$ receptor antagonist at the benzodiazepine site, flumazenil (10 mg/kg), whereas joint administration of the noneffective doses of Mg (10 mg/kg) and BZDs (diazepam (0.5 mg/kg) or chlordiazepoxide (2 mg/kg)) produced synergistic effects (increased time spent in the open arms and increased number of entries into the open arms), indicating that $GABA_A$ receptors are involved in the anxiolytic-like effects of magnesium (Poleszak, 2008). Furthermore, the anxiolytic-like effect of Mg (20 mg/kg) in the EPM test was antagonized by D-serine (100 nmol administered intracerebroventricularly in mice), which is an agonist of the glycine site of the NMDA receptor, while the noneffective dose of Mg (10 mg/kg) administered together with low, noneffective doses of NMDA receptor antagonists (the competitive NMDA antagonist, CGP 37849 (0.3 mg/kg); an antagonist of the glycine site, L-701,324 (1 mg/kg); a partial agonist of the glycine site, D-cycloserine (2.5 mg/kg); and the noncompetitive NMDA antagonist, MK-801 (0.05 mg/kg)) induced a significant increase in the percentage of time spent in the open arms of the maze, suggesting that the NMDA receptor also participates in the anxiolytic-like activity of Mg (Poleszak, Wlaz, Wrobel, Fidecka, & Nowak, 2008).

Despite preclinical evidence for the possible role of magnesium in the treatment of anxiety, clinical trials are very limited (Table 2). 200 mg Mg/day plus 50 mg vitamin B6/day were found to reduce anxiety-related premenstrual symptoms (De Souza et al., 2000). Also, a combined treatment of plant extracts of *Crataegus oxyacantha* and *Eschscholtzia californica* and

Table 2 Clinical Trials of Magnesium in Anxiety Disorders

Study Design	Participants	Intervention	Treatment Duration	Outcome	References
Randomized, double-blind, placebo-controlled	44 Elderly subjects (aged 50–70) with cognitive impairment	MMFS-01 (Mg-L-threonate)	12 weeks	Due to strong placebo effects on anxiety, the effects could not be determined	Liu, Weinger, Lu, Xue, and Sadeghpour (2015)
Randomized, double-blind, placebo-controlled	264 Patients (81% female, an average age of 44.6 years) with GAD (according to DSM-III-R criteria) of mild to moderate intensity (total Hamilton anxiety scale score: 16–28)	Sympathyl (plant extracts of *Crataegus oxyacantha* and *Eschscholtzia californica* and Mg)	3 months	Sympathyl was more effective than placebo in treating mild to moderate GAD	Hanus, Lafon, and Mathieu (2004)
Randomized, double-blind, placebo-controlled	44 Women (an average age of 32 years)	(1) 200 mg Mg*, (2) 50 mg vitamin B6, (3) 200 mg Mg* + 50 mg vitamin B6. * as Mg-oxide	one menstrual cycle	200 mg/day Mg + 50 mg/day vitamin B6 reduced anxiety-related premenstrual symptoms	De Souza, Walker, Robinson, and Bolland (2000)

DSM-III-R, Diagnostic and Statistical Manual of Psychiatric Disorders Third Edition Revision; GAD, generalized anxiety disorder.

magnesium has been shown to be more effective than placebo in patients with mild to moderate generalized anxiety disorder GAD (Hanus et al., 2004).

Although Mg administration which induces anxiolytic-like activity led to an increase in serum magnesium concentration (Poleszak et al., 2004), no differences were found in either the erythrocyte or brain Mg levels following an increase in serum Mg concentration evoked by acute and subchronic i.p. injections of Mg-chloride in rats (Wlaz, Serefko, Szopa, & Poleszak, 2016). In contrast, Mg-L-threonate, which is characterized by a higher bioavailability than Mg-chloride, -citrate, -glycinate, and -gluconate, administered chronically via the oral route, was found to increase the cerebrospinal fluid (CSF) Mg concentration (Slutsky et al., 2010). Moreover, an increase in CSF Mg concentration was associated with enhancement of synaptic plasticity in rats (Slutsky et al., 2010). The compound also enhanced fear memory extinction and prevented fear memory overgeneralization, leading to a reduction of anxiety in rats (Abumaria, Luo, Ahn, & Liu, 2013; Abumaria et al., 2011). Despite the fact that a recent clinical study involving elderly subjects with cognitive impairments could not determine the effects of Mg-L-threonate on anxiety due to strong placebo effects (Liu et al., 2015), elevating the brain magnesium level via p.o. administration of compounds with higher bioavailability than those that are commonly used may become a promising strategy for the treatment of anxiety.

The data on serum magnesium concentrations in patients with anxiety disorders are also limited (Table 3). Decreased serum Mg concentration in OCD patients compared with the control group has been observed (Shohag et al., 2012), but no differences were demonstrated between patients with GAD (Islam et al., 2013) or with panic disorder (Nahar et al., 2010) and healthy volunteers. Moreover, there was no association between anxiety scores and serum Ca/Mg ratio in adult women without psychiatric disorders (Jung et al., 2010), while university science students experienced an increase in anxiety as a consequence of exams, which was associated with increased urinary Mg concentration (Grases et al., 2006).

4.2 Magnesium Deficiency

Likewise, there is a paucity of data relating to magnesium intake in humans and anxiety scores. In a recent large, cross-sectional, population-based sample of women, no association was found between the intake of magnesium and anxiety disorders (Jacka, Maes, Pasco, Williams, & Berk, 2012).

Table 3 Levels of Magnesium in Anxiety Disorders

Participants	Mg Level	References
50 GAD (according to DSM-IV criteria) patients (39 males and 11 females with an average age of 29 ± 3 years) and 51 sex-, education-, and age-matched healthy volunteers (41 males and 10 females with an average age of 28 ± 8 years)	No differences in serum Mg concentration between the GAD patients and control group	Islam et al. (2013)
48 OCD (according to DSM-IV criteria) patients (42 males and 6 females with an average age of 24.71 ± 4.66 years) and 48 sex-, socioeconomic, and age-matched healthy volunteers (42 males and 6 females with an average age of 21.33 ± 1.01 years)	Decreased serum Mg concentration in OCD patients compared to the control group	Shohag, Ullah, Qusar, Rahman, and Hasnat (2012)
54 Panic disorder (according to DSM-IV criteria) patients (44 males and 10 females with an average age of 29 ± 7 years) and 52 sex-, education-, and age-matched healthy volunteers (42 males and 10 females with an average age of 30 ± 6 years)	No differences in serum Mg concentration between the panic disorder patients and control group	Nahar et al. (2010)
112 Adult women (with a mean of 45.11 ± 12.16 years) without psychiatric disorders	No association between serum Ca/Mg ratio and anxiety scores according to Korean version of the Hospital Anxiety and Depression Scale (Korean-HADS)	Jung, Ock, Chung, and Song (2010)
35 Volunteers (students) (12 males and 23 females aged 18–20 years)	Anxiety (according to STAI (State Treats Anxiety Inventory) questionnaire to evaluate state (STAIE) and trait (STAIR) anxiety) associated with increase in Mg in 24 h urine collected during exams	Grases et al. (2006)

DSM-IV, Diagnostic and Statistical Manual of Psychiatric Disorders Fourth Edition; GAD, generalized anxiety disorder; OCD, obsessive–compulsive disorder.

However, studies in rodents point to a relationship between a reduction of magnesium in the diet and the occurrence of anxiety-related behavior, suggesting the usefulness of an Mg-deficient diet as a model of anxiety at a preclinical level. Most of these studies were conducted on mice, for which the daily minimum magnesium requirement is 500 mg/kg of food (National Research Council (US) Subcommittee on Laboratory Animal Nutrition, 1995). The control diets contained mainly 0.2% of Mg, which is four times more than the daily minimum magnesium requirement. The amounts of magnesium present in Mg-deficient diets, the duration of the diets, as well as the outcomes of anxiety-related tests are shown in Table 4. The differences in severity and duration of the diets may account for the differences in the outcomes of the various behavioral tests.

A 3-week Mg-deficient diet containing 0.005% Mg, which corresponds to 10% of the daily minimum requirement, was shown to produce anxiety-like behavior in C57Bl/6N and BALB/c mice in a battery of established tests for anxiety (i.e., open-field, light/dark, hyponeophagia, and open-arm exposure tests) (Sartori et al., 2012; Singewald et al., 2004). Moreover, chronic administration of *H. perforatum* extract, but not of desipramine, prevented the anxiety-like behavior of the Mg-deficient mice in the open-field and light/dark tests (Singewald et al., 2004). The anxiety-like behavior of the Mg-deficient mice responded to chronic treatment with desipramine in the hyponeophagia test and to acute diazepam injection in the open-arm exposure test, while chronic paroxetine administration did not prevent enhanced anxiety-related parameters in the Mg-deficient mice in the open-field, light/dark, or hyponeophagia tests (Sartori et al., 2012). Furthermore, the anxiety-like behavior of the Mg-deficient mice was associated with enhanced transcription of the corticotropin releasing hormone in the paraventricular hypothalamic nucleus and with elevated adrenocorticotropic hormone plasma levels, whereas chronic desipramine treatment reversed these abnormalities in the stress axis (Sartori et al., 2012). The results indicate that dietary-induced Mg deficiency may represent a rodent model of anxiety, and this may support the role of Mg in the pathophysiology underlying anxiety disorders.

An Mg-deficient diet was found to change the levels of gut microbiota (Pachikian et al., 2010; Pyndt Jorgensen et al., 2015; Winther et al., 2015). Recent studies have demonstrated an anxiogenic effect (Crumeyrolle-Arias et al., 2014) as well as upregulation of synaptic plasticity-related genes and genes that promote transcription in the amygdala of germ-free mice (Stilling et al., 2015), suggesting that an absence of microbiota leads to increased

Table 4 Summary of Studies Involving an Mg-Deficient Diet and Anxiety-Like Behavior

Species	Diet	Test	Outcome	References
Male C57BL/6NBomTac mice 8-week old	MgA = 0.2% Mg MgD = 0.02% Mg 6 weeks	Light/dark box (LDB) test	Shorter latency to enter the light compartment of the LDB No differences in the number of entries into the light compartment and time spent in the light compartment	Pyndt Jorgensen et al. (2015)
Male C57BL/6NBomTac mice 8-week old	MgA = 0.2% Mg MgD = 0.02% Mg 6 weeks	Open-field test	No differences in the number of entries into the center of the field No differences in total distance traveled	Winther et al. (2015)
Male C57Bl/6N mice 8-week old	MgA = 0.2% Mg MgD = 0.005% Mg at least 3 weeks (tests were conducted in the same animals)	Stress-induced hyperthermia	Smaller stress-induced rise in body temperature	Sartori, Whittle, Hetzenauer, and Singewald (2012)
		Open-field test	Decreased number of entries into the center of an open field, decreased time spent in the center of an open field, decreased number of rearings No differences in total distance traveled	
		Light–dark test	Increased latency to enter the brightly lit compartment, no differences in the number of entries into the lit arena and time spent in the lit arena, decreased number of rearings	
		Hyponeophagia test	Increased latency to eat a preferred food placed in the center of the testing arena	
Male BALB/c mice 8-week old		Open-arm exposure test	Decreased time spent in the distal compartment	

Male C57Bl/6N mice 8-week old	MgA = 0.05% MgD = sucrose instead of MgO in the MgA diet 1, 2, and 3 weeks	Light–dark test	Increased latency to the first entry into the lit compartment after 2 and 3 weeks of the MgD diet Decreased time spent in the lit compartment after 3 weeks of the MgD diet Decreased number of transitions between the compartments after 2 weeks of the MgD diet	Muroyama et al. (2009)
Rats	MgA = ND MgD = 0.0015% Mg 7 weeks	Open-field test	Decreased number of crossed squares, number of standing on hind paws, number of visiting in central squares	Spasov, Iezhitsa, Kharitonova, and Kravchenko (2008)
		Elevated plus maze test	Decreased number of visiting open arms and residence time	
Male C57Bl/6J mice 60 days of age	MgA = 0.1% Mg MgD = 0.003% 10 days	Light–dark test	No differences in latency to entry into the light side, total time spent in the dark chamber and number of transitions	Bardgett, Schultheis, McGill, Richmond, and Wagge (2005)
Male C57Bl/6N mice 8-week old	MgA = ND MgD = 0.005% Mg 3 weeks	Open-field test	Decreased number of entries into the central zone No differences in total distance traveled	Singewald, Sinner, Hetzenauer, Sartori, and Murck (2004)
		Light–dark test	Increased latency to the first entry into the lit compartment Decreased time spent in the lit compartment	

MgA, diet containing adequate amount of Mg; MgD, Mg-deficient diet; ND, no data.

activity in the amygdala, the brain region critically involved in anxiety, highlighting the role of altered gut microbiota in the neurobiology of anxiety disorders. Interestingly, while a 6-week Mg-deficient diet reduced gut microbiota diversity, it did not produce anxiety-like behavior (Pyndt Jorgensen et al., 2015; Winther et al., 2015), but a correlation has been observed between the time spent in the light compartment of the light/dark box test and the gut microbiota profile of mice fed the control diet (Pyndt Jorgensen et al., 2015).

It is of note that the behavioral characterization of A/J and C57BL/6J mice in the modified hole-board (mHB) test was such that strain differences were revealed for anxiety-related parameters. A/J mice had a greater latency before the first board entry and a lower total number of board entries than C57BL/6J mice, which indicates that A/J mice have a high-anxiety phenotype (Laarakker, van Lith, & Ohl, 2011). Moreover, C57BL/6J mice had higher plasma Mg concentrations than A/J mice. In addition, significant positive correlations were found between the total number of board entries and the blood plasma magnesium levels after the second mHB testing; animals that made more board entries had higher plasma magnesium levels. No strain differences were found for brain Mg levels, but for C57BL/6J mice there was a significant correlation between brain magnesium level and latency before the first board entry (Laarakker et al., 2011). The results emphasize strain differences with respect to anxiety-related behavior but further support its relationship with Mg status. However, selective breeding for low and high red blood cell (RBC) magnesium values revealed contradictory results concerning the association between plasma Mg level and anxiety-like behavior. No differences in brain magnesium levels were observed between mice with low and high RBC values, but mice with a low RBC Mg level exhibited a lower plasma magnesium concentration and endured a higher number of electric shocks in the FP test, thus displaying anxiolytic behavior (Henrotte et al., 1997).

Taken together, the above-mentioned studies indicate that an Mg-deficient diet contributes to enhancement of anxiety-like behavior and may be regarded as an animal model of anxiety. However, it should be noted that as well as duration and severity of diet-induced Mg deficiency, strain differences may also have an impact on anxiety-related parameters. The data on the clinical utility of magnesium intervention as a treatment for anxiety disorders are very limited; however, increasing brain Mg levels via oral intake of salts which are effective at enhancing CSF Mg concentration seems to be a promising treatment strategy for anxiety.

5. COPPER

Increased irritability and disinhibition, cognitive impairment, anxiety, and depression are psychiatric features that are common in patients with Wilson's disease (WD), the autosomal recessive disorder of copper (Cu) metabolism (Bandmann, Weiss, & Kaler, 2015; Chan, Cheung, & Au-Yeung, 2005). Due to mutations in the ATP7B gene encoding a copper-transporting P-type ATPase, copper homeostasis undergoes disruption. Reduced biliary excretion and impaired formation of ceruloplasmin increase copper levels and provoke oxidative stress, free radical production, and mitochondrial dysfunction. Metal accumulation in the course of WD manifests itself in psychiatric symptoms, liver damage, kidney and corneal abnormalities, arthritis, and cardiomyopathy (Bandmann et al., 2015; Kaler, 2013; Kodama, Fujisawa, & Bhadhprasit, 2012; Scheiber, Mercer, & Dringen, 2014; Zimbrean & Schilsky, 2014). Compared with other inherited disorders which are, by contrast, characterized by copper deficiency, such as Menkes (kinky hair) disease and occipital horn syndrome, only in WD are psychiatric symptoms, including depression and anxiety, frequent and well documented (Dening & Berrios, 1989; Oder et al., 1991). As shown by Dening and Berrios, 51% of patients with copper overload were assessed as displaying psychopathologic features, whereas 20% had seen a psychiatrist prior to their formal WD diagnosis. According to Akil, psychiatric and behavioral abnormalities are the initial manifestations in two-thirds of WD cases and occur in 30–100% of patients (Akil & Brewer, 1995). The role of copper in anxiety and anxiety-related disorders is therefore noteworthy.

The fact that copper is essential was first demonstrated nearly a century ago in rats in which impaired erythropoiesis due to milk-based nutrition was corrected by the addition to the diet of ash containing copper sulfide (Stern et al., 2007). Since then, mammalian studies have revealed a crucial role of copper in a vast number of bodily physiological processes, such as erythropoiesis, myocardial contractility, myelination of the central nervous system, cholesterol and glucose metabolism, hormone synthesis, and immune function (Schlegel-Zawadzka & Nowak, 2000). Being an essential component of metalloenzymes and cycling between two oxidation states, copper participates in key redox reactions. Required by cuproenzymes, including cytochrome *C* oxidase, lysyl oxidase, and superoxide dismutase (SOD), it regulates cellular respiration, connective tissue formation, and free radical scavenging (Stern et al., 2007). As a component of enzymes such as monoamine oxidase and dopamine β-hydroxylase (DBH), the element also influences brain monoamine systems (Młyniec, Gaweł, Doboszewska, et al., 2015). Low copper intake may weaken neurological function, increase susceptibility to infection, and elevate risk for coronary heart disease and osteoporosis. Impaired brain development due to copper deficiency has been associated mainly with limited cytochrome-*c* oxidase, and limited SOD, DBH, and peptidyl-α-amidating monooxygenase (Prohaska, 2000; Prohaska & Smith, 1982). Reduced activity of these cuproenzymes led to decreased protein and lipid biosynthesis (decreased production of adenine nucleotides), hypo- and demyelination of nervous tissue (lipid peroxidation and membrane degeneration), and delayed maturation of bioactive neurotransmitters and growth hormones (limited amidation of peptides) (Prohaska, 2000). Studies have indicated that diminished DBH levels due to copper deficiency resulted in reduced norepinephrine (NE) and increased dopamine (DA) concentrations in the medulla/pons region of the brain (Pyatskowit & Prohaska, 2007). In mice, a 24% increase in brain DA and a 91% decrease in NE were observed after a Cu-deficient diet (Prohaska & Cox, 1983). Rat studies have demonstrated that NE levels diminished up to 35% following copper deprivation (Crayton & Walsh, 2007). However, no link confirming the catecholaminergic hypothesis related to depressive behaviors was found between copper deficiency and psychiatric symptoms in humans. On the contrary, one of the studies that was conducted postulated the relevance of an elevated serum copper level as a "trait marker" of unipolar depression (Schlegel-Zawadzka, Zieba, Dudek, Zak-Knapik, & Nowak, 1999), and another indicated it as a possible marker for postpartum depression (Crayton & Walsh, 2007). The conclusion then arises that the cause of anxiety-related disorders is more complex and involves other mechanisms besides monoaminergic disturbances.

Copper is certainly needed for human homeostatic maintenance. The human body stores slightly more than 100 mg of copper, half of which is contained in the skeleton and muscle. The highest physiological concentrations of copper are observed in the most energy-demanding organs—i.e., the kidney, liver, brain, and heart. The RDA for adults ranges from 0.9 mg/day (USA) to 1.1 mg/day (EU). It is worth mentioning that reported changes in plasma Cu may not reflect the actual condition of the body. Physiological alterations such as the acute phase of response to infection and inflammation, hormonal perturbations, carcinogenic phenotypes, and smoking may increase the amount of Cu circulating in the blood

(Collins & Klevay, 2011). Moreover, it is known that age and sex influence the serum copper. The mean values of plasma copper concentrations in healthy subjects increase with age (20–69 years), from 0.78 to 0.90 mg Cu/L for men, and from 1.01 to 1.14 mg Cu/L for women, respectively (Schlegel-Zawadzka, Zieba, Dudek, Krośniak, et al., 1999).

It is estimated that Cu concentration in the extracellular space of the brain varies between 0.2 and 1.7 μM, transiently reaching a few hundred micromolar during neurotransmission into some glutamatergic synapses. Copper ions, which are synaptically released at concentrations extending from the low nanomolar to the low micromolar range, may inhibit GABA and NMDA/AMPA receptors, respectively (Salazar-Weber & Smith, 2011). Among the first findings regarding the neuromodulatory role of copper was the study by Doreulee, which demonstrated that copper ions block long-term potentiation in the hippocampus (Doreulee, Yanovsky, & Haas, 1997). More recently, dysregulation of metal homeostasis and resulting alterations in glutamatergic and GABAergic transmission is believed to contribute to Cu toxicity in WD and to play a significant role in the etiology of anxiety and depression. During the past decade a large body of clinical and preclinical evidence supporting the GABAergic theory of anxiety and depression has emerged and overshadowed the previously popular monoamine-deficiency theory. From both human and animal studies, negative modulators of GABA receptors are commonly believed to possess anxiogenic-like effects, whereas positive modulators often exert anxiolytic activity (Kalueff & Nutt, 2007).

The recognition of genetic disorders of copper metabolism such as WD has not only spurred research into copper homeostasis and its biological role but also focused attention on the potential consequences of Cu toxicity. Exposure to high levels of copper causes oxidative damage to biological systems and mainly affects the liver, the kidneys, bone, and the central nervous and immune systems (Stern et al., 2007). Studies have demonstrated a potential association between oxidative stress and the pathogenesis of anxiety. The central nervous system is particularly sensitive to oxidative damage on account of its high metabolic activity, high content of easily oxidized membrane polyunsaturated lipids, and low level of some antioxidant enzymes such as catalase and GSH-peroxidase. Once copper ions are not effectively bound inside the cell, they generate radicals and become toxic (Carrí, Ferri, Cozzolino, Calabrese, & Rotilio, 2003). The importance of this mechanism is exemplified by one of the studies in which oxidative stress-induced anxiety in mice was reversed by inhibition of phosphodiesterase-2, an enzyme indirectly involved in oxidative stress mechanisms (Masood et al., 2008).

Excessive copper intake also leads to interactions with other nutrients. Of greatest significance to anxiety-related disorders is the interaction between copper and zinc (Zn). Most frequently, Zn supplementation, used, for example, in patients with sickle cell anemia, may lead to unwanted hypocupremia. On the other hand, Zn supplements are recommended to treat conditions associated with excessive accumulation of copper in the body such as WD (Bremner & Beattie, 1995). Furthermore, zinc deficiency is often associated with relative copper excess. Various studies have demonstrated decreased zinc and increased copper and ceruloplasmin levels in psychiatric diseases such as schizophrenia and depression (Narang, Gupta, Narang, & Singh, 1991). Russo has shown that plasma Cu levels are significantly elevated, and Zn levels are significantly lowered in individuals with anxiety (both with and without depression) compared with controls. Moreover, in patients with primary anxiety, Cu and Zn levels normalized and perceived symptoms significantly improved following Zn and antioxidant therapy. The finding that antidepressant treatment significantly reduces serum copper levels is noteworthy. It was also shown that the reduction in serum Cu during antidepressant treatment was significantly related to the reduction in anxiety scores (HARS score) (Maes et al., 1997).

It can therefore be concluded that an imbalance in the ratio between serum concentrations of several trace minerals, but not a single elemental concentration, is most likely responsible for the occurrence of anxiety and anxiety-related symptoms. Such interrelationships between elemental ratios in correlation with disease severity were investigated in patients suffering from affective disorders (González-Estecha et al., 2011; Mustak et al., 2008). Most frequently research concerned Zn/Cu (Maes et al., 1997; Russo, 2011), Fe/Cu (Collins, Prohaska, & Knutson, 2010), or Mg/Cu (Siwek, Wróbel, Dudek, Nowak, & Zieba, 2005; Zieba, Kata, Dudek, Schlegel-Zawadzka, & Nowak, 2000) ratios in serum samples. Mustak analyzed 11 trace elements (Na, K, S, Ca, Mg, P, Cu, Fe, Zn, Mn, and Al) in patients suffering from three types of bipolar disorder (I, II, and V). The obtained results revealed decreased Zn and Fe together with increased Cu and Al concentrations in all types of bipolar groups compared with controls (Mustak et al., 2008). On the basis of the studies conducted, it was affirmed that both copper deficiency and copper excess may cause adverse health effects, although high rather than low copper levels are associated with anxiety and anxiety-related disorders in humans.

6. LITHIUM

Lithium is commonly used in the treatment of bipolar disorder, although understanding its therapeutic effects remains unclear. Some studies report that Li may directly or indirectly target glycogen synthase kinase 3 (GSK3) (Freland & Beaulieu, 2012) (O'Brien et al., 2011). Inhibition of GSK3 was reported to produce antidepressive-like behavior in the mouse forced swim test (Kaidanovich-Beilin, Milman, Weizman, Pick, & Eldar-Finkelman, 2004). A study by Mines et al. which used *Fmr1* knockout mice, an animal autism model, found that deregulation of GSK3 inhibition leads to impairment of some social behaviors. In this case, lithium administration improved sociability and reduced anxiety-like-behaviors in *Fmr1* knockouts (Mines, Yuskaitis, King, Beurel, & Jope, 2010).

Among bipolar patients, comorbid anxiety disorder is a prevalent condition which indicates more severe clinical presentation and increased suicidality (Simon et al., 2004). Bipolar patients with anxiety also have a poorer response to lithium treatment (Feske et al., 2000; Freeman, Freeman, & McElroy, 2002). The existing literature on lithium's association with anxiety disorders provides a limited amount of data. The case study of Gupta et al. reported lithium carbonate to be effective in the improvement of suicidality in posttraumatic stress disorder (PTSD) patients (Gupta & Knapp, 2013). Another case study, by Forster et al., showed lithium to be a useful treatment in the regulation of impulse control in PTSD (Forster, Schoenfeld, Marmar, & Lang, 1995).

Chronic administration of lithium in a group of preadolescent rats was found to cause increases in anxiety-like behavior in the EPM and open-field test. However, in the FPS test there were no significant differences in the conditioned response between control and Li-treated animals. Lithium administration also caused no impairment of spatial memory in the Morris water maze test. The same study found that Li administration leads to changes in the expression of genes in the amygdala that are involved in neuronal rearrangements (Youngs et al., 2006). Another animal research study found that chronic lithium treatment inhibits amphetamine-induced risk-taking behavior in rats. This effect may be explained by the increased anxiety that follows lithium treatment (Zhou et al., 2015). However, human population studies do not indicate a link between lithium and treatment-induced anxiety.

7. IRON

Among the diverse biological functions of iron, there is a growing body of literature that recognizes its importance for neurological functioning. Fe plays an important role in myelination as in the brain it is mostly prevalent in oligodendrocytes (Connor & Menzies, 1996). It is also essential for the metabolism of neurotransmitters such as dopamine and GABA (Beard & Connor, 2003; Yager & Hartfield, 2002; Youdim, Ben-Shachar, & Yehuda, 1989).

Lozoff et al. in their longitudinal follow-up study reported that participants who were diagnosed and treated for iron deficiency in infancy were rated as more problematic, with higher scores for anxiety/depression as well as attention and social problems. They also presented lower scores for general mental and motor functions, as well as showing differences in some cognitive processes compared with the control (Lozoff, Jimenez, Hagen, Mollen, & Wolf, 2000). Another population-based study of Chen et al. showed a higher risk for several neurological disturbances, such as anxiety disorder, attention deficit hyperactivity disorder (ADHD), autism spectrum disorder, and mental retardation in a group of children with iron deficiency anemia. In the other group of adolescents, the risk of comorbid neurological and psychiatric disorders was increased for anxiety and depressive disorders as well as for mental retardation, delayed development, and ADHD (Chen et al., 2013).

Ceruloplasmin (Cp) is an enzyme which in its structure contains six atoms of copper and is responsible for catalyzing the oxidation of iron. Cp seems to be essential for both the influx and the efflux of iron from the brain cells (Ke & Qian, 2007). A study using Cp knockout mice has reported an anxiety phenotype in the Cp-deficient group. No deficits in terms of memory or motor function were observed nor any increase in depressive behavior. The results were correlated with decreased levels of serotonin, NE, BDNF, and trkB in the hippocampal tissue and with a heightened corticosterone plasma level (Texel et al., 2012).

Mikawa et al. in their report presented results that show lowered iron and vitamin B6 levels in a group of patients experiencing a panic attack (PA) or a hyperventilation attack (HVA). PA and HVA are psychiatric conditions considered to be caused by stress, anxiety, tension, or depression (Mikawa, Mizobuchi, Egi, & Morita, 2013).

8. MANGANESE

Manganese (Mn) is an essential element that is found mostly in the liver, bones, and kidneys. It is a cofactor for a number of ubiquitous enzymes, such as pyruvate carboxylase, arginase cholinesterase, and mitochondrial SOD (Erikson & Aschner, 2003). In the brain Mn is complexed to glutamine synthetase, an enzyme which is mostly present in astrocytes, and which is involved in the metabolic regulation of glutamate (Roth & Garrick, 2003). Mn deficiency rarely develops in the human population, but its toxic effects are known to occur in specific occupational settings. Miners, smelters, and welders are exposed to high doses of Mn through inhalation of manganese-containing dust. The oral route is less efficient at delivering Mn due to poor absorption from the gastrointestinal tract (Brenneman et al., 2000). Exposure to excessive amounts of Mn may lead to intoxication and is associated with a variety of symptoms connected with psychiatric and motor functions. The clinical picture of manganese neurotoxicity in adults, known as manganism, consists of a set of symptoms resembling idiopathic Parkinson's disease that include tremor, masked face, rigidity, bradykinesia, gait disturbance, reduced response speed, loss of postural balance, and altered behaviors (Bowler et al., 2006). It has been shown that Mn exposure leads to an accumulation of the compound in some regions of the brain that are components of the basal ganglia (Erikson & Aschner, 2003). The first stage of manganese neurotoxicity includes behavioral disturbances, deficient sleep, fatigue, and mood deregulation (Bowler et al., 2006). The mechanism of mood change is still not explained, but depression in groups of exposed workers is reported to be a common disorder (Bowler et al., 2006).

Bowler et al. in their studies reported mood and emotional alterations in a group of welders measured on the SCL-90-R (Symptom Checklist-90-Revised), POMS (The Profile of Mood States), BAI (Beck Anxiety Inventory), and BDI-II (Beck Depression Inventory-II) scales. The scores for depression and anxiety symptoms, as well as obsessive–compulsive, phobic anxiety, and hostility, were higher compared with controls (Bowler et al., 2006; Bowler et al., 2007). These results are consistent with their earlier study showing elevated levels of clinical depression and anxiety, together with signs such as dysphoric mood, lack of motivation, nervousness, tension, and panic episodes (Bowler et al., 2003). A similar observation was made in a group of

Chinese welding workers with higher scores for depression–dejection, tension–anxiety, fatigue–inertia, and confusion–bewilderment (Yuan et al., 2006). However, a study by Solís-Vivanco et al., concentrating on a population nonoccupationally exposed to manganese, presented no evidence for a higher prevalence of depression in that group (Solís-Vivanco et al., 2009).

Up to the present, there is a small amount of animal research data aimed at explaining the mechanism of mood change related to Mn exposure. It has been observed that Mn administration leads to anxiety-like or compulsive-like behaviors in the EPM test (Hogas, Ciobica, Hogas, Bild, & Hritcu, 2011). Kwieciński and Nowak in their work reported a reduced anxiolytic-like effect of diazepam after gestational Mn intoxication, with no effect on the anxiolytic-like properties of 8-OH-DPAT (5-HT_{1A} receptor agonist) (Kwieciński & Nowak, 2009). Mn was shown to affect the dopaminergic system and to be involved in the GABAergic and glutamatergic systems' function (Erikson & Aschner, 2003; Fitsanakis, Au, Erikson, & Aschner, 2006; Stanwood et al., 2009). The GABAergic system of the basal ganglia is known to be affected by Mn accumulation; however, research data remain inconsistent as to whether GABA levels decrease or increase after Mn exposure (Fitsanakis et al., 2006). Mn neurotoxic effects are also related to Glu-induced excitotoxicity as they affect astrocytic uptake of glutamate, increase activation of GluR, and disrupt mitochondrial function that leads to excessive production of reactive oxygen species (Fitsanakis et al., 2006). Zou et al. in their study showed lowered BDNF levels in a group of Mn-exposed workers, together with impairment of cognitive function, compared with the control (Zou et al., 2014).

Mn transport and homeostasis in the brain are closely related to body iron status. In a model of iron overload, hemochromatosis, which used HFE-deficient animals (high iron Fe), it was shown that Mn-exposed HFE-deficient mice expressed reduced impulsivity-like behavior compared with Mn-exposed wild-type animals. This effect may be explained by increased anxiety and/or decreased exploratory activity. The same study shows higher striatal D1 receptor and NE transporter levels in the HFE-deficient group (Ye & Kim, 2015).

9. CONCLUSIONS

In this chapter we focused on the role of zinc, magnesium, copper, lithium, iron, and manganese in anxiety disorders. As mentioned earlier, deficiency of some essential elements is linked to development of anxiety-like behavior. Moreover, presented elements show anxiolytic-like effects. Some of these elements may enhance anxiolytic-like response and reduce side effects of commonly used anxiolytics. This show how important role elements, such as zinc, magnesium, copper, lithium, iron, and manganese, play in anxiety disorders.

ACKNOWLEDGMENT

This work is supported by a grant from the National Science Centre (contract DEC-2011/03/B/NZ7/01999).

REFERENCES

Abdel-Maksoud, M. S., Abdel-hamid, M. I., & Moustafa, Y. M. (2012). NMDA/glutamate mechanism of anxiolytic-like action of zinc in Vogel conflict test in mice. *IOSR Journal of Pharmacy*, *2*, 4–12.

Abumaria, N., Luo, L., Ahn, M., & Liu, G. (2013). Magnesium supplement enhances spatial-context pattern separation and prevents fear overgeneralization. *Behavioural Pharmacology*, *24*, 255–263.

Abumaria, N., Yin, B., Zhang, L., Li, X. Y., Chen, T., & Descalzi, G. (2011). Effects of elevation of brain magnesium on fear conditioning, fear extinction, and synaptic plasticity in the infralimbic prefrontal cortex and lateral amygdala. *The Journal of Neuroscience*, *31*, 14871–14881.

Akil, M., & Brewer, G. J. (1995). Psychiatric and behavioral abnormalities in Wilson's disease. *Advances in Neurology*, *65*, 171–178.

Bandmann, O., Weiss, K. H., & Kaler, S. G. (2015). Wilson's disease and other neurological copper disorders. *The Lancet Neurology*, *14*, 103–113.

Bardgett, M. E., Schultheis, P. J., McGill, D. L., Richmond, R. E., & Wagge, J. R. (2005). Magnesium deficiency impairs fear conditioning in mice. *Brain Research*, *1038*, 100–106.

Beard, J. L., & Connor, J. R. (2003). Iron status and neural functioning. *Annual Review of Nutrition*, *23*, 41–58.

Bourin, M. (2015). Animal models for screening anxiolytic-like drugs: A perspective. *Dialogues in Clinical Neuroscience*, *17*(3), 295–303.

Bourin, M., Petit-Demoulière, B., Dhonnchadha, B. N., & Hascöet, M. (2007). Animal models of anxiety in mice. *Fundamental and Clinical Pharmacology*, *21*(6), 567–574.

Bowler, R. M., Gysens, S., Diamond, E., Booty, A., Hartney, C., & Roels, H. A. (2003). Neuropsychological sequelae of exposure to welding fumes in a group of occupationally exposed men. *International Journal of Hygiene and Environmental Health*, *206*, 517–529.

Bowler, R. M., Gysens, S., Diamond, E., Nakagawa, S., Drezgic, M., & Roels, H. a. (2006). Manganese exposure: Neuropsychological and neurological symptoms and effects in welders. *NeuroToxicology*, *27*, 315–326.

Bowler, R. M., Roels, H. A., Nakagawa, S., Drezgic, M., Diamond, E., & Park, R. (2007). Dose-effect relationships between manganese exposure and neurological, neuropsychological and pulmonary function in confined space bridge welders. *Occupational and Environmental Medicine*, *64*, 167–177.

Bremner, I., & Beattie, J. H. (1995). Copper and zinc metabolism in health and disease: Speciation and interactions. *The Proceedings of Nutrition Society*, *54*, 489–499.

Brenneman, K. A., Wong, B. A., Buccellato, M. A., Costa, E. R., Gross, E. A., & Dorman, D. C. (2000). Direct olfactory transport of inhaled manganese ((54)MnCl (2)) to the rat brain: Toxicokinetic investigations in a unilateral nasal occlusion model. *Toxicology and Applied Pharmacology*, *169*, 238–248.

Campos, A. C., Fogaça, M. V., Aguiar, D. C., & Guimarães, F. S. (2013). Animal models of anxiety disorders and stress. *Revista Brasileira de Psiquiatria*, *35*(Suppl. 2), S101–S111.
Carrí, M. T., Ferri, A., Cozzolino, M., Calabrese, L., & Rotilio, G. (2003). Neurodegeneration in amyotrophic lateral sclerosis: The role of oxidative stress and altered homeostasis of metals. *Brain Research Bulletin*, *61*, 365–374.
Chan, K., Cheung, R., & Au-Yeung, K. (2005). Wilson's disease with depression and parkinsonism. *Journal of Clinical Neuroscience*, *12*, 303–305.
Chen, M.-H., Su, T.-P., Chen, Y.-S., Hsu, J.-W., Huang, K.-L., & Chang, W.-H. (2013). Association between psychiatric disorders and iron deficiency anemia among children and adolescents: A nationwide population-based study. *BMC Psychiatry*, *13*, 161.
Chorin, E., Vinograd, O., Fleidervish, I., Gilad, D., Herrmann, S., Sekler, I., et al. (2011). Upregulation of KCC2 activity by zinc-mediated neurotransmission via the mZnR/GPR39 receptor. *The Journal of Neuroscience*, *31*, 12916–12926.
Chu, Y., Mouat, M. F., Harris, R. B. S., Coffield, J. A., & Grider, A. (2003). Water maze performance and changes in serum corticosterone levels in zinc-deprived and pair-fed rats. *Physiology & Behavior*, *78*, 569–578.
Collins, J. F., & Klevay, L. M. (2011). Copper. *Advances in Nutrition (Bethesda, Md)*, *2*, 520–522.
Collins, J., Prohaska, J., & Knutson, M. (2010). Metabolic crossroads of iron and copper. *Nutrition Reviews*, *68*, 133–147.
Connor, J. R., & Menzies, S. L. (1996). Relationship of iron to oligodendrocytes and myelination. *Glia*, *17*, 83–93.
Crayton, J. W., & Walsh, W. J. (2007). Elevated serum copper levels in women with a history of post-partum depression. *Journal of Trace Elements in Medicine and Biology*, *21*, 17–21.
Crumeyrolle-Arias, M., Jaglin, M., Bruneau, A., Vancassel, S., Cardona, A., Dauge, V., et al. (2014). Absence of the gut microbiota enhances anxiety-like behavior and neuroendocrine response to acute stress in rats. *Psychoneuroendocrinology*, *42*, 207–217.
De Souza, M. C., Walker, A. F., Robinson, P. A., & Bolland, K. (2000). A synergistic effect of a daily supplement for 1 month of 200 mg magnesium plus 50 mg vitamin B6 for the relief of anxiety-related premenstrual symptoms: A randomized, double-blind, crossover study. *Journal of Women's Health & Gender-Based Medicine*, *9*, 131–139.
Dening, T., & Berrios, G. (1989). Wilson's disease. Psychiatric symptoms in 195 cases. *Archives of General Psychiatry*, *46*, 1126–1134.
Doreulee, N., Yanovsky, Y., & Haas, H. (1997). Suppression of long-term potentiation in hippocampal slices by copper. *Hippocampus*, *7*, 666–669.
Erikson, K. M., & Aschner, M. (2003). Manganese neurotoxicity and glutamate-GABA interaction. *Neurochemistry International*, *43*, 475–480.
European Commission (1990). *Directive 90/496/EEC.*
Feske, U., Frank, E., Mallinger, A. G., Houck, P. R., Fagiolini, A., Shear, M. K., et al. (2000). Anxiety as a correlate of response to the acute treatment of bipolar I disorder. *The American Journal of Psychiatry*, *157*, 956–962.
Finn, D., Rutledge-Gorman, M. T., & Crabbe, J. C. (2003). Genetic animal models of anxiety. *Neurogenetics*, *4*(3), 109–135.
Fitsanakis, V. A., Au, C., Erikson, K. M., & Aschner, M. (2006). The effects of manganese on glutamate, dopamine and γ-aminobutyric acid regulation. *Neurochemistry International*, *48*, 426–433.
Flint, J. (2003). Animal models of anxiety and their molecular dissection. *Seminars in Cell and Developmental Biology*, *14*(1), 37–42.
Forster, P. L., Schoenfeld, F. B., Marmar, C. R., & Lang, A. J. (1995). Lithium for irritability in post-traumatic stress disorder. *Journal of Traumatic Stress*, *8*, 143–149.

Freeman, M. P., Freeman, S. a., & McElroy, S. L. (2002). The comorbidity of bipolar and anxiety disorders: Prevalence, psychobiology, and treatment issues. *Journal of Affective Disorders*, *68*, 1–23.
Freland, L., & Beaulieu, J.-M. (2012). Inhibition of GSK3 by lithium, from single molecules to signaling networks. *Frontiers in Molecular Neuroscience*, *5*, 1–7.
González-Estecha, M., Trasobares, E. M., Tajima, K., Cano, S., Fernández, C., López, J. L., et al. (2011). Trace elements in bipolar disorder. *Journal of Trace Elements in Medicine and Biology*, *25*(1), S78–S83.
Grases, G., Perez-Castello, J. A., Sanchis, P., Casero, A., Perello, J., Isern, B., et al. (2006). Anxiety and stress among science students. Study of calcium and magnesium alterations. *Magnesium Research*, *19*, 102–106.
Grober, U., Schmidt, J., & Kisters, K. (2015). Magnesium in prevention and therapy. *Nutrients*, 7, 8199–8226.
Grønli, O., Kvamme, J. M., Friborg, O., & Wynn, R. (2013). Zinc deficiency is common in several psychiatric disorders. *PloS One*, *8*, e82793.
Gupta, M. A., & Knapp, K. (2013). Lithium carbonate decreases acute suicidality in post-traumatic stress disorder. *Australian and New Zealand Journal of Psychiatry*, *47*, 1217.
Handley, S. L. (1995). 5-Hydroxytryptamine pathways in anxiety and its treatment. *Pharmacology and Therapeutics*, *66*(1), 103–148.
Hanus, M., Lafon, J., & Mathieu, M. (2004). Double-blind, randomised, placebo-controlled study to evaluate the efficacy and safety of a fixed combination containing two plant extracts (Crataegus oxyacantha and Eschscholtzia californica) and magnesium in mild-to-moderate anxiety disorders. *Current Medical Research and Opinion*, *20*, 63–71.
Henrotte, J. G., Franck, G., Santarromana, M., Frances, H., Mouton, D., & Motta, R. (1997). Mice selected for low and high blood magnesium levels: A new model for stress studies. *Physiology & Behavior*, *61*, 653–658.
Hogas, M., Ciobica, A., Hogas, S., Bild, V., & Hritcu, L. (2011). The effects of the administration of two different doses of manganese on short-term spatial memory and anxiety-like behavior in rats. *Archives of Biological Sciences*, *63*, 1031–1036.
Islam, M. R., Ahmed, M. U., Mitu, S. A., Islam, M. S., Rahman, G. K., Qusar, M. M., et al. (2013). Comparative analysis of serum zinc, copper, manganese, iron, calcium, and magnesium level and complexity of interelement relations in generalized anxiety disorder patients. *Biological Trace Element Research*, *154*, 21–27.
Jacka, F. N., Maes, M., Pasco, J. A., Williams, L. J., & Berk, M. (2012). Nutrient intakes and the common mental disorders in women. *Journal of Affective Disorders*, *141*, 79–85.
Jackson, V. R., Nothacker, H.-P., & Civelli, O. (2006). GPR39 receptor expression in the mouse brain. *Neuroreport*, *17*, 813–816.
Jahnen-Dechent, W., & Ketteler, M. (2012). Magnesium basics. *Clinical Kidney Journal*, *5*, i3–i14.
Joshi, M., Akhtar, M., Najmi, A. K., Khuroo, A. H., & Goswami, D. (2012). Effect of zinc in animal models of anxiety, depression and psychosis. *Human & Experimental Toxicology*, *31*, 1237–1243.
Jung, K. I., Ock, S. M., Chung, J. H., & Song, C. H. (2010). Associations of serum ca and mg levels with mental health in adult women without psychiatric disorders. *Biological Trace Element Research*, *133*, 153–161.
Kaidanovich-Beilin, O., Milman, A., Weizman, A., Pick, C. G., & Eldar-Finkelman, H. (2004). Rapid antidepressive-like activity of specific glycogen synthase kinase-3 inhibitor and its effect on β-catenin in mouse hippocampus. *Biological Psychiatry*, *55*, 781–784.
Kaler, S. G. (2013). Inborn errors of copper metabolism. *Handbook of Clinical Neurology*, *113*, 1745–1754.
Kalueff, A. V., & Nutt, D. J. (2007). Role of GABA in anxiety and depression. *Depression and Anxiety*, *24*, 495–517. http://dx.doi.org/10.1002/da.20262.

Kaye, P., & O'Sullivan, I. (2002). The role of magnesium in the emergency department. *Emergency Medicine Journal, 19*, 288–291.

Ke, Y., & Qian, Z. M. (2007). Brain iron metabolism: Neurobiology and neurochemistry. *Progress in Neurobiology, 83*, 149–173.

Kodama, H., Fujisawa, C., & Bhadhprasit, W. (2012). Inherited copper transport disorders: Biochemical mechanisms, diagnosis, and treatment. *Current Drug Metabolism, 13*, 237–250.

Kwieciński, A., & Nowak, P. (2009). Gestational manganese intoxication and anxiolytic-like effects of diazepam and the 5-HT1A receptor agonist 8-OH-DPAT in male Wistar rats. *Pharmacological Reports, 61*, 1061–1068.

Laarakker, M. C., van Lith, H. A., & Ohl, F. (2011). Behavioral characterization of A/J and C57BL/6J mice using a multidimensional test: Association between blood plasma and brain magnesium-ion concentration with anxiety. *Physiology & Behavior, 102*, 205–219.

Liu, G., Weinger, J. G., Lu, Z. L., Xue, F., & Sadeghpour, S. (2015). Efficacy and safety of MMFS-01, a synapse density enhancer, for treating cognitive impairment in elderly: A randomized, double-blind, placebo-controlled trial. *Journal of Alzheimer's Disease, 49*, 971–990.

Lozoff, B., Jimenez, E., Hagen, J., Mollen, E., & Wolf, A. W. (2000). Poorer behavioral and developmental outcome more than 10 years after treatment for iron deficiency in infancy. *Pediatrics, 105*, E51.

Maes, M., Vandoolaeghe, E., Neels, H., Demedts, P., Wauters, A., Meltzer, H. Y., et al. (1997). Lower serum zinc in major depression is a sensitive marker of treatment resistance and of the immune/inflammatory response in that illness. *Biological Psychiatry, 42*, 349–358.

Masood, A., Nadeem, A., Mustafa, S. J., Donnell, J. M. O., O'Donnell, J. M., & Donnell, J. M. O. (2008). Reversal of oxidative stress-induced anxiety by inhibition of phosphodiesterase-2 in mice. *Journal of Pharmacology and Experimental Therapeutics, 326*, 369–379.

Mikawa, Y., Mizobuchi, S., Egi, M., & Morita, K. (2013). Low serum concentrations of vitamin B6 and iron are related to panic attack and hyperventilation attack. *Acta Medica Okayama, 67*, 99–104.

Mines, M. A., Yuskaitis, C. J., King, M. K., Beurel, E., & Jope, R. S. (2010). GSK3 influences social preference and anxiety-related behaviors during social interaction in a mouse model of fragile X syndrome and autism. *PloS One, 5*, e9706.

Młyniec, K., Budziszewska, B., Holst, B., Ostachowicz, B., & Nowak, G. (2015). GPR39 (zinc receptor) knockout mice exhibit depression-like behavior and CREB/BDNF down-regulation in the hippocampus. *The International Journal of Neuropsychopharmacology, 18*, 1–8.

Młyniec, K., Davies, C. L., Budziszewska, B., Opoka, W., Reczyński, W., Sowa-Kućma, M., et al. (2012). Time course of zinc deprivation-induced alterations of mice behavior in the forced swim test. *Pharmacological Reports, 64*, 567–575.

Młyniec, K., Davies, C. L., de Agüero Sánchez, I. G., Pytka, K., Budziszewska, B., & Nowak, G. (2014). Essential elements in depression and anxiety. Part I. *Pharmacological Reports, 66*, 534–544.

Młyniec, K., Gaweł, M., Doboszewska, U., Starowicz, G., Pytka, K., Davies, C. L., et al. (2015). Essential elements in depression and anxiety. Part II. *Pharmacological Reports, 67*, 187–194.

Młyniec, K., Gaweł, M., Librowski, T., Reczyński, W., Bystrowska, B., & Holst, B. (2015). Investigation of the GPR39 zinc receptor following inhibition of monoaminergic neurotransmission and potentialization of glutamatergic neurotransmission. *Brain Research Bulletin, 115*, 23–29.

Młyniec, K., Singewald, N., Holst, B., & Nowak, G. (2015). GPR39 Zn(2+)-sensing receptor: A new target in antidepressant development? *Journal of Affective Disorders, 174C*, 89–100.

Muroyama, A., Inaka, M., Matsushima, H., Sugino, H., Marunaka, Y., & Mitsumoto, Y. (2009). Enhanced susceptibility to MPTP neurotoxicity in magnesium-deficient C57BL/6N mice. *Neuroscience Research, 63*, 72–75.

Mustak, M. S., Rao, T. S. S., Shanmugavelu, P., Sundar, N. M. S., Menon, R. B., Rao, R. V., et al. (2008). Assessment of serum macro and trace element homeostasis and the complexity of inter-element relations in bipolar mood disorders. *Clinica Chimica Acta; International Journal of Clinical Chemistry, 394*, 47–53.

Nahar, Z., Azad, M. A., Rahman, M. A., Rahman, M. A., Bari, W., Islam, S. N., et al. (2010). Comparative analysis of serum manganese, zinc, calcium, copper and magnesium level in panic disorder patients. *Biological Trace Element Research, 133*, 284–290.

Narang, R. L., Gupta, K. R., Narang, A. P., & Singh, R. (1991). Levels of copper and zinc in depression. *Indian Journal of Physiology and Pharmacology, 35*, 272–274.

National Research Council (US) Subcommittee on Laboratory Animal Nutrition (Ed.), (1995). *Nutrient requirements of laboratory animals* (4th revised ed.). Washington, DC: National Academy Press.

O'Brien, W. T., Huang, J., Buccafusca, R., Garskof, J., Valvezan, A. J., Berry, G. T., et al. (2011). Glycogen synthase kinase-3 is essential for β-arrestin-2 complex formation and lithium-sensitive behaviors in mice. *Journal of Clinical Investigation, 121*, 3756–3762.

Oder, W., Grimm, G., Kollegger, H., Ferenci, P., Schneider, B., & Deecke, L. (1991). Neurological and neuropsychiatric spectrum of Wilson's disease: A prospective study of 45 cases. *Journal of Neurology, 238*, 281–287.

Pachikian, B. D., Neyrinck, A. M., Deldicque, L., De Backer, F. C., Catry, E., Dewulf, E. M., et al. (2010). Changes in intestinal bifidobacteria levels are associated with the inflammatory response in magnesium-deficient mice. *The Journal of Nutrition, 140*, 509–514.

Partyka, A., Jastrzębska-Więsek, M., Szewczyk, B., Stachowicz, K., Sławińska, A., Poleszak, E., et al. (2011). Anxiolytic-like activity of zinc in rodent tests. *Pharmacological Reports, 63*, 1050–1055.

Perez-Rosello, T., Anderson, C. T., Schopfer, F. J., Zhao, Y., Gilad, D., Salvatore, S. R., et al. (2013). Synaptic Zn2+ inhibits neurotransmitter release by promoting endocannabinoid synthesis. *The Journal of Neuroscience, 33*, 9259–9272.

Poleszak, E. (2008). Benzodiazepine/GABA(A) receptors are involved in magnesium-induced anxiolytic-like behavior in mice. *Pharmacological Reports, 60*, 483–489.

Poleszak, E., Szewczyk, B., Kedzierska, E., Wlaz, P., Pilc, A., & Nowak, G. (2004). Antidepressant- and anxiolytic-like activity of magnesium in mice. *Pharmacology, Biochemistry, and Behavior, 78*, 7–12.

Poleszak, E., Wlaz, P., Wrobel, A., Fidecka, S., & Nowak, G. (2008). NMDA/glutamate mechanism of magnesium-induced anxiolytic-like behavior in mice. *Pharmacological Reports, 60*, 655–663.

Prohaska, J. (2000). Long-term functional consequences of malnutrition during brain development: Copper. *Nutrition, 16*, 502–504.

Prohaska, J. R., & Cox, D. a. (1983). Decreased brain ascorbate levels in copper-deficient mice and in brindled mice. *The Journal of Nutrition, 113*, 2623–2629.

Prohaska, J., & Smith, T. (1982). Effect of dietary or genetic copper deficiency on brain catecholamines, trace metals and enzymes in mice and rats. *The Journal of Nutrition, 112*, 1706–1717.

Pyatskowit, J., & Prohaska, J. (2007). Rodent brain and heart catecholamine levels are altered by different models of copper deficiency. *Comperative Biochemistry and Physiology. Toxicology & Pharmacology, 145*, 275–281.

Pyndt Jorgensen, B., Winther, G., Kihl, P., Nielsen, D. S., Wegener, G., Hansen, A. K., et al. (2015). Dietary magnesium deficiency affects gut microbiota and anxiety-like behaviour in C57BL/6N mice. *Acta Neuropsychiatrica, 27*, 307–311.

Razafsha, M., Behforuzi, H., Harati, H., Wafai, R. A., Khaku, A., Mondello, S., et al. (2013). An updated overview of animal models in neuropsychiatry. *Neuroscience*, *240*, 204–218.
Rodgers, R. J., Cao, B.-J., Dalvi, A., & Holmes, A. (1997). Animal models of anxiety: An ethological perspective. *Brazilian Journal of Medical and Biological Research*, *30*, 289–304.
Roth, J. A., & Garrick, M. D. (2003). Iron interactions and other biological reactions mediating the physiological and toxic actions of manganese. *Biochemical Pharmacology*, *66*, 1–13.
Russo, A. J. J. (2011). Decreased zinc and increased copper in individuals with anxiety. *Nutrition and Metabolic Insights*, *4*, 1.
Salazar-Weber, N. L., & Smith, J. P. (2011). Copper inhibits NMDA receptor-independent LTP and modulates the paired-pulse ratio after LTP in mouse hippocampal slices. *International Journal of Alzheimer's Disease*, *2011*, 864753.
Sartori, S. B., Whittle, N., Hetzenauer, A., & Singewald, N. (2012). Magnesium deficiency induces anxiety and HPA axis dysregulation: Modulation by therapeutic drug treatment. *Neuropharmacology*, *62*, 304–312.
Scheiber, I. F., Mercer, J. F. B., & Dringen, R. (2014). Metabolism and functions of copper in brain. *Progress in Neurobiology*, *116*, 33–57. http://dx.doi.org/10.1016/j.pneurobio.2014.01.002.
Schlegel-Zawadzka, M., & Nowak, G. (2000). Alterations in serum and brain trace element levels after antidepressant treatment. Part II. Copper. *Biological Trace Element Research*, *73*, 37–45.
Schlegel-Zawadzka, M., Zięba, A., Dudek, D., Krośniak, M., Szymaczek, M., & Nowak, G. (1999). Serum trace elements in animal models and human depression. Part II. Copper. *Human Psychopharmacology: Clinical and Experimental*, *14*, 447–451.
Schlegel-Zawadzka, M., Zieba, A., Dudek, D., Zak-Knapik, J., & Nowak, G. (1999). Is serum copper a 'trait marker' of unipolar depression? A preliminary clinical study. *Polish Journal of Pharmacology*, *51*, 535–538.
Schwartz, R. D., Wagner, J. P., Yu, X., & Martin, D. (1994). Bidirectional modulation of GABA-gated chloride channels by divalent cations: Inhibition by Ca2+ and enhancement by Mg2+. *Journal of Neurochemistry*, *62*, 916–922.
Seo, J. W., & Park, T. J. (2008). Magnesium metabolism. *Electrolyte & Blood Pressure*, *6*, 86–95.
Shohag, H., Ullah, A., Qusar, S., Rahman, M., & Hasnat, A. (2012). Alterations of serum zinc, copper, manganese, iron, calcium, and magnesium concentrations and the complexity of interelement relations in patients with obsessive-compulsive disorder. *Biological Trace Element Research*, *148*, 275–280.
Simon, N. M., Otto, M. W., Wisniewski, S. R., Fossey, M., Sagduyu, K., Frank, E., et al. (2004). Anxiety disorder comorbidity in bipolar disorder patients: Data from the first 500 participants in the Systematic Treatment Enhancement Program for Bipolar Disorder (STEP-BD). *American Journal of Psychiatry*, *161*, 2222–2229.
Singewald, N., Sinner, C., Hetzenauer, A., Sartori, S. B., & Murck, H. (2004). Magnesium-deficient diet alters depression- and anxiety-related behavior in mice—Influence of desipramine and hypericum perforatum extract. *Neuropharmacology*, *47*, 1189–1197.
Siwek, M., Wróbel, A., Dudek, D., Nowak, G., & Zieba, A. (2005). The role of copper and magnesium in the pathogenesis and treatment of affective disorders. *Psychiatria Polska*, *39*, 911–920.
Slutsky, I., Abumaria, N., Wu, L. J., Huang, C., Zhang, L., Li, B., et al. (2010). Enhancement of learning and memory by elevating brain magnesium. *Neuron*, *65*, 165–177.
Solís-Vivanco, R., Rodríguez-Agudelo, Y., Riojas-Rodríguez, H., Ríos, C., Rosas, I., & Montes, S. (2009). Cognitive impairment in an adult Mexican population non-occupationally exposed to manganese. *Environmental Toxicology and Pharmacology*, *28*, 172–178.

Spasov, A. A., Iezhitsa, I. N., Kharitonova, M. V., & Kravchenko, M. S. (2008). Depression-like and anxiety-related behaviour of rats fed with magnesium-deficient diet. *Zhurnal Vyssheĭ Nervnoĭ Deiatelnosti Imeni I P Pavlova, 58*, 476–485.

Stanwood, G. D., Leitch, D. B., Savchenko, V., Wu, J., Fitsanakis, V. A., Anderson, D. J., et al. (2009). Manganese exposure is cytotoxic and alters dopaminergic and GABAergic neurons within the basal ganglia. *Journal of Neurochemistry, 110*, 378–389.

Stern, B. R., Solioz, M., Krewski, D., Aggett, P., Aw, T.-C., Baker, S., et al. (2007). Copper and human health: Biochemistry, genetics, and strategies for modeling dose-response relationships. *Journal of Toxicology and Environmental Health. Part B, Critical Reviews, 10*, 157–222.

Stilling, R. M., Ryan, F. J., Hoban, A. E., Shanahan, F., Clarke, G., Claesson, M. J., et al. (2015). Microbes & neurodevelopment—Absence of microbiota during early life increases activity-related transcriptional pathways in the amygdala. *Brain, Behavior, and Immunity, 50*, 209–220.

Swardfager, W., Herrmann, N., McIntyre, R. S., Mazereeuw, G., Goldberger, K., Cha, D. S., et al. (2013). Potential roles of zinc in the pathophysiology and treatment of major depressive disorder. *Neuroscience & Biobehavioral Reviews, 37*, 911–929.

Takeda, A., Tamano, H., Itoh, H., & Oku, N. (2008). Attenuation of abnormal glutamate release in zinc deficiency by zinc and Yokukansan. *Neurochemistry International, 53*, 230–235.

Takeda, A., Tamano, H., Kan, F., Itoh, H., & Oku, N. (2007). Anxiety-like behavior of young rats after 2-week zinc deprivation. *Behavioural Brain Research, 177*, 1–6.

Texel, S. J., Camandola, S., Ladenheim, B., Rothman, S. M., Mughal, M. R., & Unger, E. L. (2012). Ceruloplasmin deficiency results in an anxiety phenotype involving deficits in hippocampal iron, serotonin, and BDNF. *Journal of Neurochemistry, 120*, 125–134.

Tong, G. M., & Rude, R. K. (2005). Magnesium deficiency in critical illness. *Journal of Intensive Care Medicine, 20*, 3–17.

Vink, R., & Nechifor, N. (2011). *Magnesium in the central nervous system*. Adelaide: University of Adelaide Press.

Whittle, N., Lubec, G., & Singewald, N. (2009). Zinc deficiency induces enhanced depression-like behaviour and altered limbic activation reversed by antidepressant treatment in mice. *Amino Acids, 36*, 147–158.

Williamson, M., Snyder, L., & Wallach, J. (2011). *Wallach's interpretation of diagnostic tests* (9th ed.). Philadelphia: Wolters Kluwer/Lippincott Williams & Wilkins Health.

Winther, G., Pyndt Jorgensen, B. M., Elfving, B., Nielsen, D. S., Kihl, P., Lund, S., et al. (2015). Dietary magnesium deficiency alters gut microbiota and leads to depressive-like behaviour. *Acta Neuropsychiatrica, 27*, 168–176.

Wlaz, P., Serefko, A., Szopa, A., & Poleszak, E. (2016). The effect of an acute and 7-day administration of magnesium chloride on magnesium concentration in the serum, erythrocytes, and brain of rats. *Pharmacological Reports, 68*, 289–291.

Yager, J. Y., & Hartfield, D. S. (2002). Neurologic manifestations of iron deficiency in childhood. *Pediatric Neurology, 27*, 85–92.

Ye, Q., & Kim, J. (2015). Effect of olfactory manganese exposure on anxiety-related behavior in a mouse model of iron overload hemochromatosis. *Environmental Toxicology and Pharmacology, 40*, 333–341.

Youdim, M. B., Ben-Shachar, D., & Yehuda, S. (1989). Putative biological mechanisms of the effect of iron deficiency on brain biochemistry and behavior. *The American Journal of Clinical Nutrition, 50*, 607.

Youngs, R. M., Chu, M. S., Meloni, E. G., Naydenov, A., Carlezon, W. A., & Konradi, C. (2006). Lithium administration to preadolescent rats causes long-lasting increases in anxiety-like behavior and has molecular consequences. *The Journal of Neuroscience, 26*, 6031–6039.

Yuan, H., He, S., He, M., Niu, Q., Wang, L., & Wang, S. (2006). A comprehensive study on neurobehavior, neurotransmitters and lymphocyte subsets alteration of Chinese manganese welding workers. *Life Sciences*, *78*, 1324–1328.

Zhou, Z., Wang, Y., Tan, H., Bharti, V., Che, Y., & Wang, J.-F. (2015). Chronic treatment with mood stabilizer lithium inhibits amphetamine-induced risk-taking manic-like behaviors. *Neuroscience Letters*, *603*, 84–88.

Zhu, Y., & Auerbach, A. (2001). Na(+) occupancy and mg(2+) block of the n-methyl-d-aspartate receptor channel. *The Journal of General Physiology*, *117*, 275–286.

Zieba, A., Kata, R., Dudek, D., Schlegel-Zawadzka, M., & Nowak, G. (2000). Serum trace elements in animal models and human depression: Part III. Magnesium. Relationship with copper. *Human Psychopharmacology*, *15*, 631–635.

Zimbrean, P. C., & Schilsky, M. L. (2014). Psychiatric aspects of Wilson disease: A review. *General Hospital Psychiatry*, *36*, 53–62.

Zou, Y., Qing, L., Zeng, X., Shen, Y., Zhong, Y., Liu, J., et al. (2014). Cognitive function and plasma BDNF levels among manganese-exposed smelters. *Occupational and Environmental Medicine*, *71*, 189–194.

CHAPTER ELEVEN

The Place of Stress and Emotions in the Irritable Bowel Syndrome

S. Pellissier*[,1], B. Bonaz[†,‡]
*Laboratoire Interuniversitaire de Psychologie, Personnalité, Cognition, Changement Social, Université Savoie Mont-Blanc, Chambéry, France
[†]Clinique Universitaire d'Hépato-Gastroentérologie, CHU de Grenoble, Grenoble 09, France
[‡]Université Grenoble Alpes, Grenoble Institut des Neurosciences, Fonctions Cérébrales et Neuromodulation, INSERM, Grenoble 09, France
[1]Corresponding author: e-mail address: sonia.pellissier@univ-smb.fr

Contents

Abstract

Our emotional state can have many consequences on our somatic health and well-being. Negative emotions such as anxiety play a major role in gut functioning due to the bidirectional communications between gut and brain, namely, the brain–gut axis. The irritable bowel syndrome (IBS), characterized by an unusual visceral hypersensitivity, is the most common disorder encountered by gastroenterologists. Among the main symptoms, the presence of current or recurrent abdominal pain or discomfort associated with bloating and altered bowel habits characterizes this syndrome that could strongly alter the quality of life. This chapter will present the physiopathology of IBS and explain how stress influences gastrointestinal functions (permeability, motility, microbiota, sensitivity, secretion) and how it could be predominantly involved in IBS.

Vitamins and Hormones, Volume 103
ISSN 0083-6729
http://dx.doi.org/10.1016/bs.vh.2016.09.005

This chapter will also describe the role of the autonomic nervous system and the hypothalamic–pituitary axis through vagal tone and cortisol homeostasis. An analysis is made about how emotions and feelings are involved in the disruption of homeostasis, and we will see to what extent the balance between vagal tone and cortisol may reflect dysfunctions of the brain–gut homeostasis. Finally, the interest of therapeutic treatments focused on stress reduction and vagal tone enforcement is discussed.

1. INTRODUCTION

Each of us, one day, could have felt the impact of stress and emotions such as anxiety or fear on the functioning of the gut. This experience often culminates in symptoms such as spasms, pain, diarrhea, bloating, and discomfort. Actually, the relationship between the emotional state and gut function, in health and disease, is the manifestation or the final output of complex bidirectional communications between the brain and the gut referred to as the concept of the brain–gut axis underpinning the biopsychosocial concept of disease[1] (Engel, 1977). According to this concept, stress may play a major influence on all attendant components of the gut including its microbiota and conversely. Recent advances revealed that gut microbiota may influence brain function and thus mood and behaviors, and in the same way, stress modifies gut microbiota. The autonomic nervous system (ANS), especially the vagus nerve, is central in this bidirectional communication. Dysfunction of homeostatic regulation at the level of the gut or the brain is mediated through the ANS that induces an adjustment of visceral functioning. In functional digestive diseases such as the irritable bowel syndrome (IBS) each level of this brain–gut axis could be involved, leading to gut dysfunctions and mood disorders. Abdominal symptoms secondarily influence anxiety and depression (bottom-up model), and psychological factors themselves influence physiological factors such as motor functions, sensory threshold, and stress reactivity of the gut via vagal and sympathetic afferents (top-down model). This bidirectional interplay between brain and gut is illustrated in population survey studies revealing a strong correlation between anxiety, depression, and functional gastrointestinal disorders. Furthermore, psychological distress can predict later onset of a functional gastrointestinal disorder and the converse (Koloski et al., 2012).

[1] Engel postulated for the first time in the late 1970s that all illnesses, but especially gastrointestinal, result from a complex reciprocal interaction between biological/genetic, psychological, and social factors.

This chapter will present the physiopathology of IBS and explain how stress may influence gastrointestinal function and how it could be involved in IBS. We will describe the role of the ANS and the hypothalamic–pituitary axis through the vagus nerve and cortisol homeostasis in relation with emotions and feelings and how this balance may reflect dysfunctions of the brain–gut pathway.

2. PATHOPHYSIOLOGY OF THE IBS

The IBS is the most common disorder encountered by gastroenterologists. It is a chronic functional gastrointestinal disorder which is remarkably prevalent in the general population affecting 5–20% people worldwide (Lovell & Ford, 2012; Park et al., 2015) and induces huge direct and indirect costs (Canavan, West, & Card, 2016). Classically the syndrome is considered as functional because of the absence of any detectable organic lesions. Generally, biological as well as morphological (e.g., colonoscopy) investigations performed in these patients are negative and if anatomical abnormalities (colonic polyps or diverticulosis) are found, they have no relation with their symptomatology. The diagnosis of this syndrome is symptom based according to Rome III criteria (Longstreth et al., 2006). Among the main symptoms, the presence of current or recurrent abdominal pain or discomfort associated with bloating and altered bowel habits characterizes the syndrome (Drossman, 1999; Mulak & Bonaz, 2004). The syndrome has been divided into subgroups based on the predominant bowel habit: diarrhea-predominant (D-IBS), constipation-predominant (C-IBS), or alternative diarrhea and constipation (M-IBS). However, each IBS patient can switch from one subgroup to another over time. Nevertheless, even if a significant variability in the clinical presentation of patients with IBS is observed, those patients usually present psychological distress comorbidities and somatic complaints (Saito, Schoenfeld, & Locke, 2002). Some studies have shown that approximately up to 60% of IBS patients have major psychosocial problems (Levy et al., 2006). According to the symptoms intensity, IBS may strongly affect quality of life with the same degree of impairment as major chronic disorder.

Among the causes that are supposed to explain this syndrome, there are two main causes that have been initially explored at the level of the digestive tract namely dysmotility and visceral hypersensitivity even if the latter is not always systematically present in IBS (Azpiroz et al., 2007). Pain is perceived by patients as the most distressing symptom and constitutes the main reason for consulting a physician (Sandler, Drossman, Nathan, & McKee, 1984). However, those consulting patients represent only a part of all the patients since a lot of IBS people who cope as possible with mild symptoms do not consult a physician. Studies have shown that patients with IBS exhibit enhanced perception of visceral sensation to esophageal, gastric, or rectal distension, and hypervigilance toward visceral sensations (Bradette et al., 1994; Elsenbruch et al., 2010). Moreover, visceral hypersensitivity is considered to play a major role in the pathophysiology of IBS since extraintestinal manifestations are also frequently described by the patients (e.g., headache, low back pain, chronic fatigue, interstitial cystitis) (Whitehead, Palsson, & Jones, 2002). The exact cause of this visceral hypersensitivity is unknown, but a number of mechanisms are evoked. One of them involves a probable dysfunction in the gastrointestinal endocrine cells. Indeed, the neuroendocrine system of the gastrointestinal tract responds to the luminal content composition, as a sensory system, by releasing a variety of hormones such as serotonin, histamine, somatostatin, ghrelin, gastrin, CCK, oxyntomodulin, motilin, and neurotensin into the lamina propria. This response could start a chain reaction that progresses along the entire neuroendocrine system to act locally and regionally on motility (Benouali-Pellissier, Roger, & Roche, 1994; Merle et al., 2000; Moro et al., 2000; Pellissier, Eribon, Chabert, Gully, & Roche, 1996; Pellissier, Sasaki, Le-Nguyen, Bataille, & Jarrousse, 2004) but also centrally via the enteric sensory neurons which in turn communicate with the central nervous system (CNS) through the parasympathetic and sympathetic afferents (El-Salhy, Gundersen, Gilja, Hatlebakk, & Hausken, 2014). This is evidenced by neuroplasticity changes observed in primary afferent terminals (peripheral sensitization) due to peripheral inflammation or infection of the gut (i.e., postinfectious IBS) but also in the spinal cord (central sensitization) and in the brain (supraspinal pain modulation) (Bonaz, 2003; Mulak & Bonaz, 2004). In addition, IBS patients are more sensitive to stress and exhibit heightened stress reactions (Dinan et al., 2008) often linked to an excess of anxiety. The everyday life stress but also the feeling of being stressed due to the burden of their symptoms is able to enhance visceral sensitivity at the central or the peripheral level or both (Larauche, Mulak, & Tache, 2012; Mulak & Bonaz, 2004).

The pathophysiology of IBS is multifactorial resulting from complex interactions between biological, psychological, and social factors, the key factors of the biopsychosocial model of illnesses (Bonaz & Sabate, 2009; Drossman, 1996; Gaynes & Drossman, 1999) (Fig. 1). This model may help

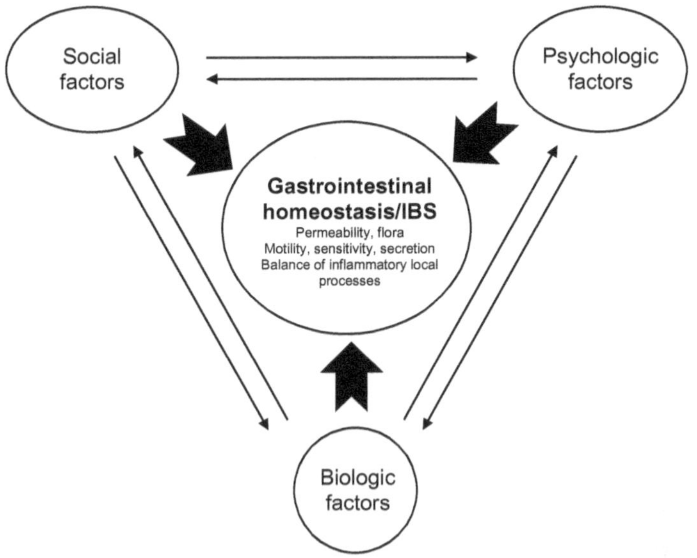

Fig. 1 The biopsychosocial model of illnesses applied to IBS.

to understand the reason why all the symptoms described in this syndrome are not always consistent and may vary over time. All these data outlined the very close link between stress, emotions, and evident dysfunctions of the brain–gut interactions.

3. THE BRAIN–GUT AXIS

The bidirectional communication between the CNS and the gastrointestinal tract, i.e., the brain–gut axis, can be viewed as a complex network that mainly includes reflex processing of sensitive and motor information playing a major role in the digestive tract function but also in the human behavior in general both in health and disease (Bonaz & Sabate, 2009). This axis is composed of receptors, afferent fibers that project to integrative central areas and efferent fibers conveying orders to smooth muscle, glands, and neuroendocrine cells (Gaman & Kuo, 2008). A bidirectional communication between the brain and the gut allows signals from the gastrointestinal tract (e.g., visceral sensation, flora) to influence the brain with regard to reflex regulation, mood states, and behavioral regulation in a bottom-up relation (Rhee, Pothoulakis, & Mayer, 2009). Conversely, signals from the brain can modify the motor, sensory, secretory, and immune functions of the gastrointestinal tract in a top-down relation. Numerous data argue for a dysfunction of this brain–gut axis in the pathophysiology of IBS (Bonaz & Sabate, 2009; Mulak & Bonaz, 2004; Pellissier et al., 2014; Tillisch, Mayer, & Labus, 2011). Behavioral, affective, and cognitive processes thus can affect brain–gut axis functioning as a vicious circle that will amplify sensitivity, motility, and anxiety (Fig. 2).

This brain–gut axis can be viewed as a network with three levels of communication and complex interplay, namely, the enteric nervous system (ENS), the ANS, and the CNS.

3.1 The ENS Level

At the bottom of the network, the ENS can control functions of the intestine even when it is completely separated from the CNS (Bayliss & Starling, 1899). The ENS contains three categories of neurons, identified as sensory, associative, and motor neurons (both excitatory and inhibitory) which are the final common pathways for the control of signals to the musculature, submucosa, mucosa, and vasculature, both blood and lymphatic, local immunity, microbiota's homeostasis. The ENS contains as many neurons as in the spinal cord (400–600 million) and confers an autonomy to the

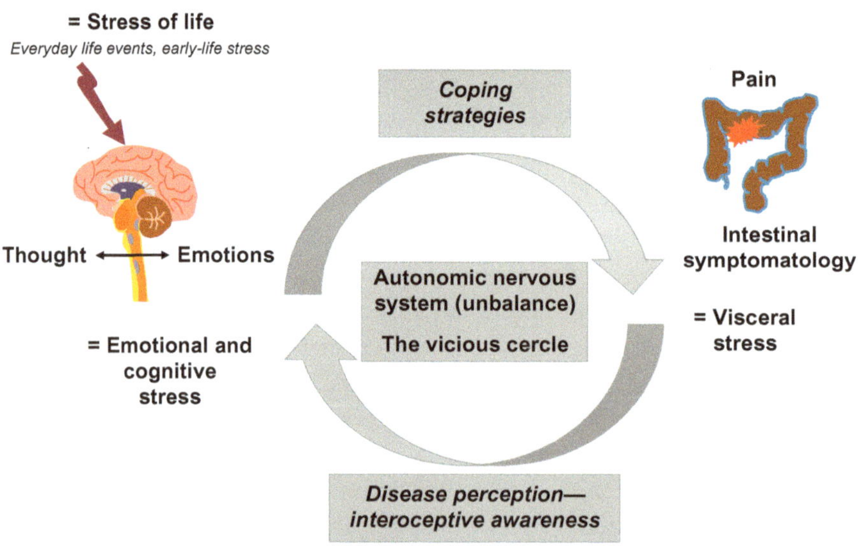

Fig. 2 Behavioral, affective, and cognitive processes can affect brain–gut axis functioning as a vicious circle that will amplify sensitivity, motility, and anxiety.

digestive tract for the programming of motility and secretion (Furness, 2012). Some neuropeptides, neurotransmitters, and receptors present in the CNS are also present in the ENS. That is the reason why this ENS has been called the "little brain." The latter communicates with the CNS or the "big brain" through afferents and the efferents of the ANS that constitute the second level of the three-stage network.

3.2 The ANS

The ANS conveys sensory, chemical, mechanical, thermic, and nociceptive information from the gut through afferent fibers that enters the CNS either at the level of the spinal cord or the medulla. The ANS is also composed of efferent fibers constituting the sympathetic (the splanchnic nerves) and the parasympathetic (vagus and pelvic nerves) branches that lead, respectively, to noradrenalin and acetylcholine release to modulate the ENS and the local activities of the gut. The nerves of the ANS are in majority mixed and convey both afferent and efferent fibers.

The vagus nerve contains essentially 80–90% of afferent fibers mediating sensorial input from the abdominal viscera to the brain (Altschuler, Bao, Bieger, Hopkins, & Miselis, 1989) with the exception of the pelvic viscera for which information is mediated to S2–S4 levels of the spinal cord by the pelvic nerves with central projections similar to other spinal visceral afferents. The vagus nerve carries mainly mechanical and chemical sensory information from the gut. If classical vagal afferents do not encode painful stimuli, they are able to modulate nociceptive processing at the levels of the spinal cord and the brain (Randich & Gebhart, 1992; Randich, Thurston, Ludwig, Robertson, & Rasmussen, 1992; Thurston & Randich, 1992). Currently, the vagus nerve is considered as the sixth sense of the body (Zagon, 2001). The sympathetic nerves contain 50% afferent fibers. Visceral afferents that enter via spinal nerves (i.e., splanchnic and pelvic nerves), at T5-L2 segments of the spinal cord, carry information concerning temperature as well as nociceptive visceral inputs related to mechanical, chemical, or thermal stimulation through C and Aδ afferents, which will reach conscious perception.

3.3 Integration of Visceral Inputs in the CNS: The Central Autonomic Network

The afferent information of the ANS reaches the CNS at both the spinal cord, for the splanchnic nerves, and the nucleus tractus solitarius (NTS) in the dorsal medulla for the vagus nerve and in the sacral (S2–S4) parasympathetic nucleus for the pelvic nerves. At the level of the spinal cord, sympathetic afferents project to the upper level through the spinoreticular and spinothalamic tracts and the dorsal column that is more particularly involved in visceral afferent pathway, and then to several cerebral cortical area (insula, cingular cortex, and prefrontal cortex) (Burstein, 1996; Cliffer, Burstein, & Giesler, 1991). At the level of the NTS, vagal afferents are integrated according to a visceral somatotopy in subnuclei (Altschuler, Escardo, Lynn, & Miselis, 1993) and then projected to the parabrachial nucleus, in the pons, which in turn sends projections to numerous structures in the brainstem, hypothalamus, basal forebrain, thalamus, and cerebral cortex (Fulwiler & Saper, 1984). Among the cerebral cortices, the insular cortex acts as a visceral (e.g., gastrointestinal) cortex through a NTS–parabrachial–thalamocortical pathway according to a viscerotopic map. The insular cortex has connections with the limbic system (bed nucleus of the *stria terminalis* and the central nucleus of the amygdala—CeA) and with the lateral frontal cortical system (Saper, 1982). Consequently, sensory messages coming from the gut are integrated through a series of areas constituting a central autonomic network. This network includes part of the frontal cortex and the insular and anterior cingulate cortices; the amygdala (CeA); several areas of the hypothalamus, in particular the paraventricular nucleus; the periaqueductal gray matter in the midbrain; the parabrachial nucleus in the pons; the *locus coeruleus*; the Barrington nucleus; and the NTS in the medulla. This central autonomic network represents a high level of sensorial integration for the interoceptive awareness that involves the autonomic, endocrine, motor, immune, and behavioral responses associated to gut inputs. After integration, these areas generate stimulus-specific patterns of autonomic responses via projections to preganglionic sympathetic neurons in the spinal cord and parasympathetic neurons of the dorsal motor nucleus of the vagus nerve to modulate the ANS. The central autonomic network appears as a complex system including many positive and negative feedback loops governing both sympathetic and parasympathetic outputs. Because of this complexity, it is difficult to predict the final results of activity of this system since prefrontal activation might induce enhancement of sympathetic activity in one situation and vagal activity in another situation. These reflexes are essentially unconscious or become conscious in pathological conditions such as visceral hypersensitivity as observed in IBS and are closely related to emotions and stress reaction.

4. THE OUTPUT OF THE CENTRAL AUTONOMIC NETWORK AND STRESS BIOLOGICAL SYSTEM

The ability of the central autonomic network to modulate gastrointestinal functions overlaps with the ability of the area involved in this system to modulate also mood, emotions, feelings, thoughts, and motor actions leading to an integrated behavioral response. The final output of the central autonomic network on the digestive tract leads to neural response mediated by the parasympathetic and sympathetic branches of the ANS and to neuroendocrine response mediated by the hypothalamic–pituitary–adrenal (HPA) axis, both axes are involved in the stress reaction.

4.1 Stress Mediators Involved in IBS

Stress is defined as the response of the organism to a solicitation of the challenging environment. The body engages a "fight or flight" response when exposed to an acute challenge with a sympathetic activation leading to an increase of heart rate and respiration, arousal, alertness, and inhibition of acutely nonadaptive vegetative functions (feeding, digestion, growth, and reproduction). The time course of the reaction corresponds to the general syndrome of adaptation defined by Selye (1950). The reaction of stress is physiological and beneficial for the body but may become pathological when an unbalance between the capacities of adaptation and the requirement of the environment appears, thus leading to functional, metabolic, and even inflammatory disorders.

4.1.1 The CRFergic System

Stress involves the CRFergic system which is composed of corticotropin-releasing factor or CRF, the principal neuromediator of stress, a 41-AA peptide, first isolated and identified in the hypothalamus (Spiess, Rivier, Rivier, & Vale, 1981). CRF and its related peptides the urocortins (Ucn 1, Ucn 2, Ucn 3) exert their biological actions on target cells through activation of two receptors known as CRF receptor type 1 (CRF1) and CRF receptor type 2 (CRF2) which are encoded by two distinct genes (Gravanis & Margioris, 2005). CRF binds more particularly to CRF1 while Ucn1 has equal affinity to CRF1 and CRF2 receptors, and Ucns 2 and 3 bind selectively to CRF2. Beside its role as a hypothalamic hypophysiotropic hormone, CRF also acts as a neurotransmitter in several brain areas and has predominantly excitatory actions on neurons in the hippocampus, cortex, *locus coeruleus*, and hypothalamic nuclei (Siggins, Gruol, Aldenhoff, & Pittman, 1985). CRF1 mediates anxiety-like behaviors, whereas CRF2 mediates anxiolytic effects (Bonaz & Tache, 1994; Heinrichs, Lapsansky, Lovenberg, De Souza, & Chalmers, 1997). The CRF system plays a critical role in coordinating the autonomic, endocrine, and behavioral responses to stress (Dunn & Berridge, 1990) appearing as a key factor in the mediation of stress effects on gut but also on mood and anxiety. The impact of stress on gut function is now more understood. Stress induces modifications of motility, secretion, visceral sensitivity, and local inflammatory responses (Tache & Million, 2015) both through a central and/or a peripheral action through CRF1,2 receptors. Alterations of this complex system in humans are linked to a variety of anxiety-related psychiatric disorders and stress-sensitive pain syndromes, including IBS. Dysfunction in the HPA axis regulation attributable to overactivation of CRF/CRF1 signaling in response to chronic stress has been implicated in the pathophysiology of IBS symptoms (Bravo, Dinan, & Cryan, 2011). Actually, under stress, both CRF receptors may be activated simultaneously and the signaling balance of CRF1 and CRF2 may determine the functional changes of gastrointestinal tract induced by stress. CRF signaling balance might be abnormally shifted toward CRF1, leading to enhanced colonic motility and visceral sensitization (Nozu & Okumura, 2015) and amygdala activation and anxiety (Tache, 2015) in IBS.

4.1.2 The Parasympathetic Vagal Tone and the Sympathetic System

During the stress reaction, the parasympathetic system is blunted as it is revealed by a decrease in vagal tone, while the sympathetic nervous system is activated and leads to the release of noradrenalin and adrenalin, two major coordinators of cardiovascular and respiratory, emotional, and cognitive adaptation to the stressful situation. Studies provide convergent evidence that measuring vagal tone in correlation with cortisol represents a window open on the homeostasis of the central autonomic network. The resting vagal tone as it could be indexed by heart rate variability (HRV) has been shown to be related to anxiety in panic disorders (Friedman & Thayer, 1998). Thus vagal tone could be seen as a fingerprint or an "endophenotype" reflecting the balance of the autonomic network (Thayer & Lane, 2009). The outputs of the central autonomic network are directly linked to HRV, and Thayer and Lane proposed HRV as an indicator of CNS–ANS integration. In this integrative interplay, the functional coupling between low cortisol levels and high vagal tone at rest would reflect the tonic inhibition of the prefrontal cortex on subcortical sympathoexcitatory circuits such as the amygdala (Bonaz et al., 2012). In functional diseases such as IBS, autonomic dysfunctions have been described with nonadapted responses to stressors (Cheng et al., 2013; Salvioli et al., 2015; Tillisch et al., 2005) or hyperactivity of the sympathetic system under resting conditions (Pellissier et al., 2014) that are specifically associated to low resting vagal tone and high score of anxiety state and trait (Pellissier, Dantzer, Canini, Mathieu, & Bonaz, 2010).

4.2 Effect of Stress on Gastrointestinal Functions

4.2.1 Stress Effect on Motility and Secretion

Stress induces several effects on gastrointestinal motility such as delayed gastric emptying, lengthened small bowel motility, and increased colonic motility and secretion. These effects are mediated by the ANS (Muth, Thayer, Stern, Friedman, & Drake, 1998) and the CRFergic system. Stress delays gastric emptying through CRF2 receptors while activating colonic motility and secretion through CRF1 receptors (Tache & Bonaz, 2007). In the small bowel, CRF-like peptides stimulate the contractile activity of the duodenum through CRF1 receptor while inhibiting phasic contractions of the ileum through CRF2 receptor (Porcher, Juhem, Peinnequin, Sinniger, & Bonaz, 2005). Stress also induces an activation of the sacral parasympathetic nucleus through the projections of the Barrington nucleus through CRF activation thus stimulating rectocolonic motility (Tache & Bonaz, 2007). Numerous data have established the involvement of peripheral CRF signaling in the modulation of secretory function under stress conditions via activation of both CRF1 and CRF2 receptors, activation of cholinergic enteric neurons, mast cells, and possibly serotonergic pathways (Kiank, Tache, & Larauche, 2010; Larauche, Kiank, & Tache, 2009).

4.2.2 Stress Effect on Visceral Sensitivity

Stress is known to increase visceral sensitivity. Stress is able to increase visceral perception and emotional response to visceral events by a perturbation of the brain–gut axis at its different gut, ANS and CNS levels. Genetic models of anxiety, such as the high-anxiety Wistar–Kyoto rats or Flinders Sensitive Line rats, exhibit increased sensitivity to colorectal distension (Overstreet & Djuric, 2001). In the same way genetic models deleting CRF1 exhibit a decreased colonic sensitivity to colonic distension (Trimble, Johnson, Foster, & Greenwood-van Meerveld, 2007), while models overexpressing CRF1 exhibit enhanced response (Million et al., 2007).

These data argue for the filiation stress–anxiety and visceral hypersensitivity. Again, the CRF signaling, both at the central and peripheral level, is a key element involved in this effect. At the peripheral level, mast cells degranulation observed in the colon following stress and peripheral administration of CRF induces visceral hypersensitivity via the release of mediators (histamine, tryptase, prostaglandin E2, nerve growth factor) that can stimulate or sensitize sensory afferents (van den Wijngaard, Klooker, de Jonge, & Boeckxstaens, 2010; Wallon et al., 2008). Intravenous administration of CRF increases GI motility and visceral pain sensitivity in IBS patients compared with healthy controls, whereas administration of a nonselective CRF receptor antagonist improved these responses (Larauche, Gourcerol, et al., 2009; Million et al., 2005).

4.2.3 Stress Effect on Intestinal Permeability

An increase of intestinal permeability is observed in the colon of IBS patients, associated with visceral or somatic hypersensitivity (Zhou & Verne, 2011). Stress is able to disrupt the intestinal epithelial barrier thus increasing the penetration of luminal antigens into the *lamina propria*, leading to nociceptors' sensitization and thus favoring the development of visceral hypersensitivity (Ait-Belgnaoui, Bradesi, Fioramonti, Theodorou, & Bueno, 2005). This increase of intestinal permeability is due to an activation of peripheral CRF signaling involving both CRF2 and CRF1 (Buckinx, Adriaensen, Nassauw, & Timmermans, 2011) as well as mast cell activation (Santos, Yang, Soderholm, Benjamin, & Perdue, 2001).

4.2.4 Stress Effect on the Human Gut Microbiota

The human microbiota is a diverse and dynamic ecosystem. The gut contains 400–1000 different bacterial species (Bercik et al., 2011). The microbiota helps safeguard the host from external pathogens and modulates the immune response, aids in the metabolism of polysaccharides and lipids, and modulates intestinal motility, in addition to modulating visceral perception (Chow, Lee, Shen, Khosravi, & Mazmanian, 2010; Lee & Mazmanian, 2010) and even social behavior (Montiel-Castro, Gonzalez-Cervantes, Bravo-Ruiseco, & Pacheco-Lopez, 2013). The microbiota also acts as a source of neurochemicals used to regulate a vast array of physiological and psychological processes. For example, the microbiota is the source of approximately 95% of the body's serotonin, a critical neurotransmitter in the modulation of gastrointestinal motility and mood regulation (Sommer & Backhed, 2013). Gut bacteria are hypothesized to cause IBS via direct effects on motility, sensitivity and inflammation, overgrowth, and changes in relative populations of phyla. Gut microbiota is included in a brain–gut biofeedback loop where local change in gut microbe enhances a local inflammation which may allow a bacterial translocation and subsequent increase in cytokine, such as TNFα, and mediator such as serotonin released from the neuroendocrine cells that induces pain sensitivity and motility changes which in turn may change gut flora. This phenomenon may be enhanced by anxiety and depression which alters ANS outflow by increasing catecholamine levels (Lyte, Vulchanova, & Brown, 2011) and increased CRF and cortisol secretion (Keightley, Koloski, & Talley, 2015). Early data have shown, for postinfectious IBS, that at the time of the initial gastroenteritis, patients who had higher scores for anxiety, depression, somatization, and neuroticism were more likely to develop symptom chronicity of IBS (Gwee et al., 1996). Exchange of gut flora between mice can alter both anxiety and expression of BDNF (neurotrophic factor involved in mood regulation) in the brain, and antibiotic treatment in these mice could transiently decrease anxious behavior and BDNF expression if given orally (active on local flora) but not if given intraperitoneally (Bercik et al., 2011). Taking evidence from both human and animal studies, Farmer, Randall, and Aziz (2014) proposed that an acute insult to the gut microbiota can cause central changes culminating in alterations in both mood and behavior in susceptible individuals. Furthermore, a body of data outlined the consequences of early-life stress on microbiota homeostasis (O'Mahony et al., 2009). Early-life stressors, represented by maternal separation in the form of limited nesting, odor attachment learning that are models of maternal neglect or abuse, and physical stressors in the form of injury such as colonic irritation, can enhance the susceptibility of individuals to develop hypersensitivity and IBS-like symptoms in the adulthood. The hypothesis from these models is that the abnormal solicitation of the HPA axis at a vulnerable stage of life and brain development can impact on microbial colonization and population, both inducing long-lasting sequelae (Anderson, Damasio, Tranel, & Damasio, 2000) on stress responses and subsequent visceral sensitivity as recently exposed in a nice review (Moloney et al., 2016).

5. EMOTIONS AND STRESS IN IBS

The GI tract is a sensitive target to stress. Numerous data argue for a role of stress in the pathophysiology of IBS. Patients with IBS report more stressful life events than medical comparison groups or healthy subjects (Drossman, 2011a; Drossman, Li, Leserman, Toomey, & Hu, 1996; Sayuk & Drossman, 2011). Stress is strongly associated with symptom onset and symptom severity in IBS patients. Illness experience, health care-seeking behavior, and treatment outcome are adversely affected by stressful life events, chronic social stress, anxiety disorders, and nonadaptive coping style. A history of emotional, sexual, or physical abuse is often found in IBS patients. For example, there is a higher prevalence (i.e., 44%) of sexual or physical abuse in patients with functional GI disorders than in controls with organic GI disorders (Drossman, 2011a; Drossman, Leserman, & Hu, 1996). Psychiatric comorbidity, especially major depression, anxiety, and somatoform disorders, occurs among 20–50% of IBS patients (Garakani et al., 2003). Stress is involved in such disorders, and psychiatric disorders precede the onset of the GI symptoms, thus suggesting a role for psychiatric disorders in functional gastrointestinal disorders development (Sykes, Blanchard, Lackner, Keefer, & Krasner, 2003).

Functional brain imaging studies have shown that there is a major influence of cognitive–affective processes on gastrointestinal sensations and its CNS correlates in health and functional digestive disorders as IBS (Naliboff & Mayer, 2006; Van Oudenhove, Coen, & Aziz, 2007; Van Oudenhove, Dupont, et al., 2008). Cognitive–affective processes including arousal, attention, conditioning, and negative affect strongly influence visceral pain perception through modulation of its neural correlates (Tillisch et al., 2011). Feeling emotions requires the participation of brain regions, such as the somatosensory cortices and the upper brainstem nuclei that are involved in the mapping and/or regulation of internal organism states (Adolphs, Damasio, Tranel, Cooper, & Damasio, 2000; Damasio et al., 2000). These data are in agreement with the role of hypervigilance in the visceral hypersensitivity observed in IBS patients (Naliboff et al., 2008). Perceived stress, anxiety, somatization, and negative illness beliefs at the time of infection have been characterized as predictors of postinfectious IBS outcome (Spence & Moss-Morris, 2007); this is in favor of a cognitive–emotional "central sensitization" model of IBS. The importance of psychosocial factors and somatization compared to gastric sensorimotor function is most pronounced in hypersensitive patients with functional dyspepsia, another functional GI disorder (Van Oudenhove, Vandenberghe, et al., 2008).

Early-life trauma (neglect, abuse, loss of caregiver, or life-threatening situation) increases susceptibility to develop later affective disorders such as depression, anxiety, and is a key factor in the development of IBS

(Bradford et al., 2012). Traumatic events, such as war, environmental disasters, physical abuse, or a bad accident in adulthood, can induce fear acquisition conditioning such as posttraumatic stress disorder where an increased prevalence of gastrointestinal symptoms, such as IBS, is found (Cohen et al., 2006). Moreover, CRF signaling system via CRF1 receptors, involved in fear acquisition and extinction learning, appears to be upregulated in patients with IBS (Labus, Hubbard, et al., 2013). The role of stress conditioning is also reproduced in preclinical studies. Adult rats previously subjected to neonatal maternal separation exhibit visceral hypersensitivity to colorectal distension in basal conditions (Ren et al., 2007). This visceral hypersensitivity is exacerbated in acute stress (e.g., water avoidance stress: WAS) conditions (Coutinho et al., 2002). Chronic exposure to repeat WAS is used to study visceral hypersensitivity and is more close to clinical conditions. However, habituation of the CRFergic system is observed in chronic conditions (Bonaz & Rivest, 1998) and may induce analgesia. It seems that these conflicting data are influenced by the basal state conditions of the animals before applying the repeated stressor (surgery and single housing) (Larauche, Gourcerol, Million, Adelson, & Tache, 2010). Surgery and housing of animals are stressful events and recovery is animal dependent; thus according to the impact of this initial stress on individuals, the final outcome of repeat stressors can vary from one extreme to another, and rather than conflicting this, results outline the priming effect of day life event on current stress reactivity and visceral sensitivity.

5.1 Central and Peripheral Autonomic Network Imbalance

The imbalance between the prefrontal cortex and the amygdala in the CNS results in an imbalance between the HPA axis and the ANS at the peripheral level (Pellissier et al., 2014). The lower the vagal tone, the less active the prefrontal cortex will be, reflecting a shifting from a homeostatic state to a stress state where the individual will have to face a situation of treat and uncertainty. This allows structures like the amygdala to be more active. This must be associated with emotional and physiological outputs representative of a stress reaction such as an increase in epinephrine and anxiety as it has been found in IBS. As an example, healthy subjects exhibiting high vagal tone at rest will have the greater decrease in salivary cortisol levels in the evening. In contrast, this inverse relationship between cortisol and HRV (vagal tone) is not found in IBS patients. This argues for an uncoupling between the HPA axis and the ANS in this syndrome. This uncoupling between the HPA axis and the ANS in IBS patients could reflect an uncoupling at the level of CNS between the prefrontal cortex and the amygdala. This is consistent with the fact that the prefrontal cortex is hypoactive in IBS in situations of uncertainty (Aizawa et al., 2012). According to the McEwen model of stress (Peters & McEwen, 2012) this uncoupling between the HPA and the ANS axis observed in IBS would be the sign of a high-cost allostatic regulation with reduced flexibility in the regulatory systems which could accelerate the relapse. Such a situation may explain why IBS patients are more reactive to stressful life events or other challenging situations and thus more probable to relapse. Indeed, inhibition, delayed response, and physiological and cognitive flexibility are important for successful adjustment and self-regulation. One could imagine that a prolonged prefrontal hypoactivity can lead to hypervigilance, anxiety, and/or depression. This is corroborated by findings showing that some IBS patients present negative psychological adjustments to their disease being perceived as an additional source of burden. Concomitantly, these patients present more negative coping strategies with low problem-focused coping (Pellissier et al., 2010) that could be related to a lack of activity of their prefrontal cortex. That is why the use of a global psychophysiological criterion of vulnerability for IBS relapse was proposed. This global criterion includes, in addition to adapted questionnaires, the evaluation of the resting vagal tone in IBS. Moreover, the prefrontal cortex, in its ventromedial part (vmPFC), is engaged in fear extinction process in relation with the amygdala (Phelps, Delgado, Nearing, & LeDoux, 2004). The amygdala is an important region in the acquisition of fear conditioning associating a conditioned stimulus with an unconditioned stimulus. The vmPFC participates in the mechanism of extinction and also in the recall of fear extinction with an active inhibition of the previous fear condition responses. This is mediated by a down-regulation of amygdala outputs with vmPFC neurons exciting (glutamate) inhibitory neurons (GABA) (for review, see Bonaz et al., 2012). The activity of intentional regulation of treat-related cues by the vmPFC is decreased in anxious patients, and the conditioned fear extinction is also less active in anxious patients and this is associated with symptoms provocations (Icenhour et al., 2015). Amygdala is also activated by uncertainty, and the capacity of the PFC to regulate attention and reappraisal modulates the level of the response of amygdala. In IBS, uncertainty plays an important role in the perception of pain. The inhibitory control of the vmPFC on amygdala would maintain a homeostatic state with an equilibrated sympathovagal balance and low glucocorticoids circulating levels. In the case of a deficit in

PFC activity with a lack of inhibitory regulatory communications with the amygdala, a chronic imbalance of the ANS with an increase sympathetic activity should appear as observed in IBS patients exhibiting a low HRV and a high score of anxiety (Pellissier et al., 2010). Moreover, there is a strong relation between the activity of the ANS and the immune system through the cholinergic antiinflammatory pathway (Huston & Tracey, 2011). Hence, when the parasympathetic system is blunted, as in anxiety for instance, it could facilitate inflammation that could be deleterious for health and well-being. The hypoactivity of the PFC and the enhancement of amygdala are strongly influenced by stress as demonstrated by a number of studies (Kavushansky & Richter-Levin, 2006; Vyas, Jadhav, & Chattarji, 2006). In IBS amygdala control is altered (Hubbard et al., 2011) along with brain structural changes of the HPA axis and other limbic structures (Blankstein, Chen, Diamant, & Davis, 2010).

5.2 Influence of Sex Hormones

Gastrointestinal disorders are strongly related to ovarian function. The prevalence of visceral pain disorders such as IBS, gastroesophageal reflux disease, gallbladder, and biliary tract diseases is significantly more frequent in women (Palomba, Di Cello, Riccio, Manguso, & La Sala, 2011). Notably, IBS appears with a higher prevalence in women in a ratio of 2:1 with the majority of them (80%) attributing psychological stress as a cause of their intestinal disorders (Drossman, 1990). Sex hormones play a significant role in the pathophysiology of IBS including regulation of motor and sensory function but also visceral perception (Chang et al., 2006). Furthermore, symptoms, such as nausea, vomiting, abdominal pain, distension, satiety, bloating, diarrhea, or constipation, frequently appear in relation with pregnancy, luteal phase of the menstrual cycle, or perimenopausal and menopausal states. Further support for the contribution of ovarian steroids to functional gastrointestinal disorders comes from studies demonstrating that pharmacological ovariectomy reduces abdominal pain symptoms (Palomba et al., 2011). Sex hormones influence gastrointestinal function by acting at each level of the brain–gut axis, i.e., centrally on the brain: on emotional-related area (Myers, Schulkin, & Greenwood-Van Meerveld, 2011), on HPA axis and on the spinal cord (Cao, Bai, Ji, & Traub, 2015), peripherally on the ANS particularly via the vagal nerve (Yan et al., 2014), and locally on the ENS and the microbiota (composition, growth, metabolism, virulence). Among numerous interactions of sex hormones with other neurotransmitters, gonadal hormones and particularly estrogens interact with serotonin and CRF signaling systems as recently reported in an exhaustive review (Mulak, Tache, & Larauche, 2014).

6. THERAPEUTIC IMPLICATIONS—TREATMENT TARGETING AMYGDALA ACTIVITY REDUCTION AND VAGAL TONE REINFORCEMENT IN IBS: MINDFULNESS AND HYPNOSIS

The effect of stress on IBS has therapeutic implications both with nonpharmacological and pharmacological treatment to reduce stress perception. Psychological mind–body interventions including psychotherapy (Zijdenbos, de Wit, van der Heijden, Rubin, & Quartero, 2009), cognitive behavioral therapy (Kearney & Brown-Chang, 2008; Labus, Gupta, et al., 2013; Toner, 2005), hypnotherapy (Flik et al., 2011; Ford et al., 2014), biofeedback (Dobbin, Dobbin, Ross, Graham, & Ford, 2013) relaxation exercises, or mindfulness meditation (Keefer & Blanchard, 2001; Shinozaki et al., 2010; Zernicke et al., 2013) have been shown to improve symptoms of IBS patients. Repetitive transcranial magnetic stimulation of the prefrontal cortex and lumbosacral area can cause changes in acute pain perception and has been recently used in IBS patients (Algladi, Harris, Whorwell, Paine, & Hamdy, 2015; Melchior et al., 2014). Modulation of the ANS for restoring the sympathovagal balance by means of vagal nerve stimulation techniques (Bonaz, Picq, Sinniger, Mayol, & Clarencon, 2013; Frokjaer et al., 2016; Zhang et al., 2013) contributes to reduce pain sensitivity and perception as well as modify coping strategies and vigilance state. This may contribute to the restoration of a functional brain–gut axis in IBS. Conventional treatment as represented by antidepressives, anxiolytics, drug targeting the central sensitization syndrome (pregabalin, gabapentin), tachykinin receptor antagonists either directly and/or indirectly is supposed to target the hyperfunctioning of the amygdala (Ghaith, El-Halabi, Hashash, & Sharara, 2010). In the context of the microbiota–brain–gut axis, probiotics, prebiotics, and antibiotics such as rifaximin, an antibacterial agent that is virtually unabsorbed after oral administration and devoid of systemic side effects, are of interest (Drossman, 2011b). If targeting CRF signaling with CRF1 receptor antagonists, based on preclinical and/or clinical data (brain imaging), has been used successfully in humans to treat depression and anxiety (Bourke, Glasper, & Neigh, 2014; Kehne & De Lombaert, 2002), their efficacy is still debated in the treatment of IBS patients (Sweetser et al., 2009).

7. CONCLUSION

The pathophysiology of IBS, mainly characterized by visceral hypersensitivity and pain, is multifactorial and is explained by the biopsychosocial model of illnesses. This model is based on the bidirectional interplay between the microbiota–gut system and the brain referred to as the brain–gut axis where the ANS and particularly the vagus nerve play a predominant role in visceroception and gut responses through multiple complex loops of regulation. Emotional states such as anxiety and depression but also stress reaction and exposition increase the risk of developing hypersensitivity through several pathways, namely, immune, neuroendocrine, and autonomic alterations. These mechanisms may alter gut microbiota's homeostasis by favoring the ability of enteric pathogens to colonize the intestine with an impact on psychiatric manifestations. Pharmacologic and nonpharmacologic therapeutic interventions that target the brain–gut axis could alleviate the debilitating symptoms of IBS.

REFERENCES

Adolphs, R., Damasio, H., Tranel, D., Cooper, G., & Damasio, A. R. (2000). A role for somatosensory cortices in the visual recognition of emotion as revealed by three-dimensional lesion mapping. *The Journal of Neuroscience*, *20*(7), 2683–2690.

Ait-Belgnaoui, A., Bradesi, S., Fioramonti, J., Theodorou, V., & Bueno, L. (2005). Acute stress-induced hypersensitivity to colonic distension depends upon increase in paracellular permeability: Role of myosin light chain kinase. *Pain*, *113*(1–2), 141–147. http://dx.doi.org/10.1016/j.pain.2004.10.002.

Aizawa, E., Sato, Y., Kochiyama, T., Saito, N., Izumiyama, M., Morishita, J., … Fukudo, S. (2012). Altered cognitive function of prefrontal cortex during error feedback in patients with irritable bowel syndrome, based on FMRI and dynamic causal modeling. *Gastroenterology*, *143*(5), 1188–1198. http://dx.doi.org/10.1053/j.gastro.2012.07.104.

Algladi, T., Harris, M., Whorwell, P. J., Paine, P., & Hamdy, S. (2015). Modulation of human visceral sensitivity by noninvasive magnetoelectrical neural stimulation in health and irritable bowel syndrome. *Pain*, *156*(7), 1348–1356. http://dx.doi.org/10.1097/j.pain.0000000000000187.

Altschuler, S. M., Bao, X. M., Bieger, D., Hopkins, D. A., & Miselis, R. R. (1989). Viscerotopic representation of the upper alimentary tract in the rat: Sensory ganglia and nuclei of the solitary and spinal trigeminal tracts. *The Journal of Comparative Neurology*, *283*(2), 248–268. http://dx.doi.org/10.1002/cne.902830207.

Altschuler, S. M., Escardo, J., Lynn, R. B., & Miselis, R. R. (1993). The central organization of the vagus nerve innervating the colon of the rat. *Gastroenterology*, *104*(2), 502–509.

Anderson, S. W., Damasio, H., Tranel, D., & Damasio, A. R. (2000). Long-term sequelae of prefrontal cortex damage acquired in early childhood. *Developmental Neuropsychology*, *18*(3), 281–296. http://dx.doi.org/10.1207/S1532694202Anderson.

Azpiroz, F., Bouin, M., Camilleri, M., Mayer, E. A., Poitras, P., Serra, J., & Spiller, R. C. (2007). Mechanisms of hypersensitivity in IBS and functional disorders. *Neurogastroenterology and Motility*, *19*(1 Suppl.), 62–88. http://dx.doi.org/10.1111/j.1365-2982.2006.00875.x.

Bayliss, W. M., & Starling, E. H. (1899). The movements and innervation of the small intestine. *The Journal of Physiology*, *24*(2), 99–143.

Benouali-Pellissier, S., Roger, T., & Roche, M. (1994). A possible interaction between serotonin and cholecystokinin-8 in the ileo-colonic motor responses to feeding in rats. *Scandinavian Journal of Gastroenterology*, *29*(2), 138–143.

Bercik, P., Denou, E., Collins, J., Jackson, W., Lu, J., Jury, J., … Collins, S. M. (2011). The intestinal microbiota affect central levels of brain-derived neurotropic factor and behavior in mice. *Gastroenterology*, *141*(2), 599–609. http://dx.doi.org/10.1053/j.gastro.2011.04.052. 609.e1–3.

Blankstein, U., Chen, J., Diamant, N. E., & Davis, K. D. (2010). Altered brain structure in irritable bowel syndrome: Potential contributions of pre-existing and disease-driven factors. *Gastroenterology*, *138*(5), 1783–1789. http://dx.doi.org/10.1053/j.gastro.2009.12.043.

Bonaz, B. (2003). Visceral sensitivity perturbation integration in the brain-gut axis in functional digestive disorders. *Journal of Physiology and Pharmacology*, *54*(Suppl. 4), 27–42.

Bonaz, B., Pellissier, S., Sinniger, V., Clarençon, D., Peinnequin, A., & Canini, F. (2012). The irritable bowel syndrome: How stress can affect the amygdala activity and the brain-gut axis. In D. B. Ferry (Ed.), *The amygdala—A discrete multitasking manager*: InTech. http://dx.doi.org/10.5772/52066. Retrieved from http://www.intechopen.com/books/the-amygdala-a-discrete-multitasking-manager/the-irritable-bowel-syndrome-how-stress-can-affect-the-amygdala-activity-and-the-brain-gut-axis.

Bonaz, B., Picq, C., Sinniger, V., Mayol, J. F., & Clarencon, D. (2013). Vagus nerve stimulation: From epilepsy to the cholinergic anti-inflammatory pathway. *Neurogastroenterology and Motility*, *25*(3), 208–221. http://dx.doi.org/10.1111/nmo.12076.

Bonaz, B., & Rivest, S. (1998). Effect of a chronic stress on CRF neuronal activity and expression of its type 1 receptor in the rat brain. *The American Journal of Physiology*, *275*(5 Pt. 2), R1438–R1449.

Bonaz, B., & Sabate, J. M. (2009). Brain-gut axis dysfunction. *Gastroentérologie Clinique et Biologique*, *33*(Suppl. 1), S48–S58. http://dx.doi.org/10.1016/S0399-8320(09)71525-8.

Bonaz, B., & Tache, Y. (1994). Water-avoidance stress-induced c-fos expression in the rat brain and stimulation of fecal output: Role of corticotropin-releasing factor. *Brain Research*, *641*(1), 21–28.

Bourke, C. H., Glasper, E. R., & Neigh, G. N. (2014). SSRI or CRF antagonism partially ameliorate depressive-like behavior after adolescent social defeat. *Behavioural Brain Research*, *270*, 295–299. http://dx.doi.org/10.1016/j.bbr.2014.05.035.

Bradette, M., Delvaux, M., Staumont, G., Fioramonti, J., Bueno, L., & Frexinos, J. (1994). Evaluation of colonic sensory thresholds in IBS patients using a barostat. Definition of optimal conditions and comparison with healthy subjects. *Digestive Diseases and Sciences*, *39*(3), 449–457.

Bradford, K., Shih, W., Videlock, E. J., Presson, A. P., Naliboff, B. D., Mayer, E. A., & Chang, L. (2012). Association between early adverse life events and irritable bowel syndrome. *Clinical Gastroenterology and Hepatology: The Official Clinical Practice Journal of the American Gastroenterological Association*, *10*(4), 385–390. http://dx.doi.org/10.1016/j.cgh.2011.12.018. e381–e383.

Bravo, J. A., Dinan, T. G., & Cryan, J. F. (2011). Alterations in the central CRF system of two different rat models of comorbid depression and functional gastrointestinal disorders. *The International Journal of Neuropsychopharmacology*, *14*(5), 666–683. http://dx.doi.org/10.1017/S1461145710000994.

Buckinx, R., Adriaensen, D., Nassauw, L. V., & Timmermans, J. P. (2011). Corticotrophin-releasing factor, related peptides, and receptors in the normal and inflamed gastrointestinal tract. *Frontiers in Neuroscience*, *5*, 54. http://dx.doi.org/10.3389/fnins.2011.00054.

Burstein, R. (1996). Somatosensory and visceral input to the hypothalamus and limbic system. *Progress in Brain Research*, *107*, 257–267.

Canavan, C., West, J., & Card, T. (2016). Calculating total health service utilisation and costs from routinely collected electronic health records using the example of patients with irritable bowel syndrome before and after their first gastroenterology appointment. *PharmacoEconomics*, *34*, 181–194. http://dx.doi.org/10.1007/s40273-015-0339-y.

Cao, D. Y., Bai, G., Ji, Y., & Traub, R. J. (2015). Epigenetic upregulation of metabotropic glutamate receptor 2 in the spinal cord attenuates oestrogen-induced visceral hypersensitivity. *Gut*, *64*(12), 1913–1920. http://dx.doi.org/10.1136/gutjnl-2014-307748.

Chang, L., Mayer, E. A., Labus, J. S., Schmulson, M., Lee, O. Y., Olivas, T. I., … Naliboff, B. D. (2006). Effect of sex on perception of rectosigmoid stimuli in irritable bowel syndrome. *American Journal of Physiology. Regulatory, Integrative and Comparative Physiology*, *291*(2), R277–R284. http://dx.doi.org/10.1152/ajpregu.00729.2005.

Cheng, P., Shih, W., Alberto, M., Presson, A. P., Licudine, A., Mayer, E. A., … Chang, L. (2013). Autonomic response to a visceral stressor is dysregulated in irritable bowel syndrome and correlates with duration of disease. *Neurogastroenterology and Motility*, *25*(10), e650–e659. http://dx.doi.org/10.1111/nmo.12177.

Chow, J., Lee, S. M., Shen, Y., Khosravi, A., & Mazmanian, S. K. (2010). Host-bacterial symbiosis in health and disease. *Advances in Immunology*, *107*, 243–274. http://dx.doi.org/10.1016/B978-0-12-381300-8.00008-3.

Cliffer, K. D., Burstein, R., & Giesler, G. J., Jr. (1991). Distributions of spinothalamic, spinohypothalamic, and spinotelencephalic fibers revealed by anterograde transport of PHA-L in rats. *The Journal of Neuroscience*, *11*(3), 852–868.

Cohen, H., Jotkowitz, A., Buskila, D., Pelles-Avraham, S., Kaplan, Z., Neumann, L., & Sperber, A. D. (2006). Post-traumatic stress disorder and other co-morbidities in a sample population of patients with irritable bowel syndrome. *European Journal of Internal Medicine*, *17*(8), 567–571. http://dx.doi.org/10.1016/j.ejim.2006.07.011.

Coutinho, S. V., Plotsky, P. M., Sablad, M., Miller, J. C., Zhou, H., Bayati, A. I., … Mayer, E. A. (2002). Neonatal maternal separation alters stress-induced responses to viscerosomatic nociceptive stimuli in rat. *American Journal of Physiology. Gastrointestinal and Liver Physiology*, *282*(2), G307–G316. http://dx.doi.org/10.1152/ajpgi.00240.2001.

Damasio, A. R., Grabowski, T. J., Bechara, A., Damasio, H., Ponto, L. L., Parvizi, J., & Hichwa, R. D. (2000). Subcortical and cortical brain activity during the feeling of self-generated emotions. *Nature Neuroscience*, *3*(10), 1049–1056. http://dx.doi.org/10.1038/79871.

Dinan, T. G., Clarke, G., Quigley, E. M., Scott, L. V., Shanahan, F., Cryan, J., … Keeling, P. W. (2008). Enhanced cholinergic-mediated increase in the pro-inflammatory cytokine IL-6 in irritable bowel syndrome: Role of muscarinic receptors. *The American Journal of Gastroenterology*, *103*(10), 2570–2576. http://dx.doi.org/10.1111/j.1572-0241.2008.01871.x.

Dobbin, A., Dobbin, J., Ross, S. C., Graham, C., & Ford, M. J. (2013). Randomised controlled trial of brief intervention with biofeedback and hypnotherapy in patients with refractory irritable bowel syndrome. *The Journal of the Royal College of Physicians of Edinburgh*, *43*(1), 15–23. http://dx.doi.org/10.4997/JRCPE.2013.104.

Drossman, D. A. (1990). Characterization of intestinal function and diagnosis of irritable bowel syndrome by surveys and questionnaires. *Gastroentérologie Clinique et Biologique*, *14*(5 (Pt. 2)), 37C–41C.

Drossman, D. A. (1996). Gastrointestinal illness and the biopsychosocial model. *Journal of Clinical Gastroenterology*, *22*(4), 252–254.

Drossman, D. A. (1999). Review article: An integrated approach to the irritable bowel syndrome. *Alimentary Pharmacology & Therapeutics*, *13*(Suppl. 2), 3–14.

Drossman, D. A. (2011a). Abuse, trauma, and GI illness: Is there a link? *The American Journal of Gastroenterology*, *106*(1), 14–25. http://dx.doi.org/10.1038/ajg.2010.453.

Drossman, D. A. (2011b). Rifaximin for treatment of irritable bowel syndrome. *Gastroenterology & Hepatology*, 7(3), 180–181.

Drossman, D. A., Leserman, J., & Hu, J. B. (1996). Gastrointestinal diagnosis, abuse history, and effects on health status. *Gastroenterology*, *111*(4), 1159–1161.

Drossman, D. A., Li, Z., Leserman, J., Toomey, T. C., & Hu, Y. J. (1996). Health status by gastrointestinal diagnosis and abuse history. *Gastroenterology*, *110*(4), 999–1007.

Dunn, A. J., & Berridge, C. W. (1990). Physiological and behavioral responses to corticotropin-releasing factor administration: Is CRF a mediator of anxiety or stress responses? *Brain Research. Brain Research Reviews*, *15*(2), 71–100.

El-Salhy, M., Gundersen, D., Gilja, O. H., Hatlebakk, J. G., & Hausken, T. (2014). Is irritable bowel syndrome an organic disorder? *World Journal of Gastroenterology*, *20*(2), 384–400. http://dx.doi.org/10.3748/wjg.v20.i2.384.

Elsenbruch, S., Rosenberger, C., Bingel, U., Forsting, M., Schedlowski, M., & Gizewski, E. R. (2010). Patients with irritable bowel syndrome have altered emotional modulation of neural responses to visceral stimuli. *Gastroenterology*, *139*(4), 1310–1319. http://dx.doi.org/10.1053/j.gastro.2010.06.054.

Engel, G. L. (1977). The need for a new medical model: A challenge for biomedicine. *Science*, *196*(4286), 129–136.

Farmer, A. D., Randall, H. A., & Aziz, Q. (2014). It's a gut feeling: How the gut microbiota affects the state of mind. *The Journal of Physiology*, *592*(14), 2981–2988. http://dx.doi.org/10.1113/jphysiol.2013.270389.

Flik, C. E., van Rood, Y. R., Laan, W., Smout, A. J., Weusten, B. L., Whorwell, P. J., & de Wit, N. J. (2011). A randomised controlled trial on hypnotherapy for irritable bowel syndrome: Design and methodological challenges (the IMAGINE study). *BMC Gastroenterology*, *11*, 137. http://dx.doi.org/10.1186/1471-230X-11-137.

Ford, A. C., Quigley, E. M., Lacy, B. E., Lembo, A. J., Saito, Y. A., Schiller, L. R., … Moayyedi, P. (2014). Effect of antidepressants and psychological therapies, including hypnotherapy, in irritable bowel syndrome: Systematic review and meta-analysis. *The American Journal of Gastroenterology*, *109*(9), 1350–1365. http://dx.doi.org/10.1038/ajg.2014.148. quiz 1366.

Friedman, B. H., & Thayer, J. F. (1998). Autonomic balance revisited: Panic anxiety and heart rate variability. *Journal of Psychosomatic Research*, *44*(1), 133–151.

Frokjaer, J. B., Bergmann, S., Brock, C., Madzak, A., Farmer, A. D., Ellrich, J., & Drewes, A. M. (2016). Modulation of vagal tone enhances gastroduodenal motility and reduces somatic pain sensitivity. *Neurogastroenterology and Motility*, *28*, 592–598. http://dx.doi.org/10.1111/nmo.12760.

Fulwiler, C. E., & Saper, C. B. (1984). Subnuclear organization of the efferent connections of the parabrachial nucleus in the rat. *Brain Research*, *319*(3), 229–259.

Furness, J. B. (2012). The enteric nervous system and neurogastroenterology. *Nature Reviews. Gastroenterology & Hepatology*, *9*(5), 286–294. http://dx.doi.org/10.1038/nrgastro.2012.32.

Gaman, A., & Kuo, B. (2008). Neuromodulatory processes of the brain-gut axis. *Neuromodulation*, *11*(4), 249–259. http://dx.doi.org/10.1111/j.1525-1403.2008.00172.x.

Garakani, A., Win, T., Virk, S., Gupta, S., Kaplan, D., & Masand, P. S. (2003). Comorbidity of irritable bowel syndrome in psychiatric patients: A review. *American Journal of Therapeutics*, *10*(1), 61–67.

Gaynes, B. N., & Drossman, D. A. (1999). The role of psychosocial factors in irritable bowel syndrome. *Baillière's Best Practice & Research. Clinical Gastroenterology, 13*(3), 437–452. http://dx.doi.org/10.1053/bega.1999.0038.

Ghaith, O., El-Halabi, M., Hashash, J. G., & Sharara, A. I. (2010). Investigational agents for the irritable bowel syndrome. *Expert Opinion on Investigational Drugs, 19*(10), 1161–1178. http://dx.doi.org/10.1517/13543784.2010.513380.

Gravanis, A., & Margioris, A. N. (2005). The corticotropin-releasing factor (CRF) family of neuropeptides in inflammation: Potential therapeutic applications. *Current Medicinal Chemistry, 12*(13), 1503–1512.

Gwee, K. A., Graham, J. C., McKendrick, M. W., Collins, S. M., Marshall, J. S., Walters, S. J., & Read, N. W. (1996). Psychometric scores and persistence of irritable bowel after infectious diarrhoea. *Lancet, 347*(8995), 150–153.

Heinrichs, S. C., Lapsansky, J., Lovenberg, T. W., De Souza, E. B., & Chalmers, D. T. (1997). Corticotropin-releasing factor CRF1, but not CRF2, receptors mediate anxiogenic-like behavior. *Regulatory Peptides, 71*(1), 15–21.

Hubbard, C. S., Labus, J. S., Bueller, J., Stains, J., Suyenobu, B., Dukes, G. E., … Mayer, E. A. (2011). Corticotropin-releasing factor receptor 1 antagonist alters regional activation and effective connectivity in an emotional-arousal circuit during expectation of abdominal pain. *The Journal of Neuroscience, 31*(35), 12491–12500. http://dx.doi.org/10.1523/JNEUROSCI.1860-11.2011.

Huston, J. M., & Tracey, K. J. (2011). The pulse of inflammation: Heart rate variability, the cholinergic anti-inflammatory pathway and implications for therapy. *Journal of Internal Medicine, 269*(1), 45–53. http://dx.doi.org/10.1111/j.1365-2796.2010.02321.x.

Icenhour, A., Langhorst, J., Benson, S., Schlamann, M., Hampel, S., Engler, H., … Elsenbruch, S. (2015). Neural circuitry of abdominal pain-related fear learning and reinstatement in irritable bowel syndrome. *Neurogastroenterology and Motility, 27*(1), 114–127. http://dx.doi.org/10.1111/nmo.12489.

Kavushansky, A., & Richter-Levin, G. (2006). Effects of stress and corticosterone on activity and plasticity in the amygdala. *Journal of Neuroscience Research, 84*(7), 1580–1587. http://dx.doi.org/10.1002/jnr.21058.

Kearney, D. J., & Brown-Chang, J. (2008). Complementary and alternative medicine for IBS in adults: Mind-body interventions. *Nature Clinical Practice. Gastroenterology & Hepatology, 5*(11), 624–636. http://dx.doi.org/10.1038/ncpgasthep1257.

Keefer, L., & Blanchard, E. B. (2001). The effects of relaxation response meditation on the symptoms of irritable bowel syndrome: Results of a controlled treatment study. *Behaviour Research and Therapy, 39*(7), 801–811.

Kehne, J., & De Lombaert, S. (2002). Non-peptidic CRF1 receptor antagonists for the treatment of anxiety, depression and stress disorders. *Current Drug Targets. CNS and Neurological Disorders, 1*(5), 467–493.

Keightley, P. C., Koloski, N. A., & Talley, N. J. (2015). Pathways in gut-brain communication: Evidence for distinct gut-to-brain and brain-to-gut syndromes. *The Australian and New Zealand Journal of Psychiatry, 49*(3), 207–214. http://dx.doi.org/10.1177/0004867415569801.

Kiank, C., Tache, Y., & Larauche, M. (2010). Stress-related modulation of inflammation in experimental models of bowel disease and post-infectious irritable bowel syndrome: Role of corticotropin-releasing factor receptors. *Brain, Behavior, and Immunity, 24*(1), 41–48. http://dx.doi.org/10.1016/j.bbi.2009.08.006.

Koloski, N. A., Jones, M., Kalantar, J., Weltman, M., Zaguirre, J., & Talley, N. J. (2012). The brain–gut pathway in functional gastrointestinal disorders is bidirectional: A 12-year prospective population-based study. *Gut, 61*(9), 1284–1290. http://dx.doi.org/10.1136/gutjnl-2011-300474.

Labus, J., Gupta, A., Gill, H. K., Posserud, I., Mayer, M., Raeen, H., … Mayer, E. A. (2013). Randomised clinical trial: Symptoms of the irritable bowel syndrome are improved by a psycho-education group intervention. *Alimentary Pharmacology & Therapeutics*, *37*(3), 304–315. http://dx.doi.org/10.1111/apt.12171.

Labus, J. S., Hubbard, C. S., Bueller, J., Ebrat, B., Tillisch, K., Chen, M., … Mayer, E. A. (2013). Impaired emotional learning and involvement of the corticotropin-releasing factor signaling system in patients with irritable bowel syndrome. *Gastroenterology*, *145*(6), 1253–1261. http://dx.doi.org/10.1053/j.gastro.2013.08.016. e1251–e1253.

Larauche, M., Gourcerol, G., Million, M., Adelson, D. W., & Tache, Y. (2010). Repeated psychological stress-induced alterations of visceral sensitivity and colonic motor functions in mice: Influence of surgery and postoperative single housing on visceromotor responses. *Stress*, *13*(4), 343–354. http://dx.doi.org/10.3109/10253891003664166.

Larauche, M., Gourcerol, G., Wang, L., Pambukchian, K., Brunnhuber, S., Adelson, D. W., … Tache, Y. (2009). Cortagine, a CRF1 agonist, induces stresslike alterations of colonic function and visceral hypersensitivity in rodents primarily through peripheral pathways. *American Journal of Physiology. Gastrointestinal and Liver Physiology*, *297*(1), G215–G227. http://dx.doi.org/10.1152/ajpgi.00072.2009.

Larauche, M., Kiank, C., & Tache, Y. (2009). Corticotropin releasing factor signaling in colon and ileum: Regulation by stress and pathophysiological implications. *Journal of Physiology and Pharmacology*, *60*(Suppl. 7), 33–46.

Larauche, M., Mulak, A., & Tache, Y. (2012). Stress and visceral pain: From animal models to clinical therapies. *Experimental Neurology*, *233*(1), 49–67. http://dx.doi.org/10.1016/j.expneurol.2011.04.020.

Lee, Y. K., & Mazmanian, S. K. (2010). Has the microbiota played a critical role in the evolution of the adaptive immune system? *Science*, *330*(6012), 1768–1773. http://dx.doi.org/10.1126/science.1195568.

Levy, R. L., Olden, K. W., Naliboff, B. D., Bradley, L. A., Francisconi, C., Drossman, D. A., & Creed, F. (2006). Psychosocial aspects of the functional gastrointestinal disorders. *Gastroenterology*, *130*(5), 1447–1458. http://dx.doi.org/10.1053/j.gastro.2005.11.057.

Longstreth, G. F., Thompson, W. G., Chey, W. D., Houghton, L. A., Mearin, F., & Spiller, R. C. (2006). Functional bowel disorders. *Gastroenterology*, *130*(5), 1480–1491. http://dx.doi.org/10.1053/j.gastro.2005.11.061.

Lovell, R. M., & Ford, A. C. (2012). Global prevalence of and risk factors for irritable bowel syndrome: A meta-analysis. *Clinical Gastroenterology and Hepatology*, *10*(7), 712–721. http://dx.doi.org/10.1016/j.cgh.2012.02.029. e714.

Lyte, M., Vulchanova, L., & Brown, D. R. (2011). Stress at the intestinal surface: Catecholamines and mucosa-bacteria interactions. *Cell and Tissue Research*, *343*(1), 23–32. http://dx.doi.org/10.1007/s00441-010-1050-0.

Melchior, C., Gourcerol, G., Chastan, N., Verin, E., Menard, J. F., Ducrotte, P., & Leroi, A. M. (2014). Effect of transcranial magnetic stimulation on rectal sensitivity in irritable bowel syndrome: A randomized, placebo-controlled pilot study. *Colorectal Disease*, *16*(3), O104–O111. http://dx.doi.org/10.1111/codi.12450.

Merle, A., Faucheron, J. L., Delagrange, P., Renard, P., Roche, M., & Pellissier, S. (2000). Nycthemeral variations of cholecystokinin action on intestinal motility in rats: Effects of melatonin and S 20928, a melatonin receptor antagonist. *Neuropeptides*, *34*(6), 385–391. http://dx.doi.org/10.1054/npep.2000.0835.

Million, M., Maillot, C., Adelson, D. A., Nozu, T., Gauthier, A., Rivier, J., … Tache, Y. (2005). Peripheral injection of sauvagine prevents repeated colorectal distension-induced visceral pain in female rats. *Peptides*, *26*(7), 1188–1195. http://dx.doi.org/10.1016/j.peptides.2005.02.004.

Million, M., Wang, L., Stenzel-Poore, M. P., Coste, S. C., Yuan, P. Q., Lamy, C., … Tache, Y. (2007). Enhanced pelvic responses to stressors in female CRF-overexpressing mice. *American Journal of Physiology. Regulatory, Integrative and Comparative Physiology*, *292*(4), R1429–R1438. http://dx.doi.org/10.1152/ajpregu.00626.2006.

Moloney, R. D., Johnson, A. C., O'Mahony, S. M., Dinan, T. G., Greenwood-Van Meerveld, B., & Cryan, J. F. (2016). Stress and the microbiota-gut-brain axis in visceral pain: Relevance to irritable bowel syndrome. *CNS Neuroscience & Therapeutics*, *22*(2), 102–117. http://dx.doi.org/10.1111/cns.12490.

Montiel-Castro, A. J., Gonzalez-Cervantes, R. M., Bravo-Ruiseco, G., & Pacheco-Lopez, G. (2013). The microbiota-gut-brain axis: Neurobehavioral correlates, health and sociality. *Frontiers in Integrative Neuroscience*, 7, 70. http://dx.doi.org/10.3389/fnint.2013.00070.

Moro, F., Levenez, F., Nemoz-Gaillard, E., Pellissier, S., Plaisancie, P., & Cuber, J. C. (2000). Release of guanylin immunoreactivity from the isolated vascularly perfused rat colon. *Endocrinology*, *141*(7), 2594–2599.

Mulak, A., & Bonaz, B. (2004). Irritable bowel syndrome: A model of the brain-gut interactions. *Medical Science Monitor*, *10*(4), RA55–RA62.

Mulak, A., Tache, Y., & Larauche, M. (2014). Sex hormones in the modulation of irritable bowel syndrome. *World Journal of Gastroenterology*, *20*(10), 2433–2448. http://dx.doi.org/10.3748/wjg.v20.i10.2433.

Muth, E. R., Thayer, J. F., Stern, R. M., Friedman, B. H., & Drake, C. (1998). The effect of autonomic nervous system activity on gastric myoelectrical activity: Does the spectral reserve hypothesis hold for the stomach? *Biological Psychology*, *47*(3), 265–278.

Myers, B., Schulkin, J., & Greenwood-Van Meerveld, B. (2011). Sex steroids localized to the amygdala increase pain responses to visceral stimulation in rats. *The Journal of Pain*, *12*(4), 486–494. http://dx.doi.org/10.1016/j.jpain.2010.10.007.

Naliboff, B. D., & Mayer, E. A. (2006). Brain imaging in IBS: Drawing the line between cognitive and non-cognitive processes. *Gastroenterology*, *130*(1), 267–270. http://dx.doi.org/10.1053/j.gastro.2005.11.034.

Naliboff, B. D., Waters, A. M., Labus, J. S., Kilpatrick, L., Craske, M. G., Chang, L., … Ornitz, E. (2008). Increased acoustic startle responses in IBS patients during abdominal and nonabdominal threat. *Psychosomatic Medicine*, *70*(8), 920–927. http://dx.doi.org/10.1097/PSY.0b013e318186d858.

Nozu, T., & Okumura, T. (2015). Corticotropin-releasing factor receptor type 1 and type 2 interaction in irritable bowel syndrome. *Journal of Gastroenterology*, *50*(8), 819–830. http://dx.doi.org/10.1007/s00535-015-1086-8.

O'Mahony, S. M., Marchesi, J. R., Scully, P., Codling, C., Ceolho, A. M., Quigley, E. M., … Dinan, T. G. (2009). Early life stress alters behavior, immunity, and microbiota in rats: Implications for irritable bowel syndrome and psychiatric illnesses. *Biological Psychiatry*, *65*(3), 263–267. http://dx.doi.org/10.1016/j.biopsych.2008.06.026.

Overstreet, D. H., & Djuric, V. (2001). A genetic rat model of cholinergic hypersensitivity: Implications for chemical intolerance, chronic fatigue, and asthma. *Annals of the New York Academy of Sciences*, *933*, 92–102.

Palomba, S., Di Cello, A., Riccio, E., Manguso, F., & La Sala, G. B. (2011). Ovarian function and gastrointestinal motor activity. *Minerva Endocrinologica*, *36*(4), 295–310.

Park, R., Mikami, S., LeClair, J., Bollom, A., Lembo, C., Sethi, S., … Nurko, S. (2015). Inpatient burden of childhood functional GI disorders in the USA: An analysis of national trends in the USA from 1997 to 2009. *Neurogastroenterology and Motility*, *27*(5), 684–692. http://dx.doi.org/10.1111/nmo.12542.

Pellissier, S., Dantzer, C., Canini, F., Mathieu, N., & Bonaz, B. (2010). Psychological adjustment and autonomic disturbances in inflammatory bowel diseases and irritable bowel syndrome. *Psychoneuroendocrinology*, *35*(5), 653–662. http://dx.doi.org/10.1016/j.psyneuen.2009.10.004.

Pellissier, S., Dantzer, C., Mondillon, L., Trocme, C., Gauchez, A. S., Ducros, V., … Bonaz, B. (2014). Relationship between vagal tone, cortisol, TNF-alpha, epinephrine and negative affects in Crohn's disease and irritable bowel syndrome. *PLoS One*, *9*(9), e105328. http://dx.doi.org/10.1371/journal.pone.0105328.

Pellissier, S., Eribon, O., Chabert, J., Gully, D., & Roche, M. (1996). Peripheral neurotensin participates in the modulation of pre- and postprandial intestinal motility in rats. *Neuropeptides*, *30*(5), 412–419.

Pellissier, S., Sasaki, K., Le-Nguyen, D., Bataille, D., & Jarrousse, C. (2004). Oxyntomodulin and glicentin are potent inhibitors of the fed motility pattern in small intestine. *Neurogastroenterology and Motility*, *16*(4), 455–463. http://dx.doi.org/10.1111/j.1365-2982.2004.00528.x.

Peters, A., & McEwen, B. S. (2012). Introduction for the allostatic load special issue. *Physiology & Behavior*, *106*(1), 1–4. http://dx.doi.org/10.1016/j.physbeh.2011.12.019.

Phelps, E. A., Delgado, M. R., Nearing, K. I., & LeDoux, J. E. (2004). Extinction learning in humans: Role of the amygdala and vmPFC. *Neuron*, *43*(6), 897–905. http://dx.doi.org/10.1016/j.neuron.2004.08.042.

Porcher, C., Juhem, A., Peinnequin, A., Sinniger, V., & Bonaz, B. (2005). Expression and effects of metabotropic CRF1 and CRF2 receptors in rat small intestine. *American Journal of Physiology. Gastrointestinal and Liver Physiology*, *288*(5), G1091–G1103. http://dx.doi.org/10.1152/ajpgi.00302.2004.

Randich, A., & Gebhart, G. F. (1992). Vagal afferent modulation of nociception. *Brain Research. Brain Research Reviews*, *17*(2), 77–99.

Randich, A., Thurston, C. L., Ludwig, P. S., Robertson, J. D., & Rasmussen, C. (1992). Intravenous morphine-induced activation of vagal afferents: Peripheral, spinal, and CNS substrates mediating inhibition of spinal nociception and cardiovascular responses. *Journal of Neurophysiology*, *68*(4), 1027–1045.

Ren, T. H., Wu, J., Yew, D., Ziea, E., Lao, L., Leung, W. K., … Sung, J. J. (2007). Effects of neonatal maternal separation on neurochemical and sensory response to colonic distension in a rat model of irritable bowel syndrome. *American Journal of Physiology Gastrointestinal and Liver Physiology*, *292*(3), G849–G856. http://dx.doi.org/10.1152/ajpgi.00400.2006.

Rhee, S. H., Pothoulakis, C., & Mayer, E. A. (2009). Principles and clinical implications of the brain-gut-enteric microbiota axis. *Nature Reviews. Gastroenterology & Hepatology*, *6*(5), 306–314. http://dx.doi.org/10.1038/nrgastro.2009.35.

Saito, Y. A., Schoenfeld, P., & Locke, G. R., 3rd. (2002). The epidemiology of irritable bowel syndrome in North America: A systematic review. *The American Journal of Gastroenterology*, *97*(8), 1910–1915. http://dx.doi.org/10.1111/j.1572-0241.2002.05913.x.

Salvioli, B., Pellegatta, G., Malacarne, M., Pace, F., Malesci, A., Pagani, M., & Lucini, D. (2015). Autonomic nervous system dysregulation in irritable bowel syndrome. *Neurogastroenterology and Motility*, *27*(3), 423–430. http://dx.doi.org/10.1111/nmo.12512.

Sandler, R. S., Drossman, D. A., Nathan, H. P., & McKee, D. C. (1984). Symptom complaints and health care seeking behavior in subjects with bowel dysfunction. *Gastroenterology*, *87*(2), 314–318.

Santos, J., Yang, P. C., Soderholm, J. D., Benjamin, M., & Perdue, M. H. (2001). Role of mast cells in chronic stress induced colonic epithelial barrier dysfunction in the rat. *Gut*, *48*(5), 630–636.

Saper, C. B. (1982). Convergence of autonomic and limbic connections in the insular cortex of the rat. *The Journal of Comparative Neurology*, *210*(2), 163–173. http://dx.doi.org/10.1002/cne.902100207.

Sayuk, G. S., & Drossman, D. A. (2011). Gastroesophageal reflux symptoms in 9/11 survivors and workers: Insights gained from tragic losses. *The American Journal of Gastroenterology*, *106*(11), 1942–1945. http://dx.doi.org/10.1038/ajg.2011.357.

Selye, H. (1950). Stress and the general adaptation syndrome. *British Medical Journal*, *1*(4667), 1383–1392.

Shinozaki, M., Kanazawa, M., Kano, M., Endo, Y., Nakaya, N., Hongo, M., & Fukudo, S. (2010). Effect of autogenic training on general improvement in patients with irritable bowel syndrome: A randomized controlled trial. *Applied Psychophysiology and Biofeedback*, *35*(3), 189–198. http://dx.doi.org/10.1007/s10484-009-9125-y.

Siggins, G. R., Gruol, D., Aldenhoff, J., & Pittman, Q. (1985). Electrophysiological actions of corticotropin-releasing factor in the central nervous system. *Federation Proceedings*, *44*(1 Pt. 2), 237–242.

Sommer, F., & Backhed, F. (2013). The gut microbiota—Masters of host development and physiology. *Nature Reviews. Microbiology*, *11*(4), 227–238. http://dx.doi.org/10.1038/nrmicro2974.

Spence, M. J., & Moss-Morris, R. (2007). The cognitive behavioural model of irritable bowel syndrome: A prospective investigation of patients with gastroenteritis. *Gut*, *56*(8), 1066–1071. http://dx.doi.org/10.1136/gut.2006.108811.

Spiess, J., Rivier, J., Rivier, C., & Vale, W. (1981). Primary structure of corticotropin-releasing factor from ovine hypothalamus. *Proceedings of the National Academy of Sciences of the United States of America*, *78*(10), 6517–6521.

Sweetser, S., Camilleri, M., Linker Nord, S. J., Burton, D. D., Castenada, L., Croop, R., … Zinsmeister, A. R. (2009). Do corticotropin releasing factor-1 receptors influence colonic transit and bowel function in women with irritable bowel syndrome? *American Journal of Physiology. Gastrointestinal and Liver Physiology*, *296*(6), G1299–G1306. http://dx.doi.org/10.1152/ajpgi.00011.2009.

Sykes, M. A., Blanchard, E. B., Lackner, J., Keefer, L., & Krasner, S. (2003). Psychopathology in irritable bowel syndrome: Support for a psychophysiological model. *Journal of Behavioral Medicine*, *26*(4), 361–372.

Tache, Y. (2015). Corticotrophin-releasing factor 1 activation in the central amygdale and visceral hyperalgesia. *Neurogastroenterology and Motility*, *27*(1), 1–6. http://dx.doi.org/10.1111/nmo.12495.

Tache, Y., & Bonaz, B. (2007). Corticotropin-releasing factor receptors and stress-related alterations of gut motor function. *The Journal of Clinical Investigation*, *117*(1), 33–40. http://dx.doi.org/10.1172/JCI30085.

Tache, Y., & Million, M. (2015). Role of corticotropin-releasing factor signaling in stress-related alterations of colonic motility and hyperalgesia. *Journal of Neurogastroenterology and Motility*, *21*(1), 8–24. http://dx.doi.org/10.5056/jnm14162.

Thayer, J. F., & Lane, R. D. (2009). Claude Bernard and the heart-brain connection: Further elaboration of a model of neurovisceral integration. *Neuroscience and Biobehavioral Reviews*, *33*(2), 81–88. http://dx.doi.org/10.1016/j.neubiorev.2008.08.004.

Thurston, C. L., & Randich, A. (1992). Effects of vagal afferent stimulation on ON and OFF cells in the rostroventral medulla: Relationships to nociception and arterial blood pressure. *Journal of Neurophysiology*, *67*(1), 180–196.

Tillisch, K., Mayer, E. A., & Labus, J. S. (2011). Quantitative meta-analysis identifies brain regions activated during rectal distension in irritable bowel syndrome. *Gastroenterology*, *140*(1), 91–100. http://dx.doi.org/10.1053/j.gastro.2010.07.053.

Tillisch, K., Mayer, E. A., Labus, J. S., Stains, J., Chang, L., & Naliboff, B. D. (2005). Sex specific alterations in autonomic function among patients with irritable bowel syndrome. *Gut*, *54*(10), 1396–1401. http://dx.doi.org/10.1136/gut.2004.058685.

Toner, B. B. (2005). Cognitive-behavioral treatment of irritable bowel syndrome. *CNS Spectrums*, *10*(11), 883–890.

Trimble, N., Johnson, A. C., Foster, A., & Greenwood-van Meerveld, B. (2007). Corticotropin-releasing factor receptor 1-deficient mice show decreased anxiety and colonic sensitivity. *Neurogastroenterology and Motility*, *19*(9), 754–760. http://dx.doi.org/10.1111/j.1365-2982.2007.00951.x.

van den Wijngaard, R. M., Klooker, T. K., de Jonge, W. J., & Boeckxstaens, G. E. (2010). Peripheral relays in stress-induced activation of visceral afferents in the gut. *Autonomic Neuroscience*, *153*(1–2), 99–105. http://dx.doi.org/10.1016/j.autneu.2009.07.004.

Van Oudenhove, L., Coen, S. J., & Aziz, Q. (2007). Functional brain imaging of gastrointestinal sensation in health and disease. *World Journal of Gastroenterology*, *13*(25), 3438–3445.

Van Oudenhove, L., Dupont, P., Vandenberghe, J., Geeraerts, B., van Laere, K., Bormans, G., … Tack, J. (2008). The role of somatosensory cortical regions in the processing of painful gastric fundic distension: An update of brain imaging findings. *Neurogastroenterology and Motility*, *20*(5), 479–487. http://dx.doi.org/10.1111/j.1365-2982.2007.01045.x.

Van Oudenhove, L., Vandenberghe, J., Geeraerts, B., Vos, R., Persoons, P., Fischler, B., … Tack, J. (2008). Determinants of symptoms in functional dyspepsia: Gastric sensorimotor function, psychosocial factors or somatisation? *Gut*, *57*(12), 1666–1673. http://dx.doi.org/10.1136/gut.2008.158162.

Vyas, A., Jadhav, S., & Chattarji, S. (2006). Prolonged behavioral stress enhances synaptic connectivity in the basolateral amygdala. *Neuroscience*, *143*(2), 387–393. http://dx.doi.org/10.1016/j.neuroscience.2006.08.003.

Wallon, C., Yang, P. C., Keita, A. V., Ericson, A. C., McKay, D. M., Sherman, P. M., … Soderholm, J. D. (2008). Corticotropin-releasing hormone (CRH) regulates macromolecular permeability via mast cells in normal human colonic biopsies in vitro. *Gut*, *57*(1), 50–58. http://dx.doi.org/10.1136/gut.2006.117549.

Whitehead, W. E., Palsson, O., & Jones, K. R. (2002). Systematic review of the comorbidity of irritable bowel syndrome with other disorders: What are the causes and implications? *Gastroenterology*, *122*(4), 1140–1156.

Yan, X. J., Feng, C. C., Liu, Q., Zhang, L. Y., Dong, X., Liu, Z. L., … Chen, S. L. (2014). Vagal afferents mediate antinociception of estrogen in a rat model of visceral pain: The involvement of intestinal mucosal mast cells and 5-hydroxytryptamine 3 signaling. *The Journal of Pain*, *15*(2), 204–217. http://dx.doi.org/10.1016/j.jpain.2013.10.012.

Zagon, A. (2001). Does the vagus nerve mediate the sixth sense? *Trends in Neurosciences*, *24*(11), 671–673.

Zernicke, K. A., Campbell, T. S., Blustein, P. K., Fung, T. S., Johnson, J. A., Bacon, S. L., & Carlson, L. E. (2013). Mindfulness-based stress reduction for the treatment of irritable bowel syndrome symptoms: A randomized wait-list controlled trial. *International Journal of Behavioral Medicine*, *20*(3), 385–396. http://dx.doi.org/10.1007/s12529-012-9241-6.

Zhang, X., Cao, B., Yan, N., Liu, J., Wang, J., Tung, V. O., & Li, Y. (2013). Vagus nerve stimulation modulates visceral pain-related affective memory. *Behavioural Brain Research*, *236*(1), 8–15. http://dx.doi.org/10.1016/j.bbr.2012.08.027.

Zhou, Q., & Verne, G. N. (2011). New insights into visceral hypersensitivity—Clinical implications in IBS. *Nature Reviews. Gastroenterology & Hepatology*, *8*(6), 349–355. http://dx.doi.org/10.1038/nrgastro.2011.83.

Zijdenbos, I. L., de Wit, N. J., van der Heijden, G. J., Rubin, G., & Quartero, A. O. (2009). Psychological treatments for the management of irritable bowel syndrome. *Cochrane Database of Systematic Reviews*, *1*, CD006442. http://dx.doi.org/10.1002/14651858.CD006442.pub2.

INDEX

Note: Page numbers followed by "*f*" indicate figures, and "*t*" indicate tables.

A

E

F

G

H

I

L

M

N

O

P

R

S

T

U

V

Kolibri 360 GmbH
Ferdinand-Jühlke-Straße 7
99095 Erfurt, Deutschland
produktsicherheit@kolibri360.de